Real-time PCR

Real-time PCR

M. Tevfik Dorak (Ed.)

School of Clinical Medical Sciences (Child Health)
Newcastle University
Newcastle-upon-Tyne, UK

Routledge
Taylor & Francis Group

LONDON AND NEW YORK

First published 2006 by Taylor & Francis

Published 2022 by Routledge
4 Park Square, Milton Park, Abingdon, Oxon OX14 4RN
605 Third Avenue, New York, NY 10017

Routledge is an imprint of the Taylor & Francis Group, an informa business

Publisher's Note

The publisher has gone to great lengths to ensure the quality of this reprint but points out that some imperfections in the original copies may be apparent.

ISBN: 978-0-415-37734-8 (pbk)

A catalog record for this book is available from the British Library.

Library of Congress Cataloging-in-Publication Data
Real-time PCR/M. Tevfik Dorak (ed.).
      p. cm.
   Includes bibliographical references and index.
   ISBN 1-415-37734-X (alk. paper)
   1.   Polymerase chain reaction. I. Dorak, M. Tevfik.
      [DNLM: 1. Polymerase Chain Reaction – methods.   QU 450
R2874 2006]
QP 606.D46R43 2006
572′.43–dc22

Editor: Elizabeth Owen
Editorial Assistant: Kirsty Lyons
Production Editor: Karin Henderson
Typeset by: Phoenix Photosetting, Chatham, Kent, UK

Contents

Contributors

Pamela Scott Adams, Trudeau Institute, 154 Algonquin Avenue, Saranac Lake, NY 12983, USA

Stephen A. Bustin, Institute of Cell and Molecular Science, Barts and the London, Queen Mary's School of Medicine and Dentistry, London, E1 1BB, UK

Burcu Cakilci, Department of Periodontology, Gazi University Dental School, Ankara, Turkey

Patrick F. Chinnery, Mitochondrial Research Group, Neurology, M4009, University of Newcastle upon Tyne, Medical School, Framlington Place, Newcastle upon Tyne, NE2 4HH, UK

Keertan Dheda, Centre for Infectious Diseases and International Health, University College London, London, W1T 4JF, UK

Steven F. Dobrowolsky, Idaho Technology, Wakara Way, Salt Lake City, UT 84108, USA

Virginie Dujols, Idaho Technology, Wakara Way, Salt Lake City, UT 84108, USA

Steve E. Durham, Mitochondrial Research Group, Neurology, M4009, University of Newcastle upon Tyne, Medical School, Framlington Place, Newcastle upon Tyne, NE2 4HH, UK

Mehmet Gunduz Department of Oral Pathology and Medicine, Okayama University School of Medicine and Dentistry, Okayama, Japan

Jim Huggett, Centre for Infectious Diseases and International Health, University College London, London, W1T 4JF, UK

Emmanuel Kanavakis, Department of Medical Genetics, National and Kapodistrian University of Athens, St. Sophia's Children's Hospital, Athens 11527, Greece

Mikael Kubista, Department of Chemistry and Bioscience, Chalmers University of Technology, and TATAA Biocenter, Box 462, 405 30 Gothenburg, Sweden

Noriko Kusukawa, Department of Pathology, University of Utah Medical School, Salt Lake City, UT 84132, USA

Kristina Lind, Department of Chemistry and Bioscience, Chalmers University of Technology, and TATAA Biocenter, Box 462, 405 30 Gothenburg, Sweden

Jason T. McKinney, Idaho Technology, Wakara Way, Salt Lake City, UT 84108, USA

Stuart N. Peirson, Department of Visual Neuroscience, Division of Neuroscience and Mental Health, Imperial College, Charing Cross Hospital, Fulham Palace Road, London, W6 8RF, UK

Michael W. Pfaffl, Institute of Physiology, Center of Life and Food Science Weihenstephan, Technical University of Munich, 85354 Freising, Weihenstephaner Berg 3, Germany

Frederique Ponchel, Molecular Medicine Unit, University of Leeds, Clinical Sciences Building, St James's University Hospital, Leeds, LS9 7TF, UK

Omaima M. Sabek, University of Tennessee Health Science Center, Memphis, TN 38103, USA

Thomas D. Schmittgen, College of Pharmacy, The Ohio State University, 500 West 12th Avenue, Columbus, OH 43210, USA

Brian Seed, Massachusetts General Hospital and Harvard Medical School, 50 Blossom Street, Boston, MA 02114, USA

Gregory L. Shipley, Department of Integrative Biology and Pharmacology, The University of Texas Health Sciences Center Houston, 6431 Fannin St, Houston, TX 77030, USA

Theo P. Sloots, Clinical Virology Research Unit, Sir Albert Sakzewski Virus Research Centre, Royal Children's Hospital and Clinical Medical Virology Centre, University of Queensland, Brisbane, Australia

Anne M. Sproul, Department of Haematology, Western General Hospital, Crewe Road South, Edinburgh, EH4 2XU, UK

Joanne Traeger-Synodinos, Department of Medical Genetics, National and Kapodistrian University of Athens, St. Sophia's Children's Hospital, Athens 11527, Greece

James (Jianming) Tang, Division of Geographic Medicine, Department of Medicine, University of Alabama at Birmingham, Birmingham, AL 35294, USA

Christina Vrettou, Department of Medical Genetics, National and Kapodistrian University of Athens, and Research Institute for the Study of Genetic and Malignant Disorders in Childhood, St. Sophia's Children's Hospital, Athens 11527, Greece

Xiaowei Wang, Ambion Inc., 2130 Woodward Street, Austin, TX 78744, USA

David M. Whiley, Clinical Virology Research Unit, Sir Albert Sakzewski Virus Research Centre, Royal Children's Hospital and Clinical Medical Virology Centre, University of Queensland, Brisbane, Australia

Carl T. Wittwer, Department of Pathology, UUMC, 5B418, 50 N. Medical Drive, Salt Lake City, UT 84105, USA

Lin Zhou, Division of Geographic Medicine, Department of Medicine, University of Alabama at Birmingham, Birmingham, AL 35294, USA

Abbreviations

ACR	Acute cellular rejection	ds	Double-stranded
ALL	Acute lymphoblastic leukemia	dTTP	Deoxythymidine triphosphate
ARMS	Amplification refractory mutation system	dUTP	2' deoxyurindine 5' triphosphate
ARP	Acidic ribosomal phosphoprotein	EAC	Europe against cancer
		EBV	Epstein–Barr virus
ASO	Allele-specific oligonucleotide	ELISA	Enzyme-linked immunosorbent assay
ATP	Adenosine triphosphate		
β2m	β2-microglobulin (B2M)	FACS	Fluorescent-assisted cell sorting
BC	Background control (in immuno-PCR)	FAM	6-carboxy fluorescein
BHQ	Black hole quencher	FDM	First derivative maximum
BKV	BK virus	FISH	Fluorescence in-situ hybridization
BLAST	Basic local alignment search tool	FITC	Fluorescein
		FRET	Fluorescence resonance energy transfer
BSA	Bovine serum albumin		
CCD	Charge-coupled device	GAPDH	glyceraldehyde-3-phosphate dehydrogenase
cDNA	Complementary DNA		
ChIP	Chromatin immunoprecipitation	gDNA	Genomic DNA
		GOI	Gene of interest
CML	Chronic myeloid leukemia	GUS	β-glucuronidase
CMV	Cytomegalovirus	HBV	Hepatitis B virus
COX	Cytochrome c oxidase	HCV	Hepatitis C virus
C_p	Crossing point (LightCycler® terminology)	HEP	Human tpigenome Project
		HEX	Hexachloro-carbonyl-fluorescein
CRS	Cambridge Reference Sequence		
C_t	Threshold cycle (ABI terminology)	HKG	House keeping gene
		HPLC	High-performance liquid chromatography
CTAB	Cetyltrimethylammonium Bromide		
CV	Coefficient of variation (standard deviation divided by the mean)	HPRT	Hypoxanthine phosphoribosyl-transferase 1
		HSV	Herpes simplex virus
CVS	Chorionic villi sampling	GOI	Gene of interest
CYC	cyclophilin	IC	Internal control (normalizer, reference)
DABCYL	4-(dimethylamino)azobenzene-4'-carboxylic acid		
		IFN	Interferon
DABSYL	4-(dimethylamino)azobenzene-4'-sulfonyl chloride	IL	Interleukin
		IPC	Internal positive control
DART	Data analysis for real-time PCR	IVF	In-vitro fertilization
DFA	Direct fluorescence assay	JOE	2, 7-dimethoxy-4, 5-dichloro-6-carboxy-fluoroscein
DGGE	Denaturing gradient gel electrophoresis		
		KOD	Kinetic outlier detection

LATE PCR	Linear after the exponential PCR
LCM	Laser capture microdissection
LED	Light emitting diode
LiCl	Lithium Chloride
LNA	Locked nucleic acid
LUX™	Light upon eXtension
^{m}C	Methylated cytosine
MDR1	Multidrug resistance protein 1
MGB	Minor groove binders
MHC	Major histocompatibility complex
MMF	Mycophenolate mofetil
MRD	Minimal residual disease
mRNA	Messenger RNA
MSE	Mean squared error (nothing to do with SEM)
MSP	Methylation-specific PCR
mtDNA	Mitochondrial DNA (see also nDNA)
NAC	No amplification control
NASBA	Nucleic acid sequence based amplification
nDNA	Nuclear DNA (as opposed to mtDNA)
NF	Normalization factor
NTC	No template control (no target control)
OD	Optical density
OD$_{260}$	Optical density at 260 nm
PAGE	Polyacrylamide gel electrophoresis
PCR	Polymerase chain reaction
PCV	Packed cell volume
PGD	Preimplantation genetic diagnosis
PGK	Phosphoglycerokinase
PMT	Photomultiplier tube
PND	Prenatal diagnosis
PTLD	Post-transplant lymphoproliferative disorders
qPCR	Quantitative PCR
r	(Pearson's) coefficient of correlation (varies between −1 and +1)

r^2	R-squared or coefficient of determination (varies between 0 and +1)
rCRS	Revised Cambridge Reference Sequence
REST	Relative expression software tool
REST-MCS	REST – multiple condition solver
REST-RG	REST Rotor-Gene
RFLP	Restriction fragment length polymorphism
ΔRn	Normalized reporter signal minus background fluorescence
−RTC	Minus reverse transcriptase control
Rn	Normalized raw fluorescence
ROS	Reactive oxygen species
ROX	6-carboxy-X-rhodamine
rRNA	Ribosomal RNA
RSV	Respiratory syncytial virus
RT	Reverse transcription
SD	Standard deviation
SDM	Second derivative maximum
SDS	Sequence Detection Software
SEM	Standard error of the mean
SMCC	Succinimidyl-4-(N-maleimidomethyl) cyclohexane-1-carboxylate
SNP	Single nucleotide polymorphism
STR	Short tandem repeat
TAMRA	6-carboxy-tetra-methyl-rhodamine
TBP	TATA-binding protein
TCA	Trichoroacetic acid
TCR	T-cell receptor
TE buffer	Tris-EDTA buffer
TfR	Transferrin receptor
Tm	Melting temperature
TREC	T-cell receptor excision circles
tRNA	Transfer RNA
UNG	Uracil-N-glycosylase
y-int	y-intercept

Preface

Polymerase chain reaction (PCR) has secured its place in biomedical history as a revolutionary method. Many techniques that have derived from PCR have come and gone. Real-time PCR is based on the conventional principles of PCR and since its beginnings about a decade ago its popularity has kept growing. With the simple shift of emphasis from the end-product to the whole course of the PCR process, real-time PCR has established itself as the most sensitive and specific quantitative PCR method. The real-time PCR concept has also contributed to the development of high-throughput allelic discrimination assays. With a variety of detection chemistries, an increasing number of platforms, multiple choices for analytical methods and the jargon emerging along with these developments, real-time PCR is facing the risk of becoming an intimidating method, especially for beginners. This book aims to provide the basics, explain how they are exploited to run a real-time PCR assay, how the assays are run and where these assays are informative in real life. The book does not intend to cover every aspect of real-time PCR in an encyclopedic fashion but instead to address the most practical aspects of the techniques with the emphasis on 'how to do it in the laboratory'. Keeping with the spirit of the Advanced Methods Series, most chapters provide an experimental protocol as an example of a specific assay. It is left with the reader to adapt the presented protocol to their individual needs.

The book is organized in two parts. The first part begins with a general introduction which is followed by chapters on the basics of data analysis, quantification, normalization and principles of primer and probe design, all contributed by leaders in the field. This part is concluded by chapters on more applied aspects of real-time PCR including an overview of high-resolution melting analysis by the most experienced users of this method. The second part of the book covers specific applications including the less recognized uses of real-time PCR: methylation detection, mitochondrial DNA analysis and immuno-PCR. The following chapters summarize many uses of real-time PCR in the clinic with examples. Clinical microbiology and virology, solid organ and bone marrow transplantation, and prenatal genetic diagnosis are among the topics covered. These chapters have been written with the laboratory and practical uses in mind and the contributors share their most valuable experiences. A comprehensive glossary and index supplement the 17 chapters and aim to make the book more accessible.

Real-time PCR is, like any other modern method in molecular genetics, expanding, with potential applications even in proteomics. This book is expected to provide the basic principles of applied real-time PCR and provide a firm grounding for those who wish to develop further applications. The selection of the chapters reflects the acknowledgment of inevitable future developments. We look forward to covering those in future editions of the book. Real-time PCR is surely here to stay and hopefully this book will help current users of this technique and also help develop further applications for it.

M. Tevfik Dorak

Glossary

This glossary has been adapted from the Editor's Real-Time PCR webpage (http://www.dorak.info/genetics/realtime.html). The online version includes references and hyperlinks.

Absolute quantification: The absolute quantitation assay is used to quantitate unknown samples by interpolating their quantity from a standard curve (as in determination of viral copy number).

Allelic discrimination assay: Assays designed to type for gene variants. Either differentially labeled probes (one for each variant) or a single probe and melting curve analysis can be used for this purpose. Alternatively, dsDNA-binding dyes can be used in combination with melting curve analysis.

Amplicon: The amplified sequence of DNA in the PCR process.

Amplification plot: The plot of cycle number versus fluorescence signal which correlates with the initial amount of target nucleic acid during the exponential phase of PCR.

Anchor & reporter probes: Two partnering LightCycler® probes that hybridize on the target sequence in close proximity. The anchor probe (donor) emits the fluorescein to excite the reporter probe (acceptor) to initiate fluorescence resonance energy (FRET). In allelic discrimination assays, it is important that the reporter probe spans the mutation and has a lower Tm than the anchor probe.

Baseline: The initial cycles of PCR during which there is little change in fluorescence signal (usually cycles 3 to 15).

Baseline value: During PCR, changing reaction conditions and environment can influence fluorescence. In general, the level of fluorescence in any one well corresponds to the amount of target present. Fluorescence levels may fluctuate due to changes in the reaction medium creating a background signal. The background signal is most evident during the initial cycles of PCR prior to significant accumulation of the target amplicon. During these early PCR cycles, the background signal in all wells is used to determine the 'baseline fluorescence' across the entire reaction plate. The goal of data analysis is to determine when target amplification is sufficiently above the background signal, facilitating more accurate measurement of fluorescence.

Calibrator: A single reference sample used as the basis for relative-fold increase in expression studies (assuming constant reaction efficiency).

Coefficient of variance (CV): Used as a measure of experimental variation. It is important that a linear value (e.g., copy numbers) is used to calculate the CV (but not C_t values which are logarithmic). Intra-assay CV quantifies the amount of error seen within the same assay (in duplicates) and inter-assay CV quantifies the error between separate assays.

C_t (threshold cycle): Threshold cycle reflects the cycle number at which the fluorescence generated within a reaction crosses the threshold. It is inversely correlated to the logarithm of the initial copy number. The C_t value assigned to a particular well thus reflects the point during the reaction at which a sufficient number of amplicons have accumulated. Also called crossing point (C_p) in LightCycler® terminology.

Derivative curve: This curve is used in Tm analysis. It has the temperature in the x axis and the negative derivative of fluorescence (F) with respect to temperature (T), shown as dF/dT, on the y axis. The reproducibility of a derivative melting curve is high with a standard deviation of only 0.1°C between runs.

dsDNA-binding agent: A molecule that emits fluorescence when bound to dsDNA. The prototype is SYBR® Green I. In real-time PCR, the fluorescence intensity increases proportionally to dsDNA (amplicon) concentration. The problem with DNA-binding agents is that they bind to all dsDNA products: specific amplicon or non-specific products (misprimed targets and primer-dimers included). For this reason, analysis using DNA-binding agents is usually coupled with melting analysis.

Dynamic range: The range of initial template concentrations over which accurate C_t values are obtained. If endogenous control is used for $\Delta\Delta C_t$ quantitation method, dynamic ranges of target and control should be comparable. In absolute quantitation, interpolation within this range is accurate but extrapolation beyond the dynamic range should be avoided. The larger the dynamic range, the greater the ability to detect samples with high and low copy number in the same run.

Efficiency of the reaction: The efficiency of the reaction can be calculated by the following equation: $E = 10^{(-1/slope)} - 1$. The efficiency of the PCR should be 90–100% meaning doubling of the amplicon at each cycle. This corresponds to a slope of 3.1 to 3.6 in the C_t vs log-template amount standard curve. In order to obtain accurate and reproducible results, reactions should have efficiency as close to 100% as possible (e.g., two-fold increase of amplicon at each cycle) and, in any case, efficiency should be similar for both target and reference (normalizer, calibrator, endogenous control, internal control). A number of variables can affect the efficiency of the PCR. These factors can include length of the amplicon, presence of inhibitors, secondary structure and primer design. Although valid data can be obtained that fall outside of the efficiency range, if it is <0.90, the quantitative real-time PCR should be further optimized or alternative amplicons designed.

Endogenous control: This is an RNA or DNA that is naturally present in each experimental sample. By using an invariant endogenous control as an active 'reference', quantitation of a messenger RNA (mRNA) target can be normalized for differences in the amount of total RNA added to each reaction and correct for sample-to-sample variations in reverse transcriptase PCR efficiency.

Exogenous control: This is a characterized RNA or DNA spiked into each sample at a known concentration. An exogenous active reference is usually an in vitro construct that can be used as an internal positive control (IPC) to distinguish true target negatives from PCR inhibition. An exogenous reference can also be used to normalize for differences in efficiency of sample extraction or complementary DNA (cDNA) synthesis by reverse transcriptase. Whether or not an active reference is used, it is important to use a passive reference dye (usually ROX–6-carboxy-X-rhodamine) in order to normalize for non-PCR-related fluctuations in fluorescence signal.

FAM: 6-carboxy fluorescein. Most commonly used reporter dye at the 5′ end of a TaqMan® probe.

Fluorescence resonance energy transfer (FRET): The interaction between the electronic excited states of two dye molecules. The excitation is transferred from one (the donor) dye molecule to the other (the acceptor) dye molecule. FRET is distance dependent and occurs when the donor and the acceptor dye are in close proximity.

Housekeeping gene: Genes that are widely expressed in abundance and are usually used as reference genes for normalization in real-time PCR with the assumption of 'constant expression'. The current trend is first to check which housekeeping genes are suitable for the target cell or tissue and then to use more than one of them in normalization.

Hybridization probe: One of the main fluorescence-monitoring systems for DNA amplification. LightCycler® probes are hybridization probes and are not hydrolyzed by Taq Polymerase. For this reason, melting curve analysis is possible with hybridization probes.

Hydrolysis probe: One of the main fluorescence-monitoring systems for DNA amplification. TaqMan® probes are an example. These kinds of probes are hydrolyzed by the 5' endonuclease activity of Taq Polymerase during PCR.

Internal positive control (IPC): An exogenous IPC can be added to a multiplex assay or run on its own to monitor the presence of inhibitors in the template. Most commonly the IPC is added to the PCR master mix to determine whether inhibitory substances are present in the mix. Alternatively, it can be added at the point of specimen collection or prior to nucleic acid extraction to monitor sample stability and extraction efficiency, respectively.

Log-dilution: Serial dilutions in powers of 10 (10, 100, 1000 etc).

LUX™ (Light upon eXtension) primers: Created by Invitrogen, LUX™ primer sets include a self-quenched fluorogenic primer and a corresponding unlabeled primer. The labeled primer has a short sequence tail of 46 nucleotides on the 5' end that is complementary to the 3' end of the primer. The resulting hairpin secondary structure provides optimal quenching of the fluorophore. When the primer is incorporated into double-stranded DNA during PCR, the fluorophore is dequenched and the signal increases by up to ten-fold. By eliminating the need for a quencher dye, the LUX™ primers reduce the cost.

Melting curve (dissociation) analysis: Every piece of dsDNA has a melting point (Tm) at which temperature 50% of the DNA is single stranded. The temperature depends on the length of the DNA, sequence order, G:C content and Watson–Crick pairing. When DNA-binding dyes are used, as the fragment is heated, a sudden decrease in fluorescence is detected when Tm is reached (due to dissociation of DNA strands and release of the dye). This point is determined from the inflection point of the melting curve or the melting peak of the derivative plot (what is meant by derivative plot is the negative first-derivative of the melting curve). The same analysis can be performed when hybridization probes are used as they are still intact after PCR. As hydrolysis probes (e.g., TaqMan®) are cleaved during the PCR reaction, no melting curve analysis possible if they are used. Mismatch between a hybridization probe and the target results in a lower Tm. Melting curve analysis can be used in known and unknown (new) mutation analysis as a new mutation will create an additional peak or change the peak area.

Minor groove binders (MGBs): These dsDNA-binding agents are attached to the 3' end of TaqMan® probes to increase the Tm value (by stabilization of hybridization) and to design shorter probes. Longer probes reduce design flexibility and are less sensitive to mismatch discrimination. MGBs also reduce background fluorescence and increase dynamic range due to increased efficiency of reporter quenching. By allowing the use of shorter probes with higher Tm values, MGBs enhances mismatch discrimination in genotyping assays.

Minus reverse transcriptase control (–RTC): A quantitative real-time PCR control sample that contains the starting RNA and all other components for one-step reaction but no reverse transcriptase. Any amplification suggests genomic DNA contamination.

Molecular beacons: These hairpin probes consist of a sequence-specific loop region flanked by two inverted repeats. Reporter and quencher dyes are attached to each end of the molecule and remain in close contact unless sequence-specific binding occurs and reporter emission (FRET) occurs.

Multiplexing: Simultaneous analysis of more than one target. Specific quantification of multiple targets that are amplified within a reaction can be performed using a differentially labeled primer or probes. Amplicon or probe melting curve analysis allows multiplexing in allelic discrimination if a dsDNA-binding dye is used as the detection chemistry.

Normalization: A control gene that is expressed at a constant level is used to normalize the gene expression results for variable template amount or template quality. If RNA quantitation can be done accurately, normalization might be done using total RNA amount used in the reaction. The use of multiple housekeeping genes that are most appropriate for the target cell or tissue is the most optimal means for normalization.

Nucleic acid sequence based amplification (NASBA): NASBA is an isothermal nucleic acid amplification procedure based on target-specific primers and probes, and the co-ordinated activity of THREE enzymes: AMV reverse transcriptase, RNase H and T7 RNA polymerase. NASBA allows direct detection of viral RNA by nucleic acid amplification.

No amplification control (NAC, a minus enzyme control): In mRNA analysis, NAC is a mock reverse transcription containing all the RT-PCR reagents, except the reverse transcriptase. If cDNA or genomic DNA is used as a template, a reaction mixture lacking Taq polymerase can be included in the assay as NAC. No product should be synthesized in the NTC or NAC. If the absolute fluorescence of the NAC is greater than that of the NTC after PCR, fluorescent contaminants may be present in the sample or in the heating block of the thermal cycler.

No template control (NTC, a minus sample control): NTC includes all of the RT-PCR reagents except the RNA template. No product should be synthesized in the NTC or NAC; if a product is amplified, this indicates contamination (fluorescent or PCR products) or presence of genomic DNA in the RNA sample.

Normalized amount of target: A unitless number that can be used to compare the relative amount of target in different samples.

Nucleic acid target (also called 'target template'): DNA or RNA sequence that is going to be amplified.

Passive reference: A dye that provides an internal reference to which the reporter dye signal can be normalized during data analysis. Normalization is necessary to correct for fluctuations from well to well caused by changes in concentration or volume. ROX is the most commonly used passive reference dye.

Quencher: The molecule that absorbs the emission of fluorescent reporter when in close vicinity. Most commonly used quenchers include 6-carboxy-tetra-methyl-rhodamine (TAMRA), (DABCYL) and black hole quencher (BHQ).

R: In illustrations of real-time PCR principles, R represents fluorescent reporter (fluorochrome).

r coefficient: Correlation coefficient, which is used to analyze a standard curve (ten-fold dilutions plotted against C_t values) obtained by linear regression analysis. It should be ≥0.99 for gene quantitation analysis. It takes values between zero and −1 for negative correlation and between zero and +1 for positive correlations.

R² coefficient: Frequently mixed up with 'r' but this is R-squared (also called coefficient of determination). This coefficient only takes values between zero and +1. R^2 is used to assess the fit of the standard curve to the data points plotted. The closer the value to 1, the better the fit.

Rapid-cycle PCR: A powerful technique for nucleic acid amplification and analysis that is completed in less than half an hour. Samples amplified by rapid-cycle PCR are immediately analyzed by melting curve analysis in the same instrument. In the presence of fluorescent hybridization probes, melting curves provide 'dynamic dot blots' for fine sequence analysis, including single nucleotide polymorphism (SNPs). Leading instruments that perform rapid-cycle PCR are RapidCycler®2 (Idaho Technology) and LightCycler® (Roche).

Real-time PCR: The continuous collection of fluorescent signal from polymerase chain reaction throughout cycles.

Reference: A passive or active signal used to normalize experimental results. Endogenous and exogenous controls are examples of active references. Active reference means the signal is generated as the result of PCR amplification.

Reference dye: Used in all reactions to obtain normalized reporter signal (Rn) adjusted for well-to-well variations by the analysis software. The most common reference dye is ROX and is usually included in the master mix.

Reporter dye (fluorophore): The fluorescent dye used to monitor amplicon accumulation. This can be attached to a specific probe or can be a dsDNA binding agent.

Relative quantification: A relative quantification assay is used to analyze changes in gene expression in a given sample relative to another reference sample (such as relative increase or decrease, compared to the baseline level, in gene expression in response to a treatment or in time, etc). Includes comparative C_t ($\Delta\Delta C_t$) and relative-fold methods.

Ribosomal RNA (rRNA): Commonly used as a normalizer in quantitative real-time RNA. It is not considered ideal due to its expression levels, transcription by a different RNA polymerase and possible imbalances in relative rRNA-to-mRNA content in different cell types.

Rn (normalized reporter signal): The fluorescence emission intensity of the reporter dye divided by the fluorescence emission intensity of the passive reference dye. Rn+ is the Rn value of a reaction containing all components, including the template and Rn is the Rn value of an unreacted sample. The Rn value can be obtained from the early cycles of a real-time PCR run (those cycles prior to a significant increase in fluorescence), or a reaction that does not contain any template.

ΔRn (delta Rn, dRn): The magnitude of the signal generated during the PCR at each time point. The ΔRn value is determined by the following formula: (Rn+) – (Rn).

ROX (6-carboxy-X-rhodamine): Most commonly used passive reference dye for normalization of reporter signal.

Scorpion: Another fluorescence detection system consists of a detection probe with the upstream primer with a fluorophore at the 5' end, followed by a complementary stem-loop structure also containing the specific probe sequence, quencher dye and a PCR primer on the 3' end. This structure makes the sequence-specific priming and probing a unimolecular event which creates enough specificity for allelic discrimination assays.

Slope: Mathematically calculated slope of standard curve, e.g., the plot of C_t values against logarithm of ten-fold dilutions of target nucleic acid. This slope is used for efficiency calculation. Ideally, the slope should be 3.3 (3.1 to 3.6), which corresponds to 100% efficiency (precisely 1.0092) or two-fold (precisely, 2.0092) amplification at each cycle. Also called gradient.

Standard: A sample of known concentration used to construct a standard curve. By running standards of varying concentrations, a standard curve is created from which the quantity of an unknown sample can be calculated.

Standard curve: Obtained by plotting C_t values against log-transformed concentrations of serial ten-fold dilutions of the target nucleic acid. Standard curve is obtained for quantitative PCR and the range of concentrations included should cover the expected unknown concentrations range.

Sunrise™ primers: Created by Oncor, Sunrise™ primers are similar to molecular beacons. They are self-complementary primers which dissociate through the synthesis of the complementary strand and produce fluorescence signals.

TAMRA (6-carboxy-tetra-methyl-rhodamine): Most commonly used quencher at the 3′ end of a TaqMan® probe.

TaqMan® probe: A dual-labeled specific hydrolysis probe designed to bind to a target sequence with a fluorescent reporter dye at one end and a quencher at the other.

Threshold: Usually 10X the standard deviation of Rn for the early PCR cycles (baseline). The threshold should be set in the region associated with an exponential growth of PCR product. It is the numerical value assigned for each run to calculate the C_t value for each amplification.

Unknown: A sample containing an unknown quantity of template. This is the sample of interest whose quantity is being determined.

Links to software and web resources cited

geNORM (download)
http://medgen.ugent.be/~jvdesomp/genorm

qBASE (download)
http://medgen.ugent.be/qbase

DART (download)
http://nar.oxfordjournals.org/cgi/content/full/31/14/e73/DC1

REST (download)
http://www.gene-quantification.de/rest.html

Best Keeper (download)
http://www.gene-quantification.de/bestkeeper.html

Mfold (nucleic acid folding and hybridization prediction)
http://www.bioinfo.rpi.edu/applications/mfold

Exiqon OligoDesign (LNA Primers)
http://lnatools.com

D-LUX Designer
http://orf.invitrogen.com/lux

PrimerBank
http://pga.mgh.harvard.edu/primerbank

RTPprimerDB
http://medgen.ugent.be/rtprimerdb

Quantitative PCR Primer Database (QPPD)
http://web.ncifcrf.gov/rtp/GEL/primerdb

Human Genome Project Database
http://genome.ucsc.edu/cgi-bin/hgGateway?org=human

NCBI Entrez GENE (Replaced Locus Link)
http://www.ncbi.nlm.nih.gov/entrez/query.fcgi?db=gene

NCBI Blast
http://www.ncbi.nlm.nih.gov/blast

MITOMAP (Human Mitochondrial Genome Database)
http://www.mitomap.org

TATAA Biocenter
www.tataa.com

Gene Quantification Web Resources
http://www.gene-quantification.info

An introduction to real-time PCR

1

Gregory L. Shipley

1.1 Introduction

There are several ways to interrogate a cell for changes induced by artificial or natural agents during a biological process. One way is to look for changes in cellular transcript levels that may indicate changes in the corresponding proteins. In another instance, the focus may be on the presence or absence of a viral or bacterial pathogen. In this case, detecting not only the presence but also the level of the pathogen provides valuable information in devising a treatment regimen. Alternatively, looking for an increase in the level of expression from a transgene or the inhibition of expression of an endogenous gene by an siRNA may be the question of interest. In all cases, quantitative real-time PCR technology can be utilized to provide the required information. Successful implementation of the technology requires users to have a basic background in the theoretical principles of real-time PCR as well as their practical application to the project at hand. The goal of this introductory chapter is to provide a basic foundation in the use of real-time PCR. Some of the following chapters will expand on topics presented here and add necessary new information required for a full understanding of this powerful technique.

1.2 A brief history of nucleic acid detection and quantification

Initially, nucleic acid quantification meant the addition of radiolabeled UTP or deoxythymidine triphosphate (dTTP) to cell cultures or one of many possible *in vitro* experimental preparations and measuring their incorporation into nucleic acids by TCA (trichoroacetic acid) precipitation. Although radioactive incorporation is a quantitative technique and gave the investigator an idea of the global changes in the nucleic acid population of their experimental system, it was not satisfactory for identifying or quantifying specific genes or transcripts. The first breakthrough in the identification of specific genes came with the development of the Southern transfer method (Southern, 1975). This was followed by the Northern blot for RNA (Alwine *et al.*, 1977). In both cases, it was now possible to specifically identify a particular gene or transcript by hybridization of a radioactively labeled probe to a membrane bearing DNA restriction fragments or RNA. Neither of these methods were quantitative techniques despite the best efforts to extract quantitative information from their use.

The next improvement in RNA transcript quantification came with the RNase-protection experiment. In this method, a short (<500 bases) highly radioactive anti-sense RNA was made from a plasmid construct utilizing T7 RNA polymerase. *In vitro* transcribed RNA was synthesized for the transcript(s) of interest along with a probe used for loading normalization. The radioactive anti-sense probes were then combined with each total RNA sample and hybridized to completion in liquid. The resulting double-stranded RNA product(s) were protected from the subsequent addition of a cocktail of single strand-specific nucleases. The protected products were then separated on a gel via denaturing polyacrylamide gel electrophoresis (PAGE) and exposed to film. This method had the advantage of hybridization in liquid rather than a solid surface and did not require the transfer of the RNAs to a membrane. However, the inherent short dynamic range when using film was the same. Again, the phosphoimager improved greatly on quantification for these experiments due to the expanded dynamic range and improved software for analysis. The down side to RNase protection experiments was the probes had to have high radioactive specific activity and required great care for safety reasons.

The procedure for performing the polymerase chain reaction (PCR) was first introduced by Kerry Mullis in 1983 (Mullis, 1990) for which he won the Nobel Prize in 1993. It is hard to think of another laboratory technique that has had a greater impact on so many different facets of biological research than PCR. In combining the reverse transcriptase (RT) reaction with the PCR, identification of a specific RNA transcript was now possible from very low copy numbers of starting material. Quantification of transcripts from sample unknowns became possible with the advent of competitive RT-PCR (Vu *et al.*, 2000). In this method, a truncated version of the target region of interest lies between the same primer-binding sites as the target transcript sequence within a plasmid clone. The easiest method for making a smaller competitive target was to digest the cloned region between the primer-binding sites with a restriction enzyme and then ligate the resulting sticky ends, dropping out a short section of sequence. The requirements for the quantification construct were that it be a similar, but different, size than the target PCR product and quantified. The plasmid contains a T7, SP6 or T3 RNA promoter sequence up stream of the cloned target sequence. Utilizing the RNA promoter, truncated *in vitro* transcribed RNA could be made and quantified. Known amounts of the RNA product were spiked into the RT reaction and converted into cDNA along with the target sequence within the unknown sample. Subsequently, both the truncated standard and unknown target sequences were amplified using the PCR. The amplified DNAs were separated using denaturing polyacrylamide gel electrophoresis. In some methods a radioactive deoxynucleotide base was added for labeling the amplified DNA and quantified using either film or a phosphoimager. In other methods, the products on the gel were imaged following staining with ethidium bromide or SYBR® Green I. Quantification of the unknown target band was determined by comparison to signal from the spiked and quantified DNA standard. Although this method was the most accurate to date it still suffered from the detection problems mentioned earlier. However, most of the criticism centered around the spiked DNA standard. The concern was that the DNA standard was competing for reagent resources and primers with the unknown target during the PCR and might, therefore, alter the final result.

In 1996, Applied Biosystems (ABI) made real-time PCR commercially available (Heid *et al.*, 1996) with the introduction of the 7700 instrument. Real-time quantitative PCR has become the most accurate and sensitive method for the detection and quantification of nucleic acids yet devised. This method has overcome most of the major shortcomings of the preceding ones. Using the specificity and sensitivity of PCR combined with direct detection of the target of choice utilizing fluorescently labeled primers, probes or dyes, the inherent problems of gels, transfer to a membrane, radioactive probe hybridization and the limitations of film as a detector have been eliminated. There are two problems for real-time PCR that still linger. They are 1) methods of quantification, i.e. kind of standard, assay quality and calculation methodology used and 2) how to properly normalize different samples to correct for differences in nucleic acid input from sample to sample. Both of these areas are the topic of much study and will be discussed in more detail in following chapters.

1.3 Real-time quantitative PCR – a definition

What exactly is real-time quantitative PCR? Some believe you have to be able to watch the growth of the amplification curves during the PCR on a computer monitor in order to be truly 'real-time'. This of course is not the case. The ABI 7700 SDS software, the original real-time program, does not allow the visualization of the amplification curves as they progress throughout the run. This was primarily due to the way the SDS software performs data analysis utilizing the final data set from the whole plate rather than analyzing each reaction individually. Thus for some instruments, the final data have to be present for data analysis to proceed. For other instruments, it does not. The latter case allows the software to follow the progress of the amplification curve in each well in real time, which can be visualized on the computer monitor. This ability does not make these instruments exclusively 'real-time instruments'.

Thus, real-time PCR is the continuous collection of fluorescent signal from one or more polymerase chain reactions over a range of cycles. Quantitative real-time PCR is the conversion of the fluorescent signals from each reaction into a numerical value for each sample.

1.4 Practical and theoretical principles underlying real-time PCR

RNA quantification begins with the making of cDNA (complementary DNA) by reverse transcriptase. There are two kinds of RT enzymes readily available on the market, AMV (Peters *et al.*, 2004) and MMLV (Gerard *et al.*, 1997). AMV is a dimeric protein from the avian myeloblastosis virus. MMLV is derived from the Moloney murine leukemia virus and is a monomeric protein. Both enzymes have RNase H activity, which is the ability to degrade RNA in an RNA–DNA hybrid. However, AMVs have higher RNase H activity compared to MMLV enzymes. In both enzymes, the RNase H activity can be separated from the RNA-dependent DNA polymerase activity by mutagenesis. More importantly, AMVs are more processive than MMLV enzymes. That is, they can incorporate more nucleotides per unit enzyme per template molecule. The temperature optimum for the native AMV enzyme is higher than that for MMLVs; 42°C versus 37°C. Cloned variants of both enzymes have pushed the temperature limits much higher: 58°C for AMVs and 55°C for MMLV modified enzymes. From this description, you most likely would deduce that engineered AMV enzymes would be best suited for making cDNAs from RNA samples prior to real-time PCR. However, in practice it is the engineered MMLVs that work best for this purpose. It is not clear why this is true but it has been proposed that although the high temperature of the PCR kills the polymerase activity of both reverse transcriptases, the DNA binding capability remains and may present a physical barrier to *Taq* polymerase during the PCR. The higher temperature stability and dimeric nature of AMV may be the reason it is less suited for use in real-time PCR. A comparison of the most commonly used reverse transcriptase enzymes is given in *Table 1.1* kindly prepared by Dr. Duncan Clark, GeneSys Ltd.

Both of these enzymes have been shown to be inhibitory to PCR in a concentration dependent manner. For this reason it is important to keep the amount of reverse transcriptase as low as possible while still achieving optimal cDNA synthesis. In practice, less reverse transcriptase is required for assay-specific primers than for assays primed by either oligo-dT or random primers because the total amount and complexity of the cDNA made is much reduced. It should be pointed out that cDNA will still be made even if no primers are added to the reverse transcriptase reaction. This is caused by self-priming due to secondary structure within the RNA template.

There are three ways to prime a reverse transcriptase reaction: oligo-dT, random primers or assay-specific primers (Winer *et al.*, 1999; Calogero *et al.*, 2000). There are pros and cons for each type and in the end, the best choice turns out to be tied to the format of the subsequent PCR. Oligo-dT has been used extensively because it will prime primarily mRNAs. However, not all messenger RNAs have poly-A tails and there are poly-A tracts within some RNA sequences. The largest problem with oligo-dT priming is that it limits the selection of the following PCR assay to a region near the poly-A tail of the mRNA. For transcripts that are very long, some are over 15 Kb in length, the assay has to be placed within the 3' UTR (3' untranslated region) of the transcript. Although this will result in high transcript fidelity for the target of interest, it means that the assay most likely will not cross an exon/exon junction. For some transcripts, the 3' UTR sequence can be very A/T rich and thus not very conducive for the design of a PCR assay. The use of random primers removes the 3' UTR bias as cDNA is made throughout the transcript sequence. However, random primers will also make cDNA from ribosomal and transfer RNAs, which makes a much more abundant and complex cDNA population thus increasing the risk of false priming during the PCR. Further, it is possible that priming will be initiated within the PCR amplicon (the sequence amplified during PCR) by one of the random primers, eliminating that molecule from possible detection. A third alternative is the use of assay-specific primers. In this case, the reverse primer for the following PCR is used to prime cDNA synthesis in the reverse transcriptase reaction. This ensures, in theory, that every potential transcript

Table 1.1 Commonly used reverse transcriptases

	MMLV	MMLV RNaseH- (equivalent to Superscript I)	MMLV RNaseH- (equivalent to Superscript II)	MMLV RNaseH- (equivalent to Superscript III)	AMV	Tth
Organism	Moloney Murine Leukemia virus	Moloney Murine Leukemia virus	Moloney Murine Leukemia virus	Moloney Murine Leukemia virus	Avian myeloblastosis virus	Thermus thermophilus HB8
Molecular weight	75 kDa	56 kDa	75 kDa	75 kDa	63 kDa + 95 kDa	94 kDa
Number of amino acids	671	503	671	671	544 + 859	832
Single chain or subunits	Single	Single	Single	Single	Two 63 kDa alpha chain and 95 kDa beta chain – heterodimer	Single
Extension rate	2 kb–4 kb/min	2 kb–4 kb/min			2 kb–4 kb/min	2 kb–4 kb/min
RNaseH activity	Yes	No (C-terminal deletion of full length enzyme)	No (single point mutation)	No (single point mutation)	Yes, strong	Yes, low
Half life at 37°C (minus template primer)	435 mins	–	390 mins	>435 mins	110 mins	–
Half life at 50°C (minus template primer)	3 mins	–	6 mins	220 mins	2 mins	20 mins at 95°C
Half life at 50°C (plus template primer)	3 mins	–	10 mins	–	15 mins	–

targeted by the PCR assay will have a corresponding cDNA synthesized. Further, the length of each cDNA has only to be the length of the amplicon. For any real-time PCR assay, the maximal length of the amplicon should be, at most 250 bases long.

From the previous discussion, the use of assay-specific primers would appear to be the best choice, and it is for many applications. However, the use of assay-specific primers has two main drawbacks. First it uses more sample RNA as only one assay can be run from each cDNA reaction and second, it cannot be used if a multiplex (more than one real-time assay run in the same PCR) reaction is contemplated. Therefore, the appropriate priming method must be determined by the investigator based on the amount of template available and the kind of real-time assay that is planned.

The business end of real-time PCR detection is, of course, the polymerase chain reaction. The heart of the modern PCR is the addition of a thermostable DNA polymerase. The most commonly used enzyme is from *Thermus aquaticus* or *Taq*. Wild type *Taq* is a 5'–3' synthetic DNA-dependent DNA polymerase with 3'–5' proof reading activity. It also has 5'-nuclease activity. Although wild-type *Taq* can be purchased commercially, the enzyme most commonly used for real-time PCR is a cloned version of the enzyme that has been mutated to remove the 3'–5' proof reading activity. *Taq* is a DNA-dependent DNA polymerase and is a very processive enzyme compared to other commercially available thermostable DNA-dependent DNA polymerases, which accounts for its popularity for use in real-time PCR. *Table 1.2*, kindly supplied by Dr Duncan Clark, GeneSys Ltd., lists comparative information for the most commonly used thermostable DNA polymerases for real-time qPCR (quantitative PCR) as well as *Pfu* from *Pyrococcus furiousus* used for standard PCR applications as a reference. The list is not intended to be totally inclusive of all available thermostable DNA polymerases.

There are two versions of *Taq* commercially available, the engineered polymerase and a hot-start enzyme. Hot-start enzymes may have added components that keep the polymerase from working until they are inactivated at a high temperature over a period of 1–15 minutes. Common additives are either one or more antibodies directed against the active site of the polymerase or one or more heat labile small molecules that block polymerase activity. Also, there are enzymes with an engineered mutation(s) that requires heating for enzymatic activation. Hot-start enzyme activation is accomplished by the high temperature (94°–95°C) of the first denaturing step in the PCR cycle. The difference among them is in how long the enzyme must stay at the high temperature for significant enzymatic activation to occur.

A hot-start enzyme is important for some PCRs as most, if not all, the false priming occurs within the first cycle while the components are being heated past the annealing temperature for the first time. Until the reaction mixture reaches the lowest temperature of the thermocycling program, usually 55° to 60°C, primers can bind and prime new strand synthesis more easily from the incorrect sequence causing false priming. Once a primer has initiated false priming, the sequence of the primer is incorporated into the rogue sequence. Should the other primer also manage to false prime from

Table 1.2 Commonly used DNA polymerases for PCR

	Taq DNA polymerase	Tth DNA polymerase	Stoffel fragment	KlenTaq fragment	Pfu DNA polymerase
Organism	Thermus aquaticus YT1	Thermus thermophilus HB8	Thermus aquaticus YT1	Thermus aquaticus YT1	Pyrococcus furiosus
Molecular weight	94 kDa	94 kDa	61 kDa	63 kDa	90 kDa
Number of amino acids	832	832	544	555	775
Single chain or subunits	Single	Single	Single	Single	Single
Extension rate	2 kb–4 kb/min	2 kb–4 kb/min	2 kb–4 kb/min	2 kb–4 kb/min	1 kb–2 kb/min
Reverse transcriptase activity	Minimal/low	Yes, Mn2+ dependent	Minimal/low	Minimal/low	No
Half life @ 95°C	40 mins	20 mins	80 mins	80 mins	>4 hrs
Processivity	50–60 bases	30–40 bases	5–10 bases	5–10 bases	15–25 bases
5′–3′ exonuclease activity	Yes	Yes	No	No	No
3′–5′ exonuclease activity	No	No	No	No	Yes
Incorporates dUTP	Yes	Yes	Yes	Yes	No
Extra A addition	Yes	Yes	Yes	Yes	No

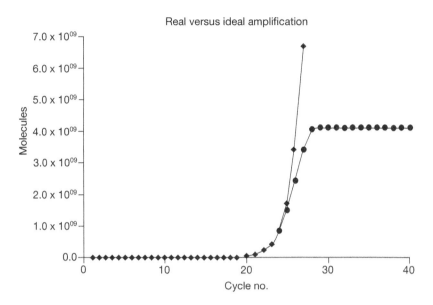

Figure 1.1

Graphical representation of the fluorescent signals from an ideal verses an actual reaction over 40 cycles of real-time PCR. In an ideal PCR, there are two phases: a baseline where the signal is below the level of instrument detection followed by a persistent geometric increase in fluorescence that continues over the remaining cycles of the experiment. However in an actual PCR, there are four phases. As in the ideal reaction, there is a baseline followed by a geometric phase. However, the amplification becomes less than ideal leading to a linear phase and finally a plateau where no further increase in signal occurs over the remaining signals. ◆ – Ideal PCR, ● – Simulated real amplification curve.

the rogue sequence, an unintended PCR product will result as both primer sequences have been incorporated into the PCR product. However, if the DNA polymerase is not active when the reaction is first heating, there will be no opportunity for false priming to occur at that time. This does not mean that there will never be false priming with hot-start enzymes but it will greatly reduce the number of falsely amplified products. Hot-start enzymes are important when the complexity of the DNA or cDNA population being amplified is very high. Examples are genomic DNA or cDNA made from oligo-dT or random primers. Members of our laboratory have empirically shown that cDNA made from assay-specific primers does not require a hot-start *Taq*.

Initially, the running of an RT-PCR or PCR experiment was akin to a black box. Reagents were put into a tube with the template, run using the appropriate program on a thermocycler and, if everything went well, a band would appear at the appropriate place on a gel following electrophoresis, at the size of the expected PCR product. With the advent of real-time PCR, it has become possible to look into the tube during the process and 'see' what is happening to the template as it is amplified through multiple PCR cycles. As a result, we have learned a lot about how the reactants, templates and primers affect the final outcome.

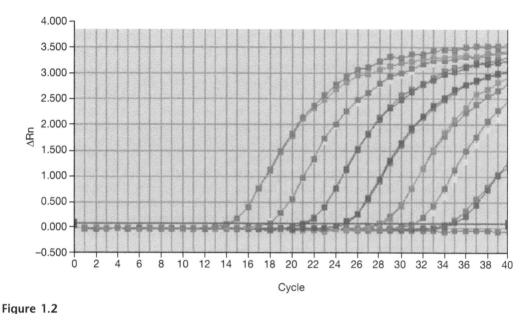

Figure 1.2

Sample dilution series of an oligonucleotide standard over a 7-log range, from a real-time PCR experiment. Each amplification curve illustrates the four phases of a polymerase chain reaction experiment: *baseline*, signal being made but not detectable by the instrument; *geometric*, detectable signal with maximal PCR efficiency; *linear*, post-geometric with slowly declining PCR efficiency; and *plateau*, no or very little new product made.

Prior to embarking on a real-time experiment, it is useful to acquire an understanding of the basic principles underlying the polymerase chain reaction. The amplification of any template is defined by four phases: 1 – baseline; 2 – exponential; 3 – linear and 4 – plateau. The baseline phase contains all the amplification that is below the level of detection of the real-time instrument. Although there is no detectible signal, exponential amplification of the template is occurring during these cycles. The exponential phase is comprised of the earliest detectible signal from a polymerase chain reaction where the amplification is proceeding at its maximal exponential rate. The length of this phase depends on the template concentration and the quality of the real-time assay. In an ideal reaction there are 2 complete molecules synthesized from every one template available in the exponential phase. This is the definition of an assay that is 100% efficient. In the linear phase, the amplification efficiency begins to taper off. *Figure 1.1* shows an ideal amplification curve that remains at 100% efficiency versus a more realistic one showing a linear and plateau phase. As can be seen, the linear phase is also a straight line but amplification is no longer 2 products from every 1 template molecule in each cycle but rather degrades to 1.95 from 1 and gradually declines in template replication efficiency further until the fourth phase, the plateau, is reached where amplification rapidly ceases for the remaining cycles of the experiment. *Figure 1.2* shows the amplification curves for a real-time PCR assay over a 7-log dilution series of synthetic oligonucleotide DNA template. Note that all the curves are parallel to one another showing that the amplification for each template dilution has the same kinetics. The four phases of amplification can clearly be seen within each individual curve.

It is not clear why the PCR goes through the linear and plateau phases. There are more than adequate reactants to sustain the PCR at an ideal rate for many more cycles than are typically seen in a 40 cycle experiment. There has been speculation in that there are 'toxic' products built up after so many cycles and these lead to a decline and cessation of product amplification. It is more likely that the highly amplified product concentration in the reaction favors association over dissociation and the denaturation of the DNA polymerase over a large number of cycles both lead to a slow down and eventual cessation of amplification. An alternative to exponential amplification, LATE PCR (linear after the exponential PCR) (Sanchez *et al.*, 2004; Pierce *et al.*, 2005), adds one primer at a reduced concentration, but comparable Tm, compared to the other primer so that the PCR proceeds at an exponential rate until it reaches a detectable level. The reaction then runs out of one primer and proceeds by copying one strand only each cycle using the remaining primer. This is called linear amplification. By slowing the reaction to a linear amplification, the occurrence of the plateau phase is greatly delayed to non-existent.

It is worth taking a moment to discuss what happens during the first two cycles of any PCR. If the template is cDNA or a single stranded DNA oligo, the first cycle of the PCR will make every molecule at least partially double stranded, depending on the length of the template and the length of time in the cycle for DNA synthesis. The main point is there is no amplification occurring during the first PCR cycle in this case. On the other hand, if the template is double stranded, like genomic DNA or a double-stranded DNA viral genome, amplification will commence during the first cycle, as both primers will have a template for synthesis. This is particularly important to keep in mind if one is using a double-stranded standard to quantify a single-stranded cDNA template.

All the real-time instruments on the market today are based on the detection of a fluorescent signal. The increase in fluorescence is directly proportional to the increase in the amplified product during the PCR. Fluorescent molecules absorb light as photons within a narrow wavelength range of light. The wavelength at which the dye absorbs light maximally is called the excitation wavelength for that molecule. Following excitation, the molecule is pushed to a higher energy state. This higher energy state is transient and short lived. The excited molecule rapidly decays, falling back to the ground energy state. When this occurs a photon of light is emitted at a longer wavelength. The light that is released is at the emission wavelength. This shift between the excitation and emission wavelengths is called a Stoke's shift (Lakowicz, 1983). For every fluorescent dye, there is an optimal excitation and emission wavelength. A fluorescent molecule can be excited or detected in a narrow range of wavelengths around these optima. Fluorescent molecules with the greatest Stoke's shift are the most desirable as they allow the cleanest separation of the excitation from the emitted wavelengths of light.

A major requirement for any fluorescent assay is that the initial and final signal intensities have as large a difference as possible. This is called the assay delta. All fluorescent assays used for real-time PCR achieve this delta by utilizing FRET (fluorescence resonance energy transfer) (Selvin, 1995). FRET requires two molecules that can interact with one another, at least one of which must be capable of fluorescence. The fluorescent component is called the donor and the second molecule is called the acceptor. During FRET, the donor fluorescent dye is excited by an external light source at or near its optimal excitation wavelength and then emits light at a shifted, longer wavelength (Stoke's shift), as described above. Instead of being detected by the instrument, the emitted light is used to excite an acceptor molecule, which is in close physical proximity. The acceptor molecule absorbs the emitted light energy from the donor dye, effectively quenching the donor signal. The wavelength emitted by the donor molecule must be near the absorbance maximum of the acceptor molecule. The acceptor molecule may or may not emit light. If light is emitted by the acceptor, it will be further shifted and a longer wavelength from that emitted by the donor. The acceptor signal will be detected by the real-time instrument, but it will not be recorded as a reporter signal by the software. FRET depends on the donor and acceptor molecules being in close proximity (10–100 Å) and falls off with the sixth power base 10 of the distance between the two molecules. The other major requirement is that the excitation wavelength of the acceptor be close to the emitted wavelength of the acceptor dye (Didenko, 2001). Some of the most common donor and acceptor (reporter and quencher) dyes currently used in real-time PCR are listed in *Table 1.3*.

There are three classes of fluorescent molecules used in real-time PCR. The three types are defined by their function within an assay. The first is the donor dye and is usually called the reporter. The fluorescent signal from the reporter is the one that is monitored during the course of the experiment. Second, is the acceptor or quencher molecule and is responsible for the initial quenching of the reporter signal. Last is the reference dye. The reference dye is common to all reactions, does not interact with the assay components and is used to normalize the signal from well to well in software.

In theory, any fluorescent dye can be a reporter in an assay. The one used most commonly is 6-FAM (6-carboxy fluorescein). This dye is efficiently excited at 488 nM, the wavelength produced by the argon-ion lasers of the original ABI real-time instrument, the now retired 7700, and the current ABI 7900. However, 6-FAM has remained the first choice for other instruments as well because it can be easily conjugated to oligonucleotides and gives a strong signal. Another reporter in wide use today is SYBR® Green I. Unlike 6-FAM, SYBR® Green I is a free dye in the PCR and works by providing a dramatic increase in fluorescence when bound to double stranded DNA. Examples of some of the many assay chemistries available for real-time PCR will be discussed in Section 1.6.

Quencher molecules can be fluorescent dyes or any molecule that can absorb light energy within the appropriate wavelength range. The original quencher dye used with the 6-FAM reporter was TAMRA (6-carboxy-tetra-methylrhodamine). When coupled to the ends of an oligonucleotide, the FAM signal is effectively quenched by the close proximity of TAMRA, due to oligo folding, while in solution. There are also dark dye quenchers that, as the name implies, quench the reporter signal but emit no light. The first of these was DABSYL (4-(dimethylamino)azobenzene-4'-sulfonyl chloride). More recently, other dark quenchers have come onto the market. Biosearch Technologies markets a series of black hole quenchers (BHQs); Integrated DNA Technologies has two forms of Iowa Black, and Nanogen (formerly Epoch Biosciences) uses its Eclipse dark quencher. It has long been recognized that fluorescein dyes are quenched by guanidine residues. This is the working principle behind the LUX system from Invitrogen (see below) and why probes with fluorescein reporters should never begin with a G residue.

The purpose of a reference dye is to monitor the fluorescence signal from each well and correct for any well-to-well differences in detection efficiency within the instrument. This was particularly important for the ABI 7700, which has a unique fiber optic cable delivering and receiving the excitation and emitted light for each of the 96 wells. The mechanism of sending and receiving light from the plate is quite different in the host of real-time instruments available today. Many still use a reference dye to ensure that the signal from each well of the plate is balanced during data analysis. Depending on the light path used by the instrument and the quality of the thermocycler used, there can be plate 'edge effects' which will require well-to-well normalization. However, if the 'edge effects' become too pronounced, differences between outer and internal wells cannot be normalized by a reference dye. The most common reference dye is ROX (6-carboxy-X-rhodamine). ROX can be found as a short 6-FAM/ROX FRET-based oligo (ABI), a free dye conjugated with glycine (Invitrogen) or as a polyethylene glycol conjugate sold as Super-ROX (Biosearch Technologies). ROX was initially used when 6-FAM was the sole reporter dye. With multiplexing becoming more common, it should be pointed out that ROX does not have to be the dye used for well-to-well normalization as ROX emits light in the middle of the valuable red spectrum. In theory, any dye with a unique spectral signature compared to the ones in the assay could be used. However, it would have to be available in a format that kept it in solution. The Bio-Rad instruments normalize the signal using a second plate with a fluorescein dye prior to the real-time run instead of an internal reference dye in the master mix on the experimental plate.

1.5 Real-time PCR instrumentation – an overview

It would be impractical to present technical information for every real-time instrument on the market today. However, it is important that users of real-time instruments have a basic understanding of how their instruments work and have some understanding of their physical capabilities and limitations. There are three major components in any real-time instrument: 1) the light source, which determines the range of reporter dyes the instrument is capable of using; 2) the detection system, where the spectral range and sensitivity of any assay will be determined; and 3) the thermocycling mechanism, the determinant of the speed at which an assay can be run, the uniformity of the temperature changes from sample to sample and the number of samples that can be accommodated at any one time.

There are currently four different light sources available in real-time instruments, argon-ion laser; LED (light emitting diode) lasers, quartz halogen tungsten lamps and xenon lamps. Each of these different light sources will determine the capabilities of the instruments that use them. The argon-ion laser, now available solely in the ABI 7900, emits powerful light primarily at 488 nm. For reporters such as 6-FAM, this is an ideal excitation wavelength. However, as you move toward the red spectrum, the excitation energy from a 488 nM light source will result in a weaker stimulation of a reporter dye with maximal excitation above 500 nM with a concomitant weaker emission signal. Weaker signals can limit the utility of a reporter dye. The ABI 7900 captures the emitted reporter signal using a CCD (charge-coupled device) camera and relies on software to parse out multiple emitted wavelengths. The ABI 7900 utilizes a Peltier-based thermocycler. There are currently 4-exchangable thermocycler blocks available for 96- and 384-well plates.

A single LED laser can be found in the LightCycler® 1 and LightCycler® 2 from Roche. Unlike classic lasers, LED lasers emit light within a 30–40 nM spectral range. The energy output of these lasers is not as strong as the argon-ion lasers but their energy use and generated heat are significantly less with a lower cost per unit. Both LightCycler®, 1 and 2, use a single blue LED laser as an excitation source. The LightCycler® have a number of photodetection diodes, each of which is specific for an increasing and narrow wavelength range of light. Unlike the plates used in the block-based ABI 7900, the LightCycler® uses glass capillary tube reaction vessels with a high surface to volume ratio. Coupled with the hot air driven thermocycler system, these machines can complete a 40 cycle experiment in about 30 minutes compared to thermocycler/plate-based instruments times of approximately 2 hours.

Currently, the most used light source for new instruments is a quartz tungsten halogen lamp. These lamps are capable of emitting a steady beam of light from 360 nM to over 1,000 nM. They are sometimes referred to as 'white light' sources as they emit light over the entire visible spectrum. Unlike either of the laser light sources described above, two sets of excitation and emission filters are required to select the desired wavelengths for multiple reporter dyes. The number of filters available determines the level of multiplexing the instrument can support. Some of these instruments use a PMT (photomultiplier tube) and others a CCD camera for signal detection depending upon how the light is captured from the plate.

One of the newest instruments coming onto the market, the Roche LightCycler® 480, has a xenon lamp, which has a higher light intensity compared to the quartz tungsten-halogen lamp and covers a similar spectral range. This machine will have 5 excitation filters coupled to 6 emission filters which will potentially allow 6 different assays in a single, multiplexed reaction. This instrument will support both 96- and 384-well plates in a Peltier-based thermocycler that is reported to have the most uniform thermal characteristics of any instrument.

Table 1.4 provides a selected overview of some of the most important features for the newest instruments available from the major real-time PCR instrument vendors. The ABI 7700, the original real-time instrument, is also listed for comparison.

Besides the hardware itself, the next most important component of a

Table 1.3 Dyes available for use in real-time PCR

Free dyes	Max. ab (nM)	Max. em (nM)
SYBR® Green I	497	525
EvaGreen™	497	525
BOXTO™	515	552

Reporter dyes	Max. ab (nM)	Max. em (nM)
Pulsar® 650	460	650
Fluorescein™	492	520
6-FAM	494	518
Alexa 488™	495	519
JOE™	520	548
TET™	521	536
Cal Fluor Gold 540™	522	544
Yakima Yellow™	530	549
HEX™	535	556
Cal Fluor Orange 560™	538	559
VIC™	538	554
Quasar® 570	548	566
Cy3™	552	570
TAMRA™	565	580
Cal Fluor Red 590™	569	591
Redmond Red™	579	595
ROX™	580	605
Cal Fluor Red 635™	618	637
LightCycler®640	625	640
Cy5™	643	667
Quasar® 670	647	667
LightCycler®705	685	705

Dark dyes	Max. ab (nM)	Max. em (nM)
DABCYL	453	None
BHQ0™	495	None
Eclipse™	522	None
Iowa Black™ FQ	531	None
BHQ1™	534	None
BHQ2™	579	None
Iowa Black™ RQ	656	None
BHQ3™	680	None

Pulsar, Cal Fluor, BHQ, Quasar and Pulsar are registered trademarks of Biosearch Technologies; Alexa is a registered trademark of Molecular Probes/Invitrogen; Iowa Black is a registered trademark of IDT; Redmond Red, Yakima Yellow and Eclipse are registered trademarks of Nanogen; LightCycler®640 and LightCycler®705 are registered trademarks of Roche; Evagreen is the registerd trademark of Biotium, Inc.

real-time instrument is the software package. This aspect of any real-time instrument should not be overlooked. An instrument with good hardware but a poor software package can diminish the overall user experience. Although many software packages offer different levels of automated data analysis, it should be pointed out that the results may not be as good as manual manipulation of the baseline and threshold settings by the investigator (see Chapter 2). Most programs offer ways to export the metadata into a file that can be opened in Microsoft Excel® and the graphical views into one or more common graphical formats. These exports can be very useful for more extensive data analysis in other programs and for incorporating data or graphics in either publications or presentations (see Chapters 2 and 3). For these reasons, it is important that the user explore all aspects of the software to ensure they are getting the most out of the data obtained by their real-time instruments.

The trends in real-time PCR instrumentation development have been in three directions. First has been an evolution in fluorescent dye detection. In the early years of real-time PCR, assays were run with one assay per reaction. This was primarily because the hardware was geared to the excitation of 6-FAM as a reporter. As more and more instruments came onto the market with more extensive capabilities in their spectral sophistication, it has become possible to excite a larger spectrum of reporters. Today, many instruments offer 5 or 6 unique excitation/emission filter combinations. In theory, this would allow up to 5 or 6 transcript or gene assays to be measured within a single multiplexed reaction. The emitted light from each reporter dye is physically isolated from the others by the filters as it is collected. This minimizes spectral spill over which was a real problem with the early instruments that relied entirely on software to differentiate signals from different reporter dyes. Thus, the instrumentation is now equipped for extensive assay multiplexing. The chemistry is just now starting to catch up and will be discussed in the next section.

The second development in real-time instrumentation has been towards faster cycling capabilities. For the Roche LightCycler® 1 and 2 and the Corbett Rotorgene, fast cycling has always been a feature because samples were not heated in a block thermocycler. However, these machines offered less than 96 sample capacities per run cycle (*Table 1.4*). The manufacturers of instruments with block thermocyclers are now offering faster cycling times coupled to new assay chemistries. This trend will undoubtedly reduce the time it takes to run each plate. A reduction in run times becomes more critical for high throughput users and will most likely lead to further reductions in run times.

The third development is the increase in sample throughput. For some time now, the ABI 7900 has been the only real-time instrument on the market capable of running 384-well plates. With the announcement of the new Roche LightCycler® 480, that will no longer be the case. The LightCycler® 480 will have the added advantage of offering true multiplexing capabilities for up to 6 dyes in each reaction. Although 384-well plates may seem overkill for the individual laboratory, for the pharmaceutical industry, even 384-well plates cannot provide the desired throughput. For true high throughput screening, 1,536-well plates are now the standard. For research core laboratories that utilize liquid handling robots to set up the reactions, however, 384-well plates are well positioned to raise the throughput bar over individual laboratories where 96-well plates have become the upper limit for hand pipetting.

The combination of multiplexed reactions, faster cycling times and higher sample throughputs per run through the use of robotics mean that the throughput potential of real-time PCR has been amplified at least 40-fold from the original instruments. What the throughput upper limit for real-time instrumentation will be is not clear at this time. What is clear is that limit has not yet been achieved.

Table 1.4 A comparison of features for the most commonly used real-time instruments

Instrument/configuration	ABI Prism® 7700	ABI Prism® 7500	ABI Prism® 7900	Bio-Rad iQ®5	Bio-Rad MJR Chromo 4™
Light source	Argon laser	Tungsten-halogen	Argon laser	Tungsten-halogen	4 LEDs
Detector	CCD camera	CCD camera	CCD camera	CCD camera	4 photodiode detectors
Scanning system	96 fiber optic cables	Whole plate	Scanning head by row	Whole plate	Scanning photonics shuttle
Excitation wavelengths (s)	488 nm	450–650 nm	488 nm	400–700 nm	450–650 nm
Emission wavelengths	500–650 nm	500–700 nm	500–660 nm	400–700 nm	515–730 nm
Multiplex capability (number)	2	5	2	5	4
Sample vessel	96-well plate	96-well plate	96/384-well plates	96-well plate	96-well plate
Sample volume	25–100 µl	20–100 µl	20–100 µl/96 5–20 µl/384	10–50 µl	20–100 µl
Temperature control	9600 Thermocycler	Peltier Thermocycler	Peltier Thermocycler	Peltier-Joule	Peltier Thermocycler
Sensitivity	9-logs	9-logs	9-logs	6-logs	10-logs

1.6 Detection chemistries used in real-time PCR

There are three basic methodologies commonly used in the detection of RNA or DNA targets by real-time PCR and all of them utilize fluorescent dyes. In each case, a low initial fluorescent signal is increased proportionally during each succeeding PCR cycle in tandem with the exponential increase in the DNA product(s) formed.

The simplest assay system involves the incorporation of a free dye into the newly formed double-stranded DNA product. The most used dye for this purpose in real-time PCR is SYBR® Green I (Molecular Probes/Invitrogen). The background fluorescence from SYBR® Green I when in solution as a free dye and stimulated by light of the appropriate wavelength is very low. The same is true for single-stranded nucleic acids at the concentrations used for real-time PCR. In contrast, as the double-stranded DNA product is formed, SYBR® Green I binds to the minor groove of the double-stranded DNA. The DNA-dye complex results in a dramatic increase in fluorescence output, when properly illuminated, of roughly 2,000 times the initial, unbound,

Cepheid SmartCycler®	Corbett Rotor-Gene™	Eppendorf Mastercycler® realplex[4]	Roche LightCycler® 2	Roche LightCycler® 480	Stratagene Mx®3005p	Techne Quantica™
4 LEDs	4 LEDs	96 LEDs	Blue LED	Xenon lamp	Tungsten-halogen	Tungsten-halogen
4 Silicon photo-detectors	4 PMTs	2 channel PMTs	Photodiodes	CCD camera	PMT	4 PMTs
All wells	Through tube	96 fiber optic cables	Through tube		Plate scanning	Whole plate
450–650 nm	470–625 nm	470 nm	480 nm	450–615 nm	350–635 nm	460–650 nm
500–650 nm	510–665 nm	520–605 nm	530–710 nm	500–670 nm	440–610 nm	500–710 nm
4	4	4	6	6	5	4
16 independent wells	36/72-well rotors	96-well plate	32 Capillaries	96/384-well plates	96-well plate	96-well plate
25 and 100 µl	100–200 µl	10–50 µl	10–20 µl/small 100 µl/large	20–100 µl/96 5–20 µl/384	25 µl	15–50 µl
Solid state heating/ forced air cooling	Forced air heating/ cooling	Peltier Thermocycler	Heater & fan for cooling thermocycler	Peltier Thermocycler	Modified Peltier	8-Peltier
Not stated	12-logs	9-logs	9-logs	8-logs	10-logs	9-logs

fluorescent signal (*Figure 1.3*). This assay system has a very good signal to noise ratio. The popularity of SYBR® Green I assays with real-time PCR users is due to three factors: 1) low cost for the dye; 2) ease of assay development, only a pair of primers is required; and 3) the same detection mechanism can be used for every assay. The down side is that every double-stranded molecule made in the reaction will generate a signal, such as primer dimers or inappropriate PCR products. This fact puts a high premium on good primer design and careful quality control during assay development. It also means that a hot-start DNA polymerase is a must to lessen or eliminate extra-assay signals. A new dye, EvaGreen™ (Biotium, Hayward, CA) has been presented for use in real-time PCR. EvaGreen™ costs more than SYBR® Green I but it is reported to show little PCR inhibition which can be a problem with high concentrations of SYBR® Green I, substantially higher fluorescence over SYBR® Green I, and greater stability at high temperatures. Whether these characteristics will be enough to entice users away from SYBR® Green I remains to be seen. A third new free dye, BOXTO (TATAA Biocenter, Gothenburg, Sweden), has come onto the market. Unlike SYBR®

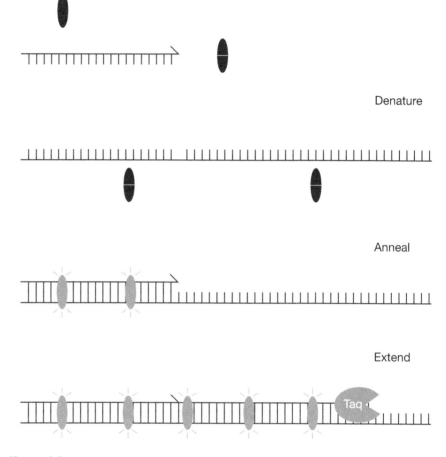

Denature

Anneal

Extend

Figure 1.3

Graphical representation of the incorporation of SYBR® Green I dye resulting in an increase in fluorescent signal during the PCR. Free dye has very low fluorescence and will not bind to single stranded or denatured DNA. During primer annealing, a double-stranded structure is formed and SYBR® Green I dye is bound resulting in a dramatic increase in fluorescent signal. During primer extension by *Taq* DNA polymerase, the fluorescent signal increases proportionally to the number of SYBR® Green I dye molecules bound per double-stranded molecule. The process is repeated in each cycle with increasing total fluorescence.

Green I or EvaGreen™, BOXTO is optimally excited and emits light at higher wavelengths than SYBR® Green I or EvaGreen™. In principle then, it could be used in the same reaction with a FAM conjugated probe for the detection of primer-dimers or to perform a melt-curve on PCR products, neither of which is possible with a probe-based assay. The most recent additions to the DNA binding dye choices are LCGreen™ I and LCGreen™ PLUS (Idaho Technology, Salt Lake City, UT) and will be discussed in

Chapter 9. More details of the use of SYBR® Green I in real-time PCR can be found in Chapter 8.

There are many quite different dye-primer based signaling systems available for real-time PCR. They range from the very simple LUX™ (light upon extension, Invitrogen) primers to more complex externally and internally quenched primers to the very complex structure of scorpion primers. It is beyond the scope of this chapter to describe each of these primer-based systems (Bustin, 2004). The template specificity for the dye-primer-based assays is the same as for SYBR® Green I. The exception to that rule is the scorpion primer, where the signal generated by the primer is dependent on a complementary match with sequence located within the PCR amplicon. Dye-primer-based assays do allow for multiplexing which is not possible with SYBR® Green I, EvaGreen™ or BOXTO and they do not require the design of another, intervening, fluorescently labeled probe (see probe-based assays below).

LUX™ primers contain a 4–6 base extension on the 5′ end of the primer that is complementary to an internal sequence of the primer near the 3′ end. The extension overlaps the position of a fluorescein dye (FAM or JOE) attached to a base near the 3′ end of the primer. The design of LUX™ primers initially was a somewhat hit or miss proposition. However, new algorithms used by the program at the Invitrogen web site now make their successful design more assured for the investigator. The reporter dye signal is quenched when in close proximity to the double stranded stem formed within the primer while in solution. Quenching of the fluorescein signal is particularly efficient when in close proximity to guanidine residues. During the first PCR cycle, the LUX™ primer is incorporated into a new strand. At this time, the stem can still be in place. However, when a new complementary second strand is made, the stem and loop are made linear. This structural change results in an increase in fluorescent signal of as much as 10-fold over background (*Figure 1.4*).

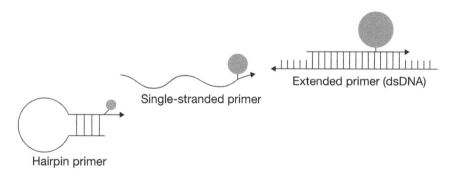

Figure 1.4

Structure and mode of action of a LUX™ (light upon eXtension) fluorescently labeled primer. LUX™ primers are inherently quenched due to a designed, short self-complementary sequence added to the 5′ end of the primer. The fluorescent moiety on the 3′ end of the primer is quenched in this confirmation. Following the melting and subsequent annealing of the primer to the template, the primer is made linear and is incorporated into the new DNA strand leading to a significant, as much as ten-fold, increase in fluorescence.

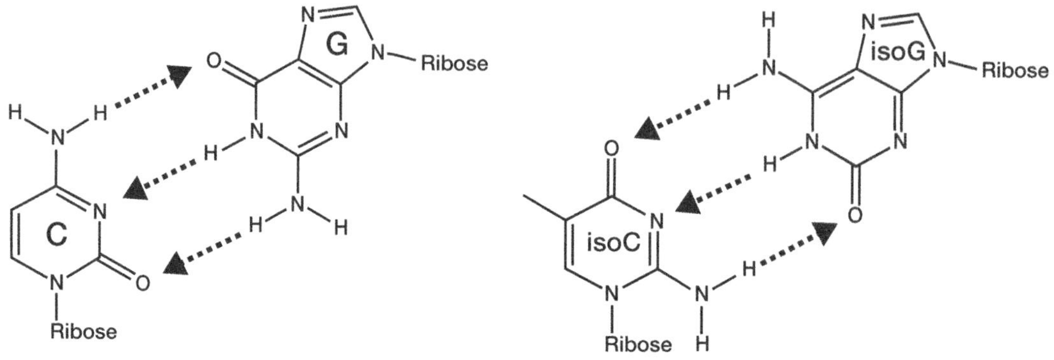

Figure 1.5

Structures of the natural guanidine and cytosine bases showing hydrogen bonding compared with the hydrogen bonded iso-base structures. The iso-bases will only base pair with themselves and not the natural dG and dC residues nor dA, dT or dU. They are recognized as dNTPs and incorporated by DNA polymerase along with natural DNA bases into newly synthesized DNA.

A new primer-based assay system that is just coming onto the market is the Plexor system from Promega. The basis for this assay is the use of two isomers of the bases cytodine and guanine, iso-C and iso-G (*Figure 1.5*) (Sherrill *et al.*, 2004). The iso-bases will only base pair with the complementary iso-base and DNA polymerases will only add an iso-base when the cognate complementary iso-base is present in the existing sequence. One primer is synthesized with a fluorescently labeled iso-C on the 5′ end. The PCR master mix contains free iso-dGTP coupled to a dark quencher dye (DABSYL). As each amplification cycle progresses, the fluorescent signal from the free fluorescently tagged iso-C primers is progressively quenched as the labeled primers are incorporated into the growing amplified PCR products (*Figure 1.6*). The quenching is accomplished following the synthesis of the complementary strand bearing the iso-G-dark dye. Thus, unlike all of the preceding assays where the fluorescent signal goes up, in this assay the reporter signal goes down with increasing PCR cycles. For this reason, special software has been written to facilitate data analysis using the exported raw data file formats from several real-time instruments. One advantage of this system is the ability to multiplex a large number of assays with only primers using a universal quencher (iso-dGTP-DABSYL). Like all primer-based assay systems, the quality of the assay is completely dependent on the specificity of the primers. Therefore, good assay (primer) design will be critical for assay specificity. On the other hand, it may be easier to multiplex five assays in a reaction without the complexity of five probe sequences along with ten primer pairs. This new assay system will have to be tried and put to the test by the real-time community to see if it will become as ubiquitous as the SYBR® Green I and TaqMan® assays are now.

The third category of signaling systems for real-time PCR are those involving a third, fluorescently labeled oligonucleotide located between the primers called a probe. Unlike the two assay systems above that consist

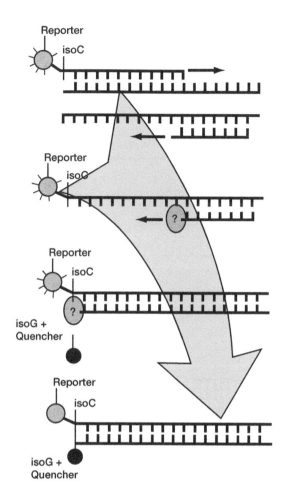

Figure 1.6

Graphic illustrating the operating principle underlying the Plexor® real-time PCR assay. One primer is labeled at the 5′ end with an iso-dC and a fluorescent reporter dye. The primer binds to the complementary template and is extended. In the next cycle, a complementary strand is made during the PCR, using a second, unlabeled primer. An iso-dG-DABCYL (4-(dimethylamino)azobenzene-4′-carboxylic acid) base, present in the PCR master mix along with the four natural dNTPs, is incorporated as a complement to the iso-dC in the new strand, The DABCYL dark dye, now physically very close to the reporter, quenches the signal from the reporter dye.

solely of a pair of primers, probe-based assays have the added specificity of an intervening third (or fourth, see below) oligo called the probe. This is because the probe is the fluorescently labeled component of these assays. A further advantage of a probe-based assay is extraneous signals from primer dimers that will be detected by free dye or dye-primer-based assays are not detected by probe-based assays. Extra-assay DNA products larger than primer dimers will also not be detected. The only PCR amplicon that can be

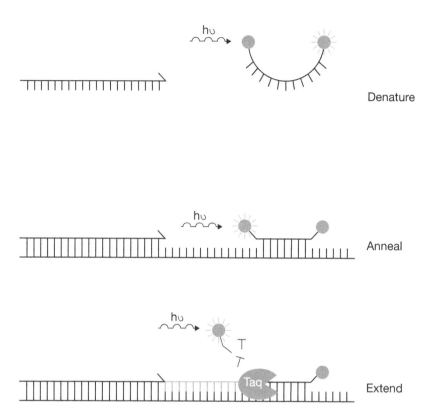

Figure 1.7

Graphical illustration showing fluorescent signal generation from a TaqMan® probe. When free in solution, the reporter fluorescent dye is quenched by a second molecule using FRET. As the reaction cools toward the annealing temperature, the probe binds to a complementary template sequence at a higher temperature than the primer on the same strand. During primer extension by *Taq* DNA polymerase, the probe is displaced and degraded by a 5′ nuclease present in *Taq* DNA polymerase. Release of the reporter dye from the probe molecule and its proximity to the quencher dye allows the full signal of the reporter dye to be realized.

detected by a probe-based assay are those to which the primers and the probe are both able to bind simultaneously. There are five distinct probe-based assay systems in use today, TaqMan® probes, Molecular Beacons®, minor groove binding (MGB) probes, Locked nucleic acid (LNA) probes and hybridization probes. In each case, multiple reporter dyes can be used with a variety of quenchers *(Table 1.3)* to form efficient FRET pairs. Thus, there are many opportunities for multiplexing with probe-based assays.

TaqMan® or hydrolysis probes are linear oligonucleotides that have historically been labeled with a reporter on the 5′ end and a quencher on the 3′ end, although the opposite orientation can and has been used. When the reporter and quencher dyes are in close proximity in solution, the reporter signal is quenched. The efficiency of the quench is determined by the probe sequence which in turn determines how efficiently the ends of the probe associate in solution. When the probe anneals to its complementary target sequence, the two dyes are maximally separated and the reporter signal detected by the instrument. There are no good predictors in any software program for how well a probe will be quenched in practice. However, there is a direct and inverse correlation between probe length and quenching efficiency. For this reason, TaqMan® probes are kept to less than 30 bases in length.

It is essential that the probe anneal to the target strand sequence prior to the primers to ensure that signal can potentially be generated from every newly synthesized amplicon. *Taq* DNA polymerase binds and begins to extend the new strand very rapidly following primer binding. For this reason, hydrolysis probes are designed to have a Tm 9–10°C higher than their matched primers. Hydrolysis probes depend on the 5′ nuclease activity of *Taq* DNA polymerase for their cleavage during new strand synthesis to generate a fluorescent signal from the newly freed reporter dye (*Figure 1.7*). The TaqMan® name comes from this hydrolysis step in an analogy to the action of the old computer game character, Pacman. New thermal stable DNA polymerases have been introduced from other genera of bacteria that do not contain 5′ nucleotidase activity. They work well for PCR but cannot cleave hydrolysis probes.

Molecular beacons are similar to TaqMan® probes in that they are labeled on each end with reporter and quencher moieties and, like TaqMan® probes, their Tm is higher than the primers used for the assay. Unlike hydrolysis probes, molecular beacons do not depend upon probe cleavage for signal generation. Molecular beacons have a 4–6 base self-complementary sequence extension on each end. Thus, in solution, the ends of molecular beacons fold into a perfect stem structure bringing the reporter and quencher dyes close together forming a close FRET association. The close proximity of the two dyes leads to very efficient quenching of the reporter signal. During the annealing step, the probe becomes unfolded, the two dyes are separated and the reporter signal detected by the real-time instrument (*Figure 1.8*).

A minor groove-binding (MGB) protein bearing probes are hydrolysis probes with shorter lengths due to the MGB moiety. As the name implies, an MGB molecule is added to one end of the nucleic acid sequence, which increases the affinity and lowers the Tm requirements of the nucleic acid portion of the probe. There are two companies that sell MGB probes. Applied Biosystems sells a probe with the MGB component and quencher on the 3′ end and the reporter on the 5′ end of the nucleic acid sequence, similar to TaqMan® probes. Like TaqMan® probes, their MGB probes can be hydrolyzed by *Taq* DNA polymerase. The second company making MGB probes is Nanogen. They put the quencher and MGB component on the 5′ end and the reporter on the 3′ end of the probe. The MGB moiety effectively blocks hydrolysis of these probes. Like molecular beacons, signal is generated during the annealing step in the thermocycle. MGB probes from either company are used primarily for SNP (single nucleotide polymorphism) and allelic discrimination assays where high sensitivity to a single base mismatch is required. They can also be used for transcript analysis,

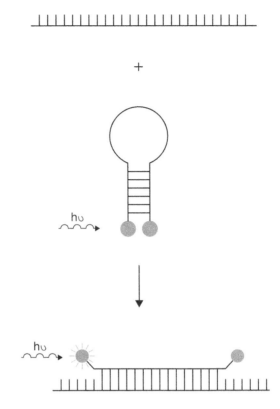

Figure 1.8

Illustration showing the mechanism of signal generation for a molecular beacon probe. Designed complementary sequences 4–6 bases in length at the 5′ and 3′ ends bring the terminally placed reporter and quencher dyes in very close proximity in solution leading to a much reduced reporter signal via FRET. Following the melting of the template and probe structures, probe molecules anneal to complementary sequences within the template as the temperature is lowered to the annealing set point, At this time, the molecular beacon is completely linear on the template, fully separating the reporter and quencher dyes and resulting in full reporter signal due to the loss of FRET. Unlike TaqMan® probes, molecular beacons do not require hydrolysis for signal generation.

particularly when the G/C content of the template is very low resulting in TaqMan® probes longer than 30 bases.

LNA probes are hydrolysis probes that are much shorter due to the higher binding affinity of locked nucleic acid bases to standard DNA (Goldenberg *et al.*, 2005). LNAs are bicyclic DNA (or RNA) analogues (*Figure 1.9*). For this reason, LNA-based probes need to be only 8–9 bases long to achieve a Tm of 68–69°C under standard assay conditions. The most extensive use of LNA probes currently is by Exiqon. It has been able to take advantage of the redundancy of nanomeric sequences in the genome to design a relatively small number of LNA probes that can be used by the investigator to design

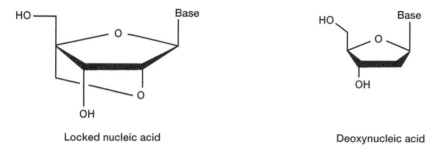

Figure 1.9

The deoxyribose ring structures of locked nucleic acids (LNAs) versus the natural bases. The Tm values for oligonucleotides with LNA bases are much higher versus those with natural bases alone.

a real-time PCR assay for every gene in many different genomes, using their web-based software. For the human genome, a set of 90 LNA probes is sufficient for complete coverage of the known transcriptome. LNA probes, due to their short length, can be used effectively in SNP and allelic discrimination assays as well.

Unlike the other probe-based assays described above, hybridization probes consist of two oligonucleotides that anneal between the primers, each with a single fluorescent dye. This probe system uses FRET to generate, rather than to quench, the fluorescent signal. The more 5′ probe oligo, closer to the forward primer sequence, has a donor dye on the 3′ end and the second probe molecule has an acceptor dye on the 5′ end. When the probes anneal, the two dyes are brought into close proximity to one another, only 1–3 bases apart. The detector dye (a fluorescein dye) absorbs energy from the light source and the resulting emitted energy is transferred to the second dye (a rhodamine dye) by FRET. It is the emitted signal from the acceptor dye that is detected by the real-time instrument and recorded (*Figure 1.10*). Hybridization probes were designed for and are used primarily in the first Roche LightCycler® although they are not limited to those instruments. This assay system works well but has the drawback of requiring an extra probe molecule, which adds to the cost and complicates assay design.

1.7 Performing a real-time RT-PCR experiment

In general, any experiment involving the PCR that begins with converting an RNA template to cDNA will be more challenging than those that begin with DNA templates. For that reason, the following discussion is centered on RT-PCR. However, the basic principles and discussion concerning PCR below would still apply to those starting with DNA.

Undoubtedly, the most important part of any RT-PCR experiment is the isolation and purification of the RNA template before the experiment is initiated. There are many methods for the isolation of nucleic acids. Most are variants on a few basic isolation themes tailored to meet the special

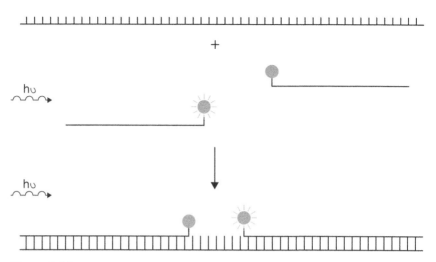

Figure 1.10

Graphical representation showing the mechanism of action for an hybridization probe pair. Two oligo probes bearing a single dye each, one with a fluorescein dye at the 3' end and the other with a rhodamine dye at the 5' end. When the two oligos anneal to a complementary template, the fluorescein dye is excited by the light source in the instrument and transfers its energy to the rhodamine dye via FRET. FRET can only occur when the two dyes are in close proximity. The instrument is set to detect the rhodamime signal.

needs of the particular organism and/or target nucleic acid of interest (Sambrook and Russell, 2001; Ausubel, 2001). There are a large number of kits on the market for RNA or DNA isolation from different organisms, tissue or cellular sources. One important factor is to minimize the amount of DNA that is carried over with the RNA during isolation. This is important because not all real-time assays cross exon junctions. Even if the assay does cross an exon junction, pseudogenes may be amplified by the assay anyway as they are spliced but imperfect copies of the original transcript reinserted into the genome. For many transcripts, there are one or more pseudogene copies in the mammalian genome. Of course for many organisms, e.g., those of viral and microbial origin, no exons exist. One sure way around this problem in all cases is to DNase I treat each total RNA preparation. Again, there are kits available for this process but it is not that hard to perform this task yourself, as described in Protocol 1.1. Be aware that DNase I treatment on a column as part of the RNA isolation procedure will not be sufficient if your real-time assay has an amplicon shorter than 100 bases. For the larger amplicons, found in SYBR® Green I or other primer-based assays that are 200–250 bases in length, treatment with DNase I on the column will suffice.

RNA from tissue culture cells can be isolated using most available methods. Isolating RNA from tissues, however, can be more problematic. Again potential complications are dependent upon the tissue involved. We have found that homogenization of the tissue using a Polytron or bead-beater device in a phenol-based guanidinium solution (Chomczynski, 1993) works best for rapid degradation of the cellular structure and inhibition of endogenous RNases. Regardless of the source, it is critical to minimize not only DNA but also protein carryover during RNA isolation. The proteins that most commonly carry over in any nucleic acid isolation procedure are nucleic acid binding proteins. These DNA or RNA binding proteins can be disruptive to the RT reaction and in particular, the following PCR. We have found that binding the RNA to a column matrix and washing away protein and other PCR contaminants results in the cleanest RNA for real-time PCR. Column-based kits are available from many manufacturers. However, we have developed a method that couples the superior disruption of a phenol-based guanidinium solution with a column-based final purification of the RNA (Protocol 1.2).

There are examples where this standard isolation method will not suffice. Many plants and some microbes contain carbohydrates that will co-purify with nucleic acids. In these instances, methods based on CTAB (Cetyltrimethylammonium Bromide) can be used to isolate RNA or DNA (Liao *et al.*, 2004). Another method for further purifying RNA is precipitation using LiCl (Lithium Chloride) instead of salt plus alcohol (Liao *et al.*, 2004). Isolation of RNA from small sample sizes, such as those from laser capture microscopy, can require special handling (Gjerdrum *et al.*, 2004). Again, there are several kits on the market for this purpose.

There is nothing about the reactants used in setting up an RT reaction and following PCR that is substantially different for analysis by real-time PCR from those reactions designed for analysis using a gel. The only differences are the concentrations of some of the reactants and the inclusion of a fluorescent moiety for detection in a real-time reaction. There are two critical elements to setting up any real-time PCR experiment, the uniformity of the components in the reaction and the accuracy and care with which each component, particularly the sample, is added to the reaction.

It is important that all the components required for the reaction be uniform from well to well if the results from multiple wells are to be compared. To accomplish this goal, the concept of a master mix is critical. Master mixes are, as the name implies, a combination of all the reactants required for the RT reaction or the PCR minus the variable in the experiment, usually the RNA, cDNA or DNA sample. Thus, all the reactants, primers and probe (if used) are combined, mixed well and dispensed in an equal volume into all the wells of a plate ensuring that each reaction is starting with the same concentration of the basic reactants.

There are many commercially available kits on the market for both the RT step and the subsequent PCR as either one-step or two-step reactions. As the names imply, a one-step kit has reactants that will perform the RT reaction and then the PCR within the same tube with no further manipulation necessary by the investigator. In a two-step reaction, the reverse transcriptase reaction is run in one tube and then all or some of the resulting cDNA is used in the polymerase chain reaction, which is performed in a second plate, tube or capillary. The advantage of a one-step method is a savings in set-up time. However, there is always a compromise between the two reactions as neither can be truly optimized. Further, all the template in a one-step reaction will be used for that one reaction. Depending on how the kit is formulated, a no-reverse transcriptase control may not possible. This important control detects DNA contamination and ensures that low-level transcript measurements are legitimate. A two-step kit takes more hands-on time, as the cDNA is set up as a separate reaction and then introduced into the PCR. In a two-step protocol, a no-reverse transcriptase control is easily performed on each sample. Both the RT and the PCR can be optimized individually and only a portion of the cDNA made need be used in any one PCR. There are times when the use of one or the other assay system makes sense and it is up to the investigator to make that decision. In general, if the same results can be obtained with a one-step method compared to the same template in a two-step procedure, the time saved and convenience using the one-step kit makes a lot of sense. If this is not the case or the template is limiting, a two-step method should be used.

In principle, kits for real-time PCR are a major convenience for the investigator. They are also a source of a readily available and consistent master mix. However, they also add substantially to the cost of performing real-time PCR. Our lab has been making master mixes by hand since 1996, when there were no kits available. We continue to do so today. Formulas for master mixes for a two-step RT and a probe-based PCR or SYBR® Green I PCR are in Protocols 1.3, 1.4, and 1.5.

For purposes of the following discussion, a two-step method will be used as an example. After distributing the RT master mix into the number of wells required for the experiment, the samples are pipetted into wells following an appropriate pattern for the experiment. This pipetting step is THE most critical part of the experiment. A slight difference in RT master mix volume from well to well will have a negligible effect on the quantitative outcome of the experiment. In contrast, a very small difference in sample volume among replicates for the same sample will have measurable effects seen in the final data set. For this reason, recently calibrated adjustable pipettors and a steady hand are the key. If the cDNA for each sample is pipetted into replicate wells for the PCR, this step is the second most critical pipetting step as errors here will be multiplied by the any errors made in the initial template pipetting step. For this reason, in the examples shown the PCR master mix is added directly to the entire cDNA reaction to speed up assay setup and to avoid further errors in pipetting.

A good way for new investigators to gauge their pipetting skills is to make a 7-log dilution series of a DNA template and then pipette the dilutions, in triplicate, into a plate bearing PCR master mix in multiple wells. There is nothing more humbling than seeing the results of this experiment for the first time. Often, the replicates are not uniform and the points of the subsequent standard curve rarely fall onto a single straight line. There are two places for pipetting error in this experiment, making the dilution series itself and placing the diluted samples into the individual wells. A mistake in either or both will be seen in the final data. With practice, the curves will become uniform and straight.

Once the PCR plate, tube or capillary sets have been filled with all the reactants, it is ready to be run on a real-time instrument. The set up of the sample positions will be dependent upon the real-time instrument and software. Although they all look different, the methodology of defining the positions and quantities for the standards, unknown samples and controls is pretty much the same for all instruments. One thing that is instrument-specific is the thermocycling conditions. Instruments that use air to heat the reactions tend to cycle much faster than instruments that use metal blocks. There are new fast-blocks for some instruments in an attempt to match the faster air heated systems. The important thing is to use the cycling conditions that match your instrument, the reagents and the assay components, mostly the primers. The first temperature is the melting step. A one-minute melt at 94–95°C is sufficient for assays with short amplicons (<300 bases) using standard *Taq*. If a hot-start *Taq* is used, up to 15 minutes will be required for enzyme activation depending on the activation system used for the enzyme. For genomic DNA, at least 3 minutes is a safe time to obtain a good melt for the PCR. Next comes the annealing temperature, which will depend primarily upon the Tm of the PCR primers, their concentration and the $MgCl_2$ concentration in the master mix. A good starting temperature range is 55–60°C. If primer-dimers or inappropriate PCR products are a problem, raise the annealing temperature in 2°C increments until the problem subsides or the reaction stops working. It may be necessary to design a better primer pair if this does not solve the problem. All probe-based assays can be run using 2-step cycles. That is from 94–95°C to 60°C and back to 94–95°C. Good cycling times for these reactions are 12–15 seconds at the high temperature followed by 30 seconds at the lower temperature, usually for 40 cycles. For dye- or primer-based assays, such as SYBR® Green 1, a 3-step cycle is generally used. Here, the cycle stays at the annealing temperature for only 2 seconds and then goes up to 72°C (the optimal temperature for *Taq* DNA polymerase) for 30 seconds for each cycle (94–95°C, 12–15 seconds; 55–60°C, 2 seconds and 72°C, 30 seconds). Collecting data at the higher temperature helps to melt out primer-dimers.

This practical introduction to running a real-time PCR experiment covered the basics of setting up a reaction. There is, of course, much more to know about running and analyzing the data from a real-time experiment. Those topics will be covered in succeeding chapters of this book.

1.8 What lies ahead

In this introduction, an attempt has been made to discuss some of the key aspects that underlie the theory and practice of real-time PCR. It is far from complete in covering all aspects necessary for the completion of a successful real-time PCR experiment. In the chapters immediately following this one, important topics critical for performing a real-time PCR experiment are discussed in detail such as primer and probe design (Chapter 5), a discussion of SYBR® Green I assay design and utilization (Chapters 7 and 8), post-run data analysis (Chapter 2), different strategies for quantifying data (Chapter 3) and data normalization (Chapter 4). With this knowledge base, the new investigator should be well armed to initiate their own studies using this most robust and sensitive technique. Chapters 10–17 discuss, in a more focused fashion, how real-time PCR is being utilized as an important tool in different ways and areas of biological research.

References

Alwine JC, Kemp DJ, Stark GR (1977) Method for detection of specific RNAs in agarose gels by transfer to diazobenzyloxymethyl-paper and hybridization with DNA probes. *Proc Natl Acad Sci USA* **74**: 5350–4.

Ausubel FM (2001) *Current protocols in molecular biology.* Wiley InterScience, New York.

Bustin SA (2004) *A–Z of quantitative PCR.* International University Line, La Jolla, CA.

Calogero A, Hospers GA, Timmer-Bosscha H, Koops HS, Mulder NH (2000) Effect of specific or random c-DNA priming on sensitivity of tyrosinase nested RT-PCR: potential clinical relevance. *Anticancer Res* **20**: 3545–3548.

Chomczynski P (1993) A reagent for the single-step simultaneous isolation of RNA, DNA and proteins from cell and tissue samples. *Biotechniques* **15**: 532–534, 536–537.

Didenko VV (2001) DNA probes using fluorescence resonance energy transfer (FRET): designs and applications. *Biotechniques* **31**: 1106–1116, 1118, 1120–1121.

Gerard GF, Fox DK, Nathan M, D'Alessio JM (1997) Reverse transcriptase. The use of cloned Moloney murine leukemia virus reverse transcriptase to synthesize DNA from RNA. *Mol Biotechnol* **8**: 61–77.

Gjerdrum LM, Sorensen BS, Kjeldsen E, Sorensen FB, Nexo E, Hamilton-Dutoit S (2004) Real-time quantitative PCR of microdissected paraffin-embedded breast carcinoma: an alternative method for HER-2/neu analysis. *J Mol Diagn* **6**: 42–51.

Goldenberg O, Landt O, Schumann RR, Gobel UB, Hamann L (2005) Use of locked nucleic acid oligonucleotides as hybridization/FRET probes for quantification of 16S rDNA by real-time PCR. *Biotechniques* **38**: 29–30, 32.

Heid CA, Stevens J, Livak KJ, Williams PM (1996) Real time quantitative PCR. *Genome Res* **6**: 986–994.

Lakowicz JR (1983) *Principles of fluorescence spectroscopy.* Plenum Press, New York.

Liao Z, Chen M, Guo L, Gong Y, Tang F, Sun X, Tang K (2004) Rapid isolation of high-quality total RNA from taxus and ginkgo. *Prep Biochem Biotechnol* **34**: 209–214.

Mullis KB (1990) Target amplification for DNA analysis by the polymerase chain reaction. *Ann Biol Clin (Paris)* **48**: 579–582.

Peters IR, Helps CR, Hall EJ, Day MJ (2004) Real-time RT-PCR: considerations for efficient and sensitive assay design. *J Immunol Methods* **286**: 203–217.

Pierce KE, Sanchez JA, Rice JE, Wangh LJ (2005) Linear-After-The-Exponential (LATE)-PCR: Primer design criteria for high yields of specific single-stranded DNA and improved real-time detection. *Proc Natl Acad Sci USA* **102**: 8609–8614.

Sambrook J, Russell DW (2001) *Molecular cloning: a laboratory manual.* Cold Spring Harbor Laboratory Press, Cold Spring Harbor, NY.

Sanchez JA, Pierce KE, Rice JE, Wangh LJ (2004) Linear-after-the-exponential (LATE)-PCR: an advanced method of asymmetric PCR and its uses in quantitative real-time analysis. *Proc Natl Acad Sci USA* **101**: 1933–1938.

Selvin PR (1995) Fluorescence resonance energy transfer. *Methods Enzymol* **246**: 300–334.

Sherrill CB, Marshall DJ, Moser MJ, Larsen CA, Daude-Snow L, Prudent JR (2004) Nucleic acid analysis using an expanded genetic alphabet to quench fluorescence. *J Am Chem Soc* **126**: 4550–4556.

Southern EM (1975) Detection of specific sequences among DNA fragments separated by gel electrophoresis. *J Mol Biol* **98**: 503–517.

Vu HL, Troubetzkoy S, Nguyen HH, Russell MW, Mestecky J (2000) A method for quantification of absolute amounts of nucleic acids by (RT)-PCR and a new mathematical model for data analysis. *Nucleic Acids Res* **28**: E18.

Winer J, Jung CK, Shackel I, Williams PM (1999) Development and validation of real-time quantitative reverse transcriptase-polymerase chain reaction for monitoring gene expression in cardiac myocytes in vitro. *Anal Biochem* **270**: 41–49.

Protocol 1.1: Treatment of total RNA with RNase-free DNase I

1) Determine the RNA concentration of your samples
2) Dilute the total RNA to 100–500 ng/µl in DEPC-H_2O
3) Prepare the following dilution of RNase-free DNase I

1/10 Dilution of RNase-free DNase I

Stock	Volume
H_2O	8 µl
10X PCR Bfr*	1 µl
RNase-free DNase I (10 U/µl)	1 µl
	10 µl

4) Using the following table as a guide, make enough DNase I master mix to add to all your RNA samples + 1 extra sample for overage
5) Add the appropriate volume of DNase I master mix to each RNA sample tube

Treatment of total RNA

Stock	Volume
Total RNA	120 µl
50 mM MgCl2 (1 mM final)	2.5 µl
1/10 DNase I (1 U/µl)	2.5 µl
	125 µl

6) Incubate at 37°C for 30 min
7) Immediately transfer to 75°C for 10 min (exactly) to kill the DNase I; put sample tubes on ice immediately after heating
8) Store the treated RNA at –80°C (good for at least 3 years)

*Any commercial nuclease-free 10X PCR buffer

Standard DNase I procedure from the Quantitative Genomics Core Laboratory; UTHSC-Houston, Houston, TX 77030, USA

Protocol 1.2: Hybrid protocol for the isolation of total RNA

Follow the phenol/chloroform method (Tri-Reagent, Trizol, RNAsol, etc) with the following modifications:

Step 1: Homogenize tissue samples

1) Homogenize tissue samples in a phenol-based reagent in the amounts indicated:
   The amount of reagent required may be greater for certain tissues; e.g., brain. If you don't get a clear aqueous phase, add another ml of homogenization reagent, vortex and spin. Repeat if necessary.

Tissue (mg)	Phenol Reagent (ml)
100	1 ml
500	5 ml
1000	10 ml

2) Add tissue to the phenol reagent. Use a power homogenizer to homogenize tissue for 1 min or until there are no tissue chunks (settings should be determined and optimized for the specific instrument). If tissue culture cells are used, the phenol-based reagent can be added directly to the plates after the media is removed. Cells removed by scraping.
3) Separate insoluble material from the homogenate by centrifugation at 12,000 × g for 10 min at 2–4°C. The supernatant contains the RNA while the pellet contains extracellular membranes. If working with adipose tissue, remove the supernatant from under fat that collects as a top layer.
4) Transfer the cleared supernatant to a fresh tube being careful not to carry-over any of the interface. Add 0.2 ml of chloroform for every 1.0 ml of phenol reagent.

Step 2: Separate phases

1) Securely cap the tube from 4 above. Shake the tube vigorously for 2 min (end to end).
2) Incubate at room temperature for 2 min.
3) Centrifuge at 12,000 × g for 15 min, 2–4°C.
   The resulting sample should show 3 phases:
   1–Lower red (Tri-Reagent) phenol-chloroform phase
   2–White interphase
   3–Upper colorless aqueous phase

The RNA remains exclusively in the upper aqueous phase. If the aqueous phase appears cloudy, you may perform the chloroform extraction step up to 3 times.

Step 3: Purify RNA on a Qiagen spin column

1) Transfer the clear aqueous phase into a new RNase-free tube. Add 0.5 ml isopropyl alcohol per 600 µl of isolated aqueous phase. The correct final alcohol concentration is critical for maximal binding to the column matrix.
2) Place solution on a Qiagen RNeasy mini or midi column (750 µl at a time). Spin 30 sec at 13,000 RPM reloading with remaining solution until all has passed through the column.
3) Continue with the washing and elution of the RNA as outlined in the Qiagen instructions.
4) Elute the final RNA in DEPC-H$_2$O.

Note: other column types can be used but the alcohol concentration has to be compatible, otherwise the RNA may not stick to the column matrix.

Step 4: Check the quality of the RNA

1) Read a sample of your RNA solution (1/50 dilution) on a spectrophotometer. The ratio of absorbance at 260 nm to that at 280 nm (A260/280 ratio) should have values of 1.8–2.0 (2.2 is max for RNA).
2) Calculate the concentration of total RNA in the sample from the 260 nm reading. An optical density (OD) of 1.0 corresponds to approximately 40 µg/ml of single-stranded RNA.

Standard Protocol for the isolation of RNA from the Quantitative Genomics Core Laboratory; UTHSC-Houston, Houston, TX 77030, USA

Protocol 1.3: Two-step master mix

RT-1 Stock	Component	Final	A – 10 µl Rxn Volume (µl)	B – 5 µl Rxn Volume (µl)
*****	DEPC-H$_2$O	*****	2.60	1.30
10X	Rev Transcriptase Bfr	1X	1.00	0.50
2.5 mM	dNTP mix	500 µM	2.00	1.00
20 µM	Assay-specific reverse primer	400 nM	0.20	0.10
50 U/µl	MMLV reverse transcriptase	10 U/10 µl	0.20	0.10
			6.00	3.00

RT-2 Stock	(–RT or NAC control) Component	Final conc.	10 µl Rxn Volume (µl)	5 µl Rxn Volume (µl)
*****	DEPC-H$_2$O	*****	2.80	1.40
10X	Rev transcriptase Bfr	1X	1.00	0.50
2.5 mM	dNTP mix*	500 µM	2.00	1.00
20 µM	Assay-specific reverse primer	400 nM	0.20	0.10
			6.00	3.00

1) Add 6.0 µl (A) or 3 µl (B) RT-1 to 3-wells/sample
2) Add 6.0 µl (A) or 3 µl (B) RT-2 to 1-well/sample
3) Add 4.0 µl (A) or 2 µl (B) total RNA/well (2.5–50 ng/µl)
4) Incubate using RT program = 50°C – 30 min; 72°C – 5 min; 20°C – hold

*–2.5 mM dNTP mix – equal volumes of the 4 dNTPs at 10 mM

Standard RT master mix from the Quantitative Genomics Core Laboratory; UTHSC-Houston, Houston, TX 77030, USA

Protocol 1.4: Two-step PCR master mix for probe-based assays

PCR-1 Stock	Component	Final conc.	50 µl Rxn Volume (µl)	25 µl Rxn Volume (µl)
*****	PCR-H$_2$O	*****	23.20	11.60
10X	PCR Bfr	1X	5.00	2.50
15 µM	SuperROX™ (Biosearch)	90 nM	0.30	0.15
50 mM	MgCl$_2$	5 mM	5.00	2.50
2.5 mM	dNTP mix*	200 µM	4.00	2.00
20 µM	Forward primer (+)	400 nM	1.00	0.50
20 µM	Reverse primer (–)	400 nM	1.00	0.50
20 µM	Probe	100 nM	0.25	0.13
5 U/µl	*Taq* (regular or hot-start)	1.25 U/50 µl	0.25	0.13
			40.00	20.00

1) Add 40 µl (A) or 20 µl (B) PCR-1 to each of 4 wells cDNA/assay
2) Use the following cycle program:
   95°C – 1–15 min; 40 cycles of 95°C – 12–15 sec, 60°C – 30–60 sec

*–2.5 mM dNTP mix – equal volumes of the 4 dNTPs at 10 mM

Standard RT master mix from the Quantitative Genomics Core Laboratory; UTHSC-Houston, Houston, TX 77030, USA

Protocol 1.5: Two-step PCR master mix – primer or SYBR® Green I based assay

PCR-1 Stock	Component	Final conc.	50 µl Rxn Volume (µl)	25 µl Rxn Volume (µl)
*****	PCR-H$_2$O	*****	12.70	6.35
10X	PCR Bfr	1X	5.00	2.50
15 µM	SuperROX™ (Biosearch)	90 nM	0.30	0.15
40%	Glycerol	8%	10.00	5.00
10%	Tween 20	0.10%	0.50	0.25
50 mM	MgCl$_2$	5 mM	5.00	2.50
2.5 mM	dNTP mix*	200 µM	4.00	2.00
20 µM	Forward primer (+)	400 nM	1.00	0.50
20 µM	Reverse primer (–)	400 nM	1.00	0.50
20 µM	Probe	100 nM	0.25	0.13
5 U/µl	*Taq* (regular or hot-start)	1.25 U/50 µl	0.25	0.13
			40.00	20.00

1. Add 40 µl (A) or 20 µl (B) PCR-1 to each of 4 wells cDNA/assay
2. Use the following cycle program:
   95°C – 1–15 min; 40 cycles of 95°C – 12–15 sec, 60°C – 2 sec, 72°C – 30–60 sec

*–2.5 mM dNTP mix – equal volumes of th 4 dNTPs at 10 mM

Standard RT master mix from the Quantitative Genomics Core Laboratory; UTHSC-Houston, Houston, TX 77030, USA

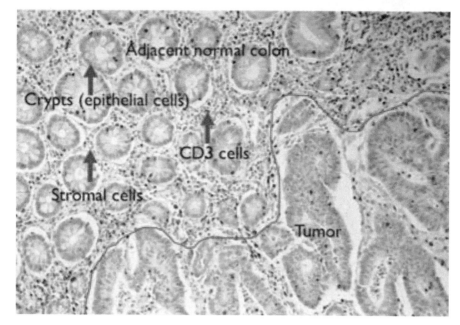

Figure 4.3

Colorectal cancer biopsy. Just over one third of the section is made up of tumour, the rest is adjacent normal colon. At least three cell types (epithelial, stromal and T cells) are clearly distinguishable.

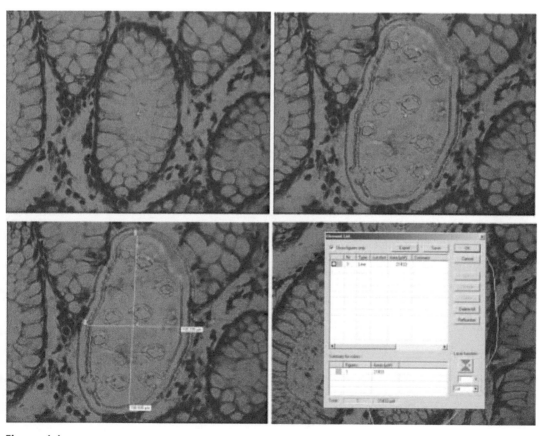

Figure 4.4

Laser microdissected colorectal crypt. The crypt highlighted by an arrow in the first picture has been ablated in the second picture. The software measures the dimensions of the dissected sample and displays its area in μm².

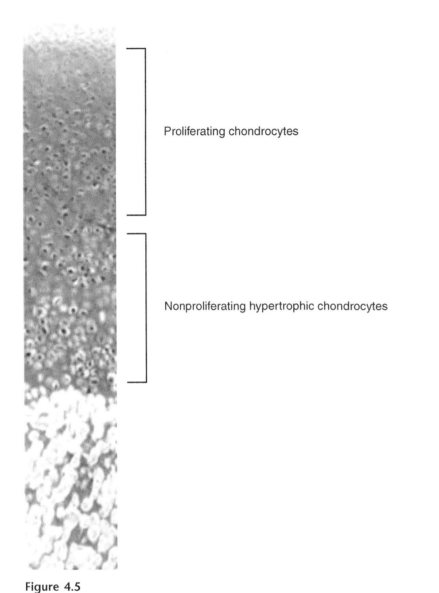

Proliferating chondrocytes

Nonproliferating hypertrophic chondrocytes

Figure 4.5

Longitudinal section of an epiphysis growth plate. Sample sizing comparisons between the larger nonproliferating cells with the proliferating cells are extremely difficult with this tissue. Any result must also consider the difference in the size of the active small cells and the less active larger cells.

Data analysis and reporting

2

Pamela Scott Adams

2.1 Introduction

Proper data analysis is crucial to obtaining valid and relevant results from any experimental system, but is especially critical when assessing variations in mRNA expression of genes, the so-called transcriptome. Unlike DNA which is present in each cell throughout the life of the organism, RNA is transiently expressed and levels vary according to cell type, developmental stage, physiology and pathology. Therefore, quantification of RNA is context dependent and inherently variable. Extra vigilance must be exercised to ensure the data are technically accurate before they can be examined for biological relevance. It should always be remembered that mRNA expression does not necessarily correlate with protein expression and protein expression may not correlate with function. Regardless of the instrument used to perform quantitative real-time PCR, certain basic conditions should be met. The assay should be robust and fulfill all criteria for a good assay. A standard curve using a defined template should result in a slope, coefficient of determination (r^2) and y-intercept that demonstrate good efficiency, accuracy and sensitivity. The baseline and threshold should be properly set. The amplification curves should demonstrate exponential amplification and be within the detection limits of the instrument. Proper controls should be included to monitor contamination and sensitivity of the assay. Genomic contamination should be assessed, if applicable. Data should be reported in a manner that allows the variability of results to be seen. Proper statistical analyses should be applied to the data.

What will be covered in this chapter? There are many applications utilizing real-time PCR, but the most common is the 5′ exonuclease or TaqMan® assay used to quantify gene expression using a set of primers and a dual-labeled fluorescent probe. This chapter will focus on the preliminary analysis of data for this popular utilization of real-time PCR. All subsequent analyses will be based on the accuracy of the post-run preliminary analysis. Therefore, the samples in a real-time PCR run must be verified as valid before any further interpretation of the results is possible. Assay design, SYBR® Green I assays, normalization, relative quantification, along with more specialized uses and analyses will be covered in other chapters. Starting with the questions: are the primer/probes giving a good assay and what constitutes a good assay, standard curves, how to interpret them and use them to improve your assays, will be discussed. When you take the first look at the results of a real-time PCR run, what to look for and how to look will be described: how to set the baseline, the threshold, assess the controls and how to interpret the wide variety of amplification curves (or no curves) which might be encountered. Are the amplifications curves real? Are the no template controls (NTCs) negative? Examples are shown for specific instruments, but the principles should be applicable to all instruments. Illustrations of common problems that can arise will be described, so that they will be recognized if they occur. Data comparing the results obtained from nine different instruments using default instrument values or operator-adjusted values will be shown. An example of the basic principles of data reporting and statistics will be shown. This chapter should be especially helpful to researchers, such as core facility personnel, who have the opportunity to deal with the real-time PCR preliminary data analysis of numerous users.

A quick review of terminology used in preliminary data analysis is in order to promote understanding of why the analysis settings are so important. The C_t is defined as the cycle when sample fluorescence exceeds a chosen threshold above background fluorescence. This is also known as the C_p or crossing point. A positive C_t results from genuine amplification, but some C_t values are not due to genuine amplification and some genuine amplification does not result in a C_t value. Therefore, regardless of the terminology, the key word is 'chosen'. The C_t is determined by both the **chosen** baseline setting and the **chosen** threshold setting. These settings may be selected by the user or the user may accept the default values of the software. Values can be fixed or they can be adaptive. In most cases the reporter signal is normalized to a reference dye by dividing the raw fluorescence of the reporter by the fluorescence of the passive reference, to compensate for minor well-to-well variation. This value is referred to as the Rn, the normalized reporter signal. When the background value has been subtracted from the Rn, then this value is referred to as delta (Δ) Rn, the normalized fluorescence value corrected for the background.

2.2 Standard curves

Before discussing the actual preliminary analysis procedure, it is wise to have a good understanding of what constitutes the best possible assay. Information on primer and probe design will help with this task if you are designing your own assay. Often it is more convenient to use a published assay, however, keep in mind that just because a primer/probe set is published, it is no guarantee that an optimal assay will result. Also, typographical errors in published DNA sequences of any type abound, so BLAST (basic local assignment tool) all sequences before using them to insure they are correct. If the information is not included in the publication, it is a good idea to check the sequence and melting temperatures (Tm) of the primers and the probe just to assure that the assay meets basic design requirements, before going to the expense of purchasing the primer/probe set. Checking the relationship of the primers/probe to intron/exon boundaries will make you aware of the importance of having controls that determine genomic DNA contamination in RNA preparations. Similarly, checking amplicon size will prevent any unpleasant surprises. Perhaps these measures are 'overkill' but they can save time and money in the long run. The assay should always be validated using the conditions in your laboratory.

How to determine if an assay is optimal? The standard curve is a very useful tool for determining the qualities of an assay. Using a defined template, such as a plasmid containing a relevant portion of the gene of interest, PCR product, synthetic oligonucleotide or transcribed RNA to perform a standard curve, will allow determination of the PCR efficiency of the assay along with the sensitivity and dynamic range independently of any variables associated with the sample preparation and/or reverse transcription.

How is a standard curve generated? Begin with determining the OD_{260} of the template sample.

The amount of the template should be expressed in molecules. Convert the mass to molecules using the formula:

$$\frac{\text{Mass (in grams)} \times \text{Avogadro's Number}}{\text{Average mol. wt. of a base} \times \text{template length}} = \text{molecules of DNA}$$

For example, if a synthetic 75-mer oligonucleotide (single-stranded DNA) is used as the template, 0.8 pg would be equal to 2×107 molecules:

$$\frac{0.8 \times 10^{-12} \text{ gm} \times 6.023 \times 10^{23} \text{ molecules/mole}}{330 \text{ gm/mole/base} \times 75 \text{ bases}}$$

$$= 2.0 \times 10^7 \text{ molecules (copies) of DNA}$$

For double-stranded templates, use 660 gm/mole/base. To generate a standard curve, prepare seven 10-fold dilutions, starting with 1×10^7 template copies and ending with 10 copies. Be sure to use a carrier such as tRNA to minimize loss of template at low concentrations. Use a previously determined optimized primer/probe concentration that gives the highest ΔRn and the lowest C_t.

Most platform software will plot a standard curve using designated wells or Microsoft Excel® can be used to draw xy plots with the log template amount as the x value and the cycle threshold (C_t) as the y value. A line representing the best fit is calculated for a standard curve using the least squares method of linear regression:

$$y = m\,x + b \text{ where } y = C_t, \text{ m = slope, } x = \log_{10} \text{ template amount,}$$
$$\text{and b = y-intercept.}$$

The coefficient of determination, r^2 should also be noted.

What does each parameter tell us? Assay efficiency is based on the slope of the line, calculated by the formula:

$$\text{Efficiency} = [10^{(-1/\text{slope})}] - 1$$

Efficiency is primarily an indication of how well the PCR reaction has proceeded.

The integrity of the data fit to the theoretical line is described by the r^2. This is a measure of the accuracy of the dilutions and precision of pipetting. The y-intercept is an indication of the sensitivity of the assay and how accurately the template has been quantified.

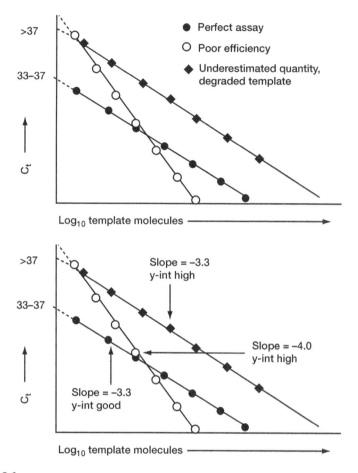

Figure 2.1

Standard curve cartoons. Exaggerated depictions of a 'perfect' standard curve (closed circles) and the effect that low efficiency (open circles) or inaccurate template quantification or degraded template has on the y-intercept (closed diamonds).

What constitutes a good assay? *Figure 2.1* is an exaggerated depiction of the standard curve of a 'perfect' assay and what influence an alteration of the parameters will have on the outcome. By expressing the template amount in molecules, the formula below can be used to determine how many cycles are necessary to produce a specified number of molecules.

$n = Log(N_n) - log(N_o)/log(1+E)$ Where N_n = number of copies of the template after n cycles, N_0 = number of copies of the template at cycle 0, E = efficiency and n = number of cycles.

Taking into account that the signal detection limit of free FAM is $10^{10} - 10^{11}$ molecules on most platforms, it is possible to determine that at 100% PCR efficiency, a single copy of template should be detected between 33.3–36.5 cycles. Therefore, a perfect assay would have a slope of –3.32

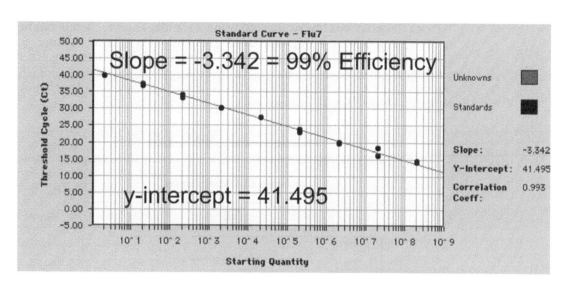

Figure 2.2

Standard curve of degraded template. Real example of a standard curve using a template that has degraded during storage.

(100% efficiency), a y-intercept between 33 and 37 cycles and an r^2 of 1.00. *Figure 2.1* demonstrates that as the slope of the line increases (becomes more negative), the efficiency decreases, the y-intercept increases and the sensitivity of the assay suffers because more starting copies are needed before the detection limit is reached. Or, more cycles are required to detect the same number of template molecules. This may be a distinct disadvantage if dealing with genes that have low expression. Make a spreadsheet using the formulas above and see for yourself what effect alteration of the efficiency has on detectable copy number or the cycle at which the sample will be detected.

An example where efficiency may not be optimal, even though the primers and probes have been properly designed, is the case of a non-linearized plasmid being used as a template for the standard curve. Because of super-coiling, 100% of the copies of the plasmid are not available for initial PCR resulting in low efficiency at the start, but once the new amplicons outnumber the plasmid, efficiency is 100%. The net result is an under-estimation of copy number. This is why plasmids should always be linearized before using them as templates for standard curves. However, efficiency is not necessarily the only measure of assay quality. An assay can have an apparently acceptable efficiency of 95–100% but if the y-intercept is substantially higher than 37 or lower than 33, it indicates that the amount of template has not been correctly determined. In the case of a high y-intercept, deterioration of the template is likely (*Figure 2.2*). Degradation of template is a common result of too many freeze–thaw cycles of the standard or storing the template at too low a concentration without a carrier.

Once the assay has been validated using a standard curve, then it can be used with cDNA or a universal RNA standard in a similar manner to optimize sample preparation and reverse transcription by monitoring assay efficiency. Of course, in reality, not every assay will be perfect, but an attempt should be made to develop the best assay possible or at least know what the weaknesses of an assay are.

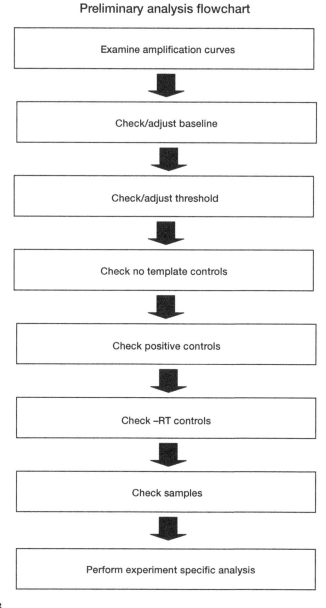

Preliminary analysis flowchart

Examine amplification curves

Check/adjust baseline

Check/adjust threshold

Check no template controls

Check positive controls

Check –RT controls

Check samples

Perform experiment specific analysis

Figure 2.3

Flow chart for preliminary analysis. Parameters to check during preliminary analysis of data obtained from a real-time PCR run.

2.3 Preliminary assay analysis

The experiment has been carefully planned, set up and run. The run is finished and now it is time to analyze the data. What to look at first? Do we really know what the analysis settings do? Does it matter? The most common place to start preliminary analysis of a real-time PCR run is using the Amplification view that allows adjustment of the baseline and threshold values. The default view for most software plots cycle number on the x axis versus the logarithm of the ΔRn on the y axis. *Figure 2.3* shows a flowchart with suggested steps of parameters to check and in what order. The steps in the flowchart are meant to be a guideline and it may be necessary to adjust some parameters, such as the threshold, more than once. Some instruments require that a baseline and threshold be accepted before further manipulations of the software can be executed, but all instruments allow the user to change the baseline and threshold and reanalyze at will. All analysis scenarios presented in this chapter assume that the instrument is working properly.

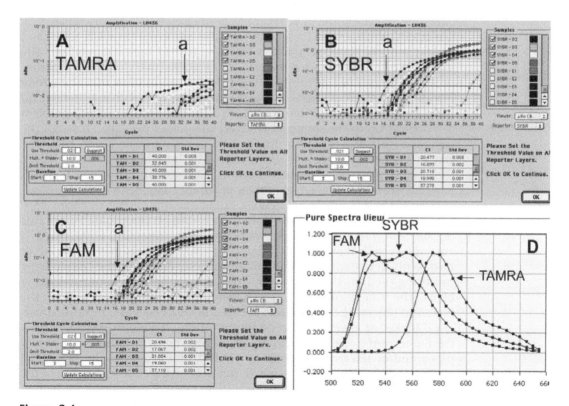

Figure 2.4

Effect of incorrect dye layer settings. **A.** Amplification view showing the incorrect dye layer (TAMRA). **B.** Amplification view showing the incorrect dye layer (SYBR®). **C.** Amplification view showing the correct dye layer (FAM). **D.** Curves showing the relationship among spectra for the dyes FAM, SYBR® (Green I) and TAMRA. Arrow at 'a' identifies C_t of the same amplification curve under different analysis conditions.

2.3.1 Amplification curves

What if there are no curves or curves that look unusual? This situation must be dealt with before any further analysis is possible. The first parameter that should be checked is that the correct dye layer or reporter has been designated. This seems like a very simple proposition, but if there are no amplification curves or strange curves this is the most likely explanation. This problem occurs most often on multi-user instruments and when templates are routinely used. The result of choosing the wrong template with the incorrect dye layer can be as drastic, as very few amplification curves above the threshold as shown in *Figure 2.4A* (C_t of a = 33) when TAMRA is chosen for the dye layer for a FAM probe or very subtle as the instance when a SYBR® Green I template is chosen when using a FAM probe, as shown in *Figure 2.4B*, (C_t of a = 17). If the spectral calibration curves are viewed (*Figure 2.4D*), it is easy to see that the TAMRA spectrum is quite different from FAM. However, because the FAM and SYBR® Green I have overlapping spectra, the amplification curves will appear quite respectable and the C_t values may be indistinguishable (C_t of curve 'a' is 17 in *Figure 2.4B and 2.4C*). Only by checking that the correct dye layer has been chosen will the discrepancy be detected. One other way to detect/verify this problem is to select a well and look at it in the Raw Spectra view with the best fit option turned on. The curve of the raw spectra for the well and the fit curve should be superimposed on each other. If they are not this is an indication that the wrong reporter is being monitored (or there is contamination with another dye on the plate or in the sample block). What can be done to correct this problem if it occurs? Go to the set-up view, select the proper dye layer, reassign the wells and reanalyze to get the proper C_t values as shown in *Figure 2.4C*. One way to minimize the chances of the wrong dye layer being used is to reboot the data collection computer between individual runs, which resets template/software parameters to their previously assigned values.

What if the Amplification view looks like the one shown in *Figure 2.5A*, with very erratic amplification plots and an unusual threshold? This occurs on the ABI 7700 (Applied Biosystems, Foster City, CA, USA) when it encounters wells that have no fluorescence in them. This may be a result of mistakenly designating wells as containing a sample when they are actually not being used, or the over zealous newbie who wants to include a water blank. The troublesome wells can be identified by selecting all wells on the plate and observing them in the Raw Spectra view. They will be most apparent if looked at after 40 cycles of amplification. The empty wells will have flat fluorescent traces, as shown in *Figure 2.5B*. The lack of fluorescence from probe and/or reference dye can be verified by examining the individual wells in the Multicomponent view (*Figure 2.5C*) which allows viewing of the raw fluorescence in an individual well. The problem is resolved by switching to the set-up view and designating the offending wells as 'not in use' and reanalyzing the data without the blank wells. Now the curves look quite respectable, as shown in *Figure 2.5D* and an example C_t (b) is corrected from 29 to 17.

One other anomaly that can generate very weird curves is having ROX designated as the reference dye when in fact there is no ROX in the reaction. The amplification curves will look similar to Figure 2.5A. Turning

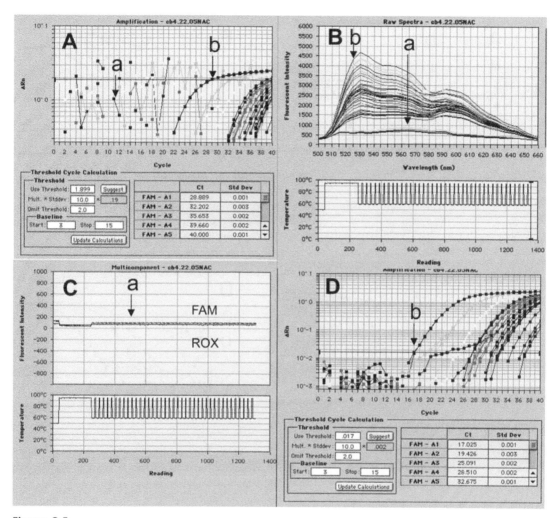

Figure 2.5

Effect of including wells with no fluorescence. **A.** Appearance of the amplification view if wells have been selected that contain no fluorescence. (a) artifact, (b) real amplification. **B.** Raw data view of all samples. **C.** Multicomponent view of (a) verifying the absence of fluorescence. **D.** Appearance of the amplification view after removing the wells with no fluorescence and reanalyzing. Sample b has a C_t of 17 compared to a C_t of 29 in panel A.

off ROX as a reference and reanalyzing, restores the curves to their proper configuration.

2.3.2 Baseline

If there are some reasonable amplification curves, the next parameter that should be examined is the baseline setting. Each real-time PCR instrument

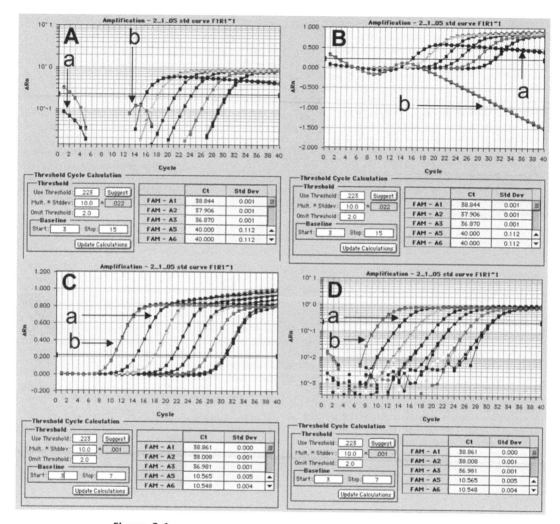

Figure 2.6

Effect of baseline settings. **A.** Amplification view showing an incorrect baseline setting (3–15 cycles) and the resulting artifacts: (a) half rainbows and (b) a full rainbow, plotted using a logarithmic scale for the y axis (ΔRn). **B.** The same curves in the amplification view plotted using a linear scale for the y axis. **C.** The same curves corrected to a baseline of 3–7 cycles in the amplification view plotted using a linear scale for the y axis. **D.** The same curves corrected to a baseline of 3–7 cycles in the amplification view plotted using a logarithmic scale for the y axis.

has baseline-setting algorithms that determine the background noise of the detector and reagents using the fluorescence generated in a designated number of cycles at the beginning of the run. This may be a fixed number of cycles for all samples or adaptive for each sample, depending on the type of instrument that is being used. Baseline anomalies may be slight or severe. Look at the left hand corner of the Amplification plot. Are there 'half-

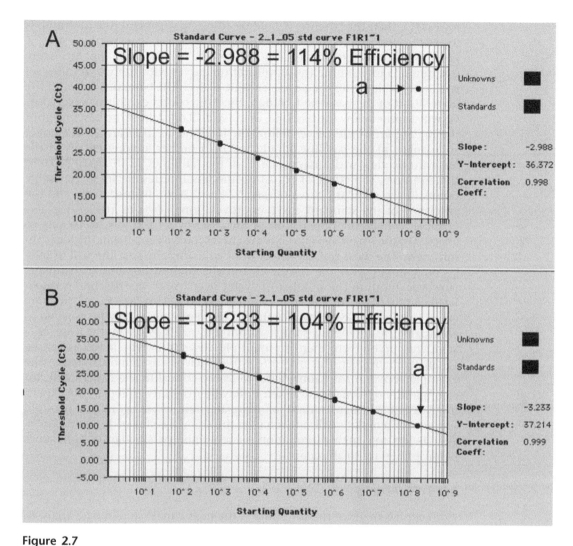

Figure 2.7

Effect of baseline settings on a standard curve. **A.** Standard curve of the data shown in Figure 2.4 before adjusting the baseline. **B.** Standard curve of the data shown in Figure 2.4 after adjusting the baseline.

rainbows' or what looks like half an arc or hump between cycles 1 and 10? These arcs frequently start above the threshold, as shown in *Figure 2.6A.a*. In extreme cases, complete 'rainbows' in which the total curve is affected, may be seen. The 'rainbows' may appear under the threshold and the C_t may register as 40 (*Figure 2.6A.b*) or the curve may cross the threshold in two places (not shown). Examination of the Raw Spectra view or the Multicomponent view, which allows viewing of the raw fluorescence in an individual well, will clarify if there is real amplification or not. At what cycle does the lowest C_t appear? If the lowest C_t is less than the upper limit of the baseline setting (the upper limit of the baseline setting is greater than the lowest C_t), then the baseline should be adjusted. A general rule of thumb is to set the upper limit of the baseline 2–3 cycles less than the lowest C_t. A good way to determine exactly where to set the baseline on many machines, is to examine the amplification curves with the y axis (ΔRn) set to a linear scale rather than a logarithmic scale. If the baseline is set with the upper limit of the baseline too high, then the curves exhibiting a very low C_t, will appear below the baseline, as in *Figure 2.6B*. Adjust the baseline until the linear part of the curves follow the baseline, as in *Figure 2.6C*. Correct adjustment of the baseline should result in proper looking curves in the logarithmic view with disappearance of the rainbows, as in *Figure 2.6D*. In this example, the effect of adjusting the baseline results in seven usable points in the standard curve *(Figure 2.7A* and *2.7B)*, instead of six and the efficiency of the standard curve improves from 114% to 104%. It is also possible to have the upper limit of the baseline set too low, so be sure to check it in the linear view, even if the curves look acceptable in the logarithmic view. The most notable effect from adjusting the baseline will usually be on samples with the lowest C_t and the highest amount of template. However, limiting the number of cycles over which the baseline is calculated can interfere with the detection of less abundant targets, in some rare cases. In cases of severe overloading of the template, it may be possible to eliminate the offending samples from the analysis as in the case of a standard curve where there are multiple doses from which values can be calculated. However, this may affect the efficiency as demonstrated in the example above. If the problematic wells are critical to the experiment, for example the reference gene, then it may be necessary to repeat the experiment using a lesser amount of template, remembering that a two-fold dilution of template will only change the C_t by a factor of 1. Gene expression being too high is common to reference or standardization genes because they are so abundant.

2.3.3 Threshold

Correct placement of the threshold is the next crucial step in data analysis. As for the baseline, each instrument uses variations of a threshold-setting algorithm to set this value. For instance, the Applied Biosystems 7700 SDS (Sequence Detection Software) calculates the threshold value as 10 standard deviations from the baseline. For this reason, the baseline must be set before adjusting the threshold value. Keep in mind that an amplification curve is composed of four components: background, exponential phase, linear phase and the plateau (Chapter 1). To adjust the threshold properly, set the threshold value within the exponential phase of all amplification plots when viewed using the logarithmic scale for the y axis. If it is not possible to place the threshold in the proper place for all curves due to variations in the abundance of the various genes and/or the amount of fluorescence generated by a particular probe, then it is permissible to analyze sets of curves with different thresholds. Multiple thresholds are the exception rather than the rule, but can be very helpful in situations where there is a low abundance mRNA that generates very low ΔRn values (Bustin and Nolan, 2004a). How do you know exactly where to set the threshold in the exponential phase? This is a matter of debate, with some believing that the

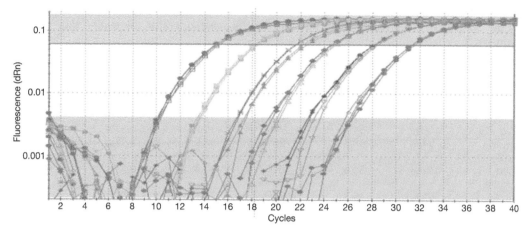

Threshold	Calculated RNA copy #			Slope	y-Int	r^2
0.004	2.26E7	2.92E4	3.01E2	3.23	33.86	0.999
0.012	2.51E7	2.46E4	2.72E2	3.25	35.45	0.999
0.020	2.46E7	2.44E4	2.52E2	3.26	36.32	0.999
0.028	2.33E7	2.57E4	2.41E2	3.26	36.98	1.000
0.036	2.30E7	2.58E4	2.40E2	3.27	37.56	1.000
0.044	2.23E7	2.57E4	2.32E2	3.27	38.02	1.000
0.052	2.27E7	2.53E4	2.25E2	3.27	38.54	1.000
0.060	2.30E7	2.51E4	2.17E2	3.26	38.86	0.999
Mean	2.33	2.57	2.48			
SD	0.10	0.15	0.27			

Figure 2.8

Comparison of parameters relative to adjustment of the threshold. A standard curve was run using a synthetic oligonucleotide. Ten-fold dilutions of a known amount of *in vitro* transcribed RNA were used as 'unknowns'. Representative values of the calculated copy number of RNA at each threshold setting in the unshaded (exponential phase) area of the amplification plot figure are shown. Slope, y-Intercept and r^2 for the RNA curves are calculated values. Data were provided by Brian P. Holloway and Karen McCaustland from the DNA Core Facility at the Centers for Disease Control and Prevention, Atlanta, GA, USA.

lowest possible threshold is best. Some analysis software adjusts the threshold so that a standard curve will have the highest r^2. Still others believe that the most accurate C_t is obtained by selecting the point of maximum curvature on a 2nd derivative curve (second derivative maximum (SDM)). *Figure 2.8* shows the effect of adjusting the threshold within the exponential phase of the curves on the calculation of values from a standard curve. Little effect on the calculated number of known template copies is seen in this example where the threshold setting is varied, demonstrating that there is not a single optimal threshold value. Setting the threshold as low as possible results in lower C_t values, which may be beneficial under some

Table 2.1 Comparison of instrumentation efficiency using default versus manually adjusted settings. Standard curves were set up and run by the same operator using nine different instruments*

Instrument	Default analysis settings		Adjusted analysis settings		Difference	
	Efficiency	y-Intercept	Efficiency	y-Intercept	Δ Efficiency	Δ y-Intercept
ABI Prism® 7500	128%	33.3	99%	35.9	−29.0%	−2.67
ABI Prism® 7700	106%	35.8	103%	35.9	−3.0%	−0.10
BioRad iCycler®	100%	34.8	98%	35.9	−1.3%	−1.12
Cepheid SmartCycler I®	94%	38.5	96%	36.9	2.7%	1.62
Corbett Rotorgene™	101%	36.3	101%	37.4	0.3%	−1.04
Roche LC 1.2®	95%	36.8	94%	38.1	−0.8%	−1.32
Roche LC 2®	99%	34.5	98%	35.9	−1.7%	−1.41
Stratagene Mx3000p®	102%	37.8	103%	35.9	0.6%	1.83
Stratagene Mx4000®	101%	36.9	102%	35.9	0.5%	0.97
Mean	**103%**	**36.1**	**99%**	**35.9**	**−3.5%**	**0.21**
Range	**94%–128%**	**33.3–38.5**	**94%–103%**	**35.90–38.11**	**0.3%–29%**	**0.21–2.67**

*Data was provided by Brian P. Holloway and Karen McCaustland from the DNA Core Facility at the Centers for Disease Control and Prevention, Atlanta, GA, USA.

circumstances. However, a two-fold difference in copy number should have one C_t difference no matter where the threshold is set within the exponential phase of amplification, providing the efficiency is close to 100%. Curve fitting, baseline and threshold algorithms are not universal across platforms, however the end result can be similar. A comparison of PCR efficiency and y-intercept values on nine different platforms using the instrument default settings versus manually optimizing baseline settings and thresholds is shown in *Table 2.1*. In the majority of cases, the default settings provide results quite similar to those that were manually adjusted. Adjusting the baseline was especially beneficial for some instruments in this example because a high dose of standard was used and the default baseline was set too high. Because calculated concentrations will vary somewhat across different platforms, a standard curve should be used when comparing quantitative data between platforms. In the end, the data is relative. Probably the most important factor is to be consistent with subsequent runs. If large differences exist between test samples, then fine-tuning the baseline and threshold may not be so important, but if detecting two-fold or single C_t differences are desired, then proper adjustment becomes critical.

2.3.4 Proper controls

No template controls

Part of deciding where to set the threshold depends upon the negative controls. No template controls (NTCs) should always be included on every plate in every experiment! These controls assure that what is being

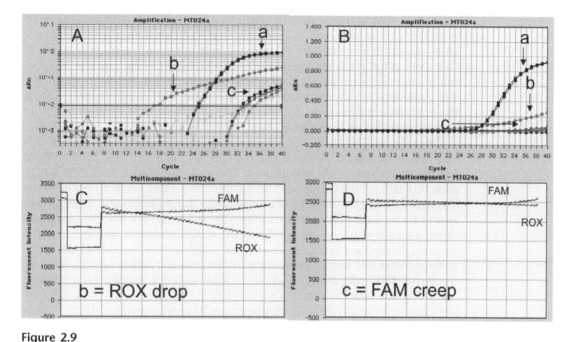

Figure 2.9

False amplification. **A.** Amplification view showing an acceptable amplification curve (a) and unacceptable curves (b, c) plotted using a logarithmic scale for the y axis (ΔRn). **B.** The same curves plotted using a linear scale for the y axis. **C.** Multicomponent view of sample b. **D.** Multicomponent view representative of one of the c sample wells.

measured is in the sample and not a contaminant. A good recommendation (Bustin and Nolan, 2004a) is to disperse NTCs over the plate. For example, prepare 2–3 NTCs in the first few wells on the plate and seal these wells before adding template to any of the other wells. Then prepare 2–3 NTC wells after all the templates have been added. This technique should help determine how the contamination is being introduced if positive NTCs become an issue. What if the NTC is reported as a C_t value less than 40? Inspection of the shape of the amplification curves can be informative. Sometimes a curve is seen that is gradually increasing, but not exponentially, and crosses the threshold at some point, as shown in *Figure 2.9A.b*. In other cases one might see curves that look like very low amplification curves (*Figure 2.9A.c*). Observing the shape of the curves with ΔRn expressed on a linear scale (*Figure 2.9B*) emphasizes that these curves are far from optimal. One cause is a decreasing ROX signal as the FAM signal remains the same, aptly called ROX drop. Another cause is a slight increase in FAM signal while the ROX remains the same, known as FAM creep. Both can occur to varying degrees and at the same time. Since most software divides the FAM signal by the ROX signal to normalize each well, the software interprets this change as amplification. Examination of the Multicomponent view allows viewing of all the fluorescent reagents in an individual well so the relationship between the reporter signal and the reference dye can be observed as demonstrated in *Figure 2.8C* and *8D*. Extreme cases can often be caused by evaporation resulting from improper sealing of that well. What is the solution to non-exponential or false positive NTC? In many cases, upward adjustment of the threshold is sufficient to eliminate the positive NTC. In some instances, the rogue well can be removed from the analysis if it is technically faulty. Careful attention to technique should minimize this type of problem.

What if the NTC show real exponential amplification? NTC contamination can be of several varieties. The contaminant can be a PCR product from a previous real-time PCR experiment. Theoretically, if the tubes from a real-time PCR run are never opened and disposed in an area far removed from the set-up area, then this type of contamination should not occur. However, in reality there might be occasions when the tubes are opened. For example, when an assay is being validated and the amplicon is run on a gel to check for a single product or purified to sequence the amplicon to check for specificity. At least one manufacturer of master mixes has addressed this problem by using dUTP (2′ deoxyurindine 5′ triphosphate) instead of dTTP and including the enzyme uracil-N-glycosylase (UNG). The UNG should digest any carry-over real-time PCR products that have incorporated dUTP. This is why there is a 2 minute step at 50°C step included in the thermocycling protocol if there is UNG in the master mix. The 2 minute step at 50°C inactivates the UNG so it does not affect subsequent PCR products being generated in the current run. A down side of using this enzyme is that if a little bit of residual activity is left, the UNG will begin to digest the amplicons in the current PCR reaction. This only becomes important if it is necessary to sequence the amplicon or run it on a gel. In this case, remove the samples from the machine immediately after the run and store them in the freezer until time to run them on a gel. How to determine if contamination is from carryover from previous real-time PCR runs? If a master mix containing dUTP but no UNG is used, add UNG to the master mix and see if this eliminates the positive NTC. Consider using UNG routinely. Of course, the ideal situation is to prevent the carryover contamination from happening, but this is not always practical, especially in multi-user situations. If dTTP is used in the real-time PCR reactions, then it will not be possible to determine the source of contamination by this method. What if dUTP and UNG are used in the reactions and the NTCs are still less than 40? Examination of the Raw Spectra and/or Multicomponent views shows the beginning of genuine exponential amplification. Primary contamination (as opposed to carry over contamination) is especially common when plasmids, PCR products and/or synthetic templates are used for standard curves. The standard becomes a contaminant unless stringent technique is exercised. Replacing all reagents, cleaning preparation sites, and moving to another building are all methods by which to escape from contamination. These measures are not always totally successful and the positive NTC situation must be dealt with. Likewise, if the samples are small and irreplaceable, there may be no choice but to deal with a positive NTC. Here is where some judgment much be carefully exercised. Here are some recommended suggestions (Bustin and Nolan, 2004a) that should be considered. If the NTC is between 35 and 40 cycles and the sample C_t is 25 or greater than 10 cycles lower than the NTC, then it is reasonable to think that the sample amplification is real. If the sample C_t is 30 and the NTC is 35 this should give serious pause. If the NTC is less than 30 cycles, then there is most likely a serious contamination problem. Another example of when NTCs need to be carefully examined is when the sample C_t values fall in the 35–39 cycle range. In this case, it is very important to verify that the sample is truly above background. One way to accomplish this is to add more template to the reaction. The NTCs should remain the same and the sample C_t should decrease 1 C_t for every two-fold increase in template. If the template is a limiting factor, as is frequently the case when using a small number of cells or micro-dissected tissue, then it is important to increase the number of cycles in the run to verify that there is greater than 10 C_t difference between the sample and the NTC.

No reverse transcription control

If using a real-time PCR assay to quantify mRNA, then an important control is a reaction that allows one to assess the amount of genomic contamination in the sample. This control is commonly referred to as the minus reverse transcription (–RT) control. It is a sample of starting RNA that is treated identically to all the samples, but has no reverse transcriptase added during the reverse transcription step. How important this control is to the experiment and whether it is needed to include this control in every experiment is dependent upon the genes being examined and how the primer/probes were designed. What if the –RT control is positive? There are several strategies for eliminating or minimizing genomic detection. Some advocate designing primer/probed sets that span introns. Try to span a large intron rather than a small one. Some introns are less than 100 base pairs and could compete with the designated amplicon. Spanning an intron-exon border is not always possible, as with single exon genes and those with pseudogenes. Treating the samples with DNase is a good method to remove the majority of genomic DNA contamination from a RNA sample and is generally recommended. DNase treatment, however, usually reduces RNA yields, and this may be undesirable in the case where small numbers of cells are analyzed. If DNase treatment is not possible and the primer/probe set does not span an intron, a –RT control should always be included.

Positive control

There are several types of positive controls, but one of the best is a standard curve. The merits of using a standard curve have been discussed earlier in this chapter. If using a standard curve for quantification, check the slope and the y-intercept to be sure that the standard is not deteriorating. If it is not possible to include an artificial standard curve for each primer/probe set, then a standard curve over a smaller range using genomic DNA or a universal RNA or a known positive sample as a template should be included to monitor the efficiency of the assay. Include a single positive sample if a standard curve isn't possible. If the situation arises where there is no amplification in the test samples, then having a positive control will aid in troubleshooting the problem and narrowing down the possibilities to reagent versus template problems.

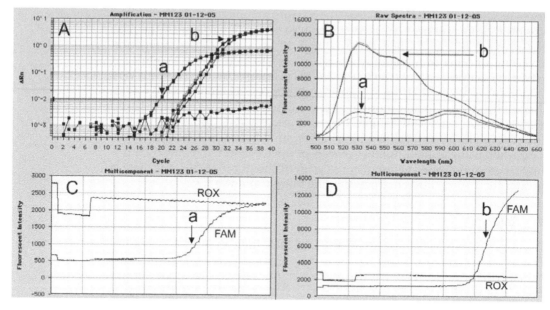

Figure 2.10

Two differently labeled probes. **A.** Amplification view showing a probe labeled with FAM/BHQ1 (a) and a MGB probe from an Applied Biosystems Gene Expression Assay (b). **B.** The same in the Raw Spectra view. **C.** Multicomponent view of sample a. **D.** Multicomponent view representative of one of the b sample wells.

2.3.5 Experimental samples

The baseline is acceptable, the threshold is adjusted properly, the NTCs are negative, the positive control is good – now you are ready to look at the samples. First, look at the amplification curves. What is the ΔRn at the plateau? If the ΔRn is not greater than 0.5, then there may not be real amplification. The amplitude of the ΔRn depends upon several factors. One important factor is the probe. Each probe may perform differently and have different maximum ΔRn values. The type of probe and what it is labeled with can affect the ΔRn substantially as can the type of master mix that is being used. *Figure 2.10A*, shows an example of the two different ΔRn plateaus that are obtained using a 5' FAM, 3' BHQ1 probe (Biosearch Technologies, Novato, CA, USA) and a MGB probe from a commercial Gene Expression assay (Applied Biosystems, Foster City, CA, USA). Checking the Raw Spectra view after 40 cycles of amplification (*Figure 2.10B*) reveals that there is a substantial difference in the raw fluorescence generated by each probe. Further investigation into the Multicomponent view (*Figure 2.10C* and *2.10D*) demonstrates that even though the raw fluorescence values are quite different, there is exponential amplification using both probes. As long as the threshold is set so that it is in the exponential phase of both sets of curves, analysis should be valid.

Look at the samples' curves again. Are all the curves exponential or are there some curves that run parallel to the threshold before becoming exponential? This can happen when the ROX is gradually dropping, as discussed in the NTC section and demonstrated in *Figure 2.9A. Curve b*. It is also possible to have a run with no real amplification that looks like there are amplification curves. The software will do its best to make curves. However, if the ΔRn is much less than 1, then it should immediately be suspected that no real amplification has taken place even though it looks like curves have been generated. *Figure 2.9A. Curve c*, which has already been described in the NTC section, demonstrates this situation. If a good positive control was included, then it should be easy to distinguish real amplification from machine artifacts.

Check the replicates next. They should be within 0.5 C_t of each other. Determine the coefficient of variance. What is the coefficient of variance (CV)? The CV is the standard deviation (SD) divided by the arithmetic mean and is used to measure intra-assay reproducibility from well to well and is also useful to measure inter-assay variation from assay to assay. If the CV for the samples is small, then there is no problem. What if one well is different from the others? Hopefully, triplicates were run. Here again, the Multicomponent view or Raw Spectra view is useful. If there was no amplification, check for the presence of the probe fluorescence. Maybe the probe did not get added? If there is no sign of probe fluorescence, then this would be a legitimate reason to exclude this data. If the probe is present, but there is no sign of amplification, perhaps no template was added. Similarly, if the aberrant well has a lower C_t, then perhaps template was added twice. Of course, adding template to the same well twice would only decrease the C_t by one, so if the CV is large, this phenomena would not be detected. Another scenario is that the plate was not sealed tightly and there has been evaporation in one or more of the wells/tubes. The Multicomponent view might show all fluorescent compounds increasing over the course of the run due to evaporation. Using genomic DNA as a template can be especially problematic for getting identical replicates and careful attention needs to be paid to mixing between every aliquot. If dealing with samples that have low expression and therefore a very high C_t, then the Monte Carlo effect should be considered. Try to run triplicates and run the same experiment three times if possible, to gain statistical significance for the results (Bustin and Nolan, 2004b). As a general rule, the more replicate reactions there are, and the lower the CV of the replicates, the better the ability to discriminate between samples.

2.3.6 Quantifying data

Once the initial examination and analysis of the preliminary data have been completed, the next step is to decide how to compare the data in a meaningful manner. *Figure 2.11* illustrates the general concepts of secondary analysis. Normalization and relative quantification are extensive topics that are covered in other chapters. Using a standard curve for quantifying mRNA or DNA is sometimes referred to as absolute quantification, but the term absolute is debatable since it is only as accurate as the measurement of the material used for the standard curve. This method is better

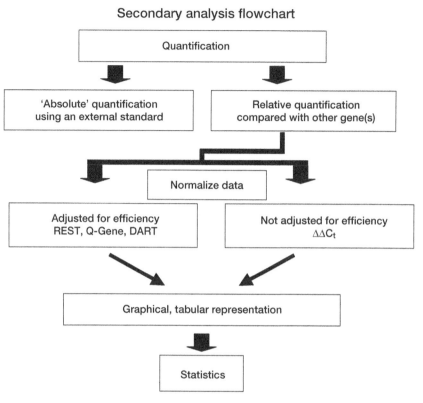

Figure 2.11

Flow chart for secondary analysis. Suggested steps for analysis of real-time PCR data after the preliminary data has been verified.

referred to as quantification using a standard curve. The standard curve allows the amount of unknown samples to be computed on a per cell or unit mass basis. But, no matter how accurately the concentration of the standard material has been determined, the final result is relative to a defined unit of interest (Pfaffl, 2004).

The characteristics of a standard curve are shown in *Figure 2.1* and described in the section on standard curves. Most real-time instruments have software that will calculate the amount of unknown values in the same units designated in the standard curve, but unknowns can be calculated in a spreadsheet using the formula:

$$\text{Log}_{10} \text{ copy number} = C_t - \text{y-intercept/slope}$$

It is important to remember that verifying your reagents are capable of giving a 100% efficient reaction, as described in the section on standard curves earlier in this chapter, does not necessarily mean that particular template will give an accurate measure of gene expression in your sample. Using a defined template, such as a plasmid, PCR product or synthetic oligonucleotide, while easy to prepare, does not take into account the sample preparation and reverse transcription step. Transcribed RNA can be used to control for the reverse transcription step, but it does not control for template preparation and is quite labor intensive. Genomic DNA can be used for the standard curve and may be a more realistic control for sample preparation, but is only useful if the primer/probe set are designed within one exon. Obviously, there is not one perfect method for quantification. Be aware of the pitfalls and choose the method which is best suited to your goals.

2.4 Data reporting and statistics

The experiment is complete, the data have been analyzed, now what should be done?

There are a huge number of variables associated with real-time PCR assays. Variation is introduced by harvesting procedures, nucleic acid extraction techniques, reverse transcription, PCR conditions, reagents, etc. For this reason, data reporting and statistics are important steps in preparing your results for peer review. Data presentation is dependent upon the type of experiment. For example, experimental protocols might examine relative gene expression before and after treatment, normal versus tumor, time courses, responses to inflammation or disease. Other experiments might measure the quantity or strain of organisms in food, water, or the environment. Validations of microarray and siRNA results are other common uses for real-time PCR technology. Regardless of the type of experiment, data should be presented in a manner which allows the reader to observe the amount of variation inherent to the experiment; for example, mean, standard deviation and confidence intervals. Some sort of statistical analysis should be performed to apprise the reader of probabilities of differences being significant. Most real-time PCR experiments are based on hypothesis testing. What is the probability that randomly selected samples have a difference larger than those observed? In some situations, the differences in data are obvious and statistics are a formality. But because biological systems are subject to variation and experimental imprecision, sometimes statistics can reveal differences that are not otherwise discernible, especially if there are a large amount of data. Ideally, an experiment should be planned with a statistical analysis in mind. Large studies should be planned with the advise of a statistician.

If statistical and bioinformatics resources are limited, then there are several choices. Freely available software packages for this purpose include Q-Gene (Muller *et al.*, 2002), DART-PCR (Peirson *et al*, 2003) and REST (relative expression software) (or REST-XL) (Pfaffl *et al.*, 2002). These and other software are presented in detail in Chapter 3. For a more generic approach, a wise investment is to purchase some type of software for reporting data and performing statistical analyses that allows both functions to be performed using the same software.

One such package is Graphpad Prism® (Graphpad Software Inc., San Diego, CA, USA). This software provides many data presentation choices, handbooks to explain basic statistics (Statistics Guide: *Statistical analyses for Laboratory and Clinical Researchers, Fitting Models to Biological Data using Linar*

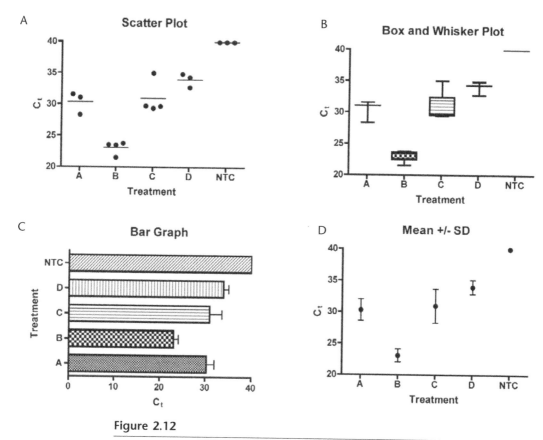

Figure 2.12

Examples of data reporting. The data for each panel is the same representing the real-time PCR results for four different treatments. NTC = No template control. **A.** Scatter plot of individual data points with the mean shown as a bar. **B.** Box and whisker plot: box extends from the 25th to 75th percentile with a line at the median. Whiskers extend to show the highest and lowest value. **C.** Bar graph of the mean and standard deviation for each group. **D.** Mean of each group plotted as a single point +/– standard deviation.

and Nonlinear Regression) and automatic performance of a wide variety of statistical analyses. Prism® provides checklists to help the user decide which is the correct statistical analysis and even describes the best way to cite statistical analyses. With a few basic concepts understood, the mathematics can be left to the software. A good general review of statistics and how they might apply to real-time PCR can be found in *A–Z of Quantitative PCR* edited by Stephen Bustin. *Figure 2.12* shows several different methods of presenting the same data. Scatter plots (*Figure 2.12A*) allow the reader to see the individual data points and the number of samples in each group (n) is readily apparent. A box and whiskers plot (*Figure 2.12B*) conveys data about the mean, the 25th and 75th quartile values and the highest and lowest value in the group. Some prefer bar graphs or single means with the standard deviation (*Figure 2.12C* and *12D*). Be sure to use the standard deviation (SD)

Table 2.2 One-way analysis of variance of the data graphically represented in *Figure 2.11* demonstrating the calculation of P values between groups

P value	P<0.0001
Are means significantly different? (P <0.05)	Yes
Number of groups	5
F	46.4
R squared	0.9391

ANOVA table	SS	df	MS
Treatment (between columns)	436.7	4	109.2
Residual (within columns)	8.712	10	0.8712
Total	445.4	14	

Tukey's multiple comparisons test	MD	q	P value	99% CI of difference
A vs. B	7.218	7.961	P <0.001	1.93 to 12.51
A vs. C	−0.6567	0.7242	P >0.05	−5.95 to 4.64
A vs. D	−3.67	3.786	P >0.05	−9.38 to 1.99
A vs. NTC	−9.697	10	P <0.001	−15.35 to −4.04
B vs. C	−7.875	9.381	P <0.001	−12.77 to −2.98
B vs. D	−10.89	12.01	P <0.001	−16.18 to −5.60
B vs. NTC	−16.92	18.66	P <0.001	−22.21 to −11.62
C vs. D	−3.013	3.323	P >0.05	−8.31 to 2.28
C vs. NTC	−9.04	9.97	P <0.001	−14.33 to −3.75
D vs. NTC	−6.027	6.217	P <0.01	−11.68 to −0.37

and not the standard error of the mean (SEM) because the standard deviation is a better indicator of how much variability there is in the data. A statistical analysis using a one-factor, analysis of variance (ANOVA) and Tukey's multiple comparison test was performed on the data in graphically represented in *Figure 2.12* and the results are shown in *Table 2.2*.

Statistics are important but must be tempered with scientific and/or clinical experience. Remember that statistical significance and biological relevance may not be directly correlated.

Acknowledgments

Many thanks to Dr Deborah Grove, Brian Holloway, Karen McCaustland, Dr Greg Shipley, Dr Tony Yeung and Dr Stephen Bustin.

References

Bustin SA, Nolan T (2004a) Pitfalls of quantitative real-time reverse-transcription polymerase chain reaction. *J Biomol Tech* **15**: 155–166.

Bustin SA, Nolan T (2004b) Data analysis and interpretation. In: *A–Z of Quantitative PCR* (ed Bustin, SA) pp 441–492. International University Line, La Jolla, CA.

Muller PY, Janovjak H, Miserez AR, Dobbie Z (2002) Processing of gene expression data generated by quantitative real-time RT-PCR. *Biotechniques* **32**(6): 1372–1378.

Peirson SN, Butler JB, Foster RG (2003). Experimental validation of novel and conventional approaches to quantitative real-time PCR data analysis. *Nucleic Acids Res* **31**(14): e73.

Pfaffl M (2004) Quantification strategies in real-time PCR. In: *A–Z of Quantitative PCR* (ed Bustin SA) pp 441–492. International University Line, La Jolla, CA.

Pfaffl M, Horgan GW, Dempfle L (2002) Relative expression software tool (REST) for group-wise comparison and statistical analysis of relative expression results in real-time PCR. *Nucleic Acids Res* **30**(9): E36.

Relative quantification

3

Michael W. Pfaffl

3.1 Introduction

Reverse transcription (RT) followed by a polymerase chain reaction (PCR) represents the most powerful technology to amplify and detect trace amounts of mRNA (Heid *et al.*, 1996; Lockey, 1998). To quantify these low abundant expressed genes in any biological matrix the real-time quantitative RT-PCR (qRT-PCR) is the method of choice. Real-time qRT-PCR has advantages compared with conventionally performed 'semi-quantitative end point' RT-PCR, because of its high sensitivity, high specificity, good reproducibility, and wide dynamic quantification range (Higuchi *et al.*, 1993; Gibson *et al.*, 1996; Orland *et al.*, 1998; Freeman *et al.*, 1999; Schmittgen *et al.*, 2000; Bustin, 2000). qRT-PCR is the most sensitive and most reliable method, in particular for low abundant transcripts in tissues with low RNA concentrations, partly degraded RNA, and from limited tissue sample (Freeman *et al.*, 1999; Steuerwald *et al.*, 1999; Mackay *et al.*, 2002). While real-time RT-PCR has a tremendous potential for analytical and quantitative applications in transcriptome analysis, a comprehensive understanding of its underlying quantification principles is important. High reaction fidelity and reliable results of the performed mRNA quantification process is associated with standardized pre-analytical steps (tissue sampling and storage, RNA extraction and storage, RNA quantity and quality control), optimized RT and PCR performance (in terms of specificity, sensitivity, reproducibility, and robustness) and exact post-PCT data procession (data acquisition, evaluation, calculation and statistics) (Bustin, 2004; Pfaffl, 2004; Burkardt, 2000).

The question which might be the 'best RT-PCR quantification strategy' to express the exact mRNA content in a sample has still not been answered to universal satisfaction. Numerous papers have been published, proposing various terms, like 'absolute', 'relative', or 'comparative' quantification. Two general types of quantification strategies can be performed in qRT-PCR. The levels of expressed genes may be measured by an 'absolute' quantification or by a relative or comparative real-time qRT-PCR (Pfaffl, 2004). The 'absolute' quantification approach relates the PCR signal to input copy number using a calibration curve (Bustin, 2000; Pfaffl and Hageleit, 2001; Fronhoffs *et al.*, 2002). Calibration curves can be derived from diluted PCR products, recombinant DNA or RNA, linearized plasmids, or spiked tissue samples. The reliability of such a an absolute real-time RT-PCR assay depends on the condition of 'identical' amplification efficiencies for both the native mRNA target and the target RNA or DNA used in the calibration curve (Souaze et al., 1996; Pfaffl, 2001). The so-called 'absolute' quantification is misleading, because the quantification is shown relative to the used calibration curve. The mRNA copy numbers must be correlated to some biological parameters, like mass of tissue, amount of total RNA or DNA, a defined amount of cells, or compared with a reference gene copy number (e.g. ribosomal RNA, or commonly used house keeping genes (HKG)). The 'absolute' quantification strategy using various calibration curves and applications are summarized elsewhere in detail (Pfaffl and Hageleit, 2001; Donald et al., 2005; Lai et al., 2005; Pfaffl et al., 2002).

This chapter describes the relative quantification strategies in quantitative real-time RT-PCR with a special focus of relative quantification models and newly developed relative quantification software tools.

3.2 Relative quantification: The quantification is relative to what?

Relative quantification or comparative quantification measures the relative change in mRNA expression levels. It determines the changes in steady-state mRNA levels of a gene across multiple samples and expresses it relative to the levels of another RNA. Relative quantification does not require a calibration curve or standards with known concentrations and the reference can be any transcript, as long as its sequence is known (Bustin, 2002). The units used to express relative quantities are irrelevant, and the relative quantities can be compared across multiple real-time RT-PCR experiments (Orlando et al., 1998; Vandesompele et al., 2002; Hellemans et al., 2006). It is the adequate tool to investigate small physiological changes in gene expression levels. Often constant expressed reference genes are chosen as reference genes, which can be co-amplified in the same tube in a multiplex assay (as endogenous controls) or can be amplified in a separate tube (as exogenous controls) (Wittwer et al., 2001; Livak, 1997, 2001; Morse et al., 2005). Multiple possibilities are obvious to compare a gene of interest (GOI) mRNA expression to one of the following parameters. A gene expression can be relative to:

- an endogenous control, e.g. a constant expressed reference gene or another GOI
- an exogenous control, e.g. an universal and/or artificial control RNA or DNA
- an reference gene index, e.g. consisting of multiple averaged endogenous controls
- a target gene index, e.g. consisting of averaged GOIs analyzed in the study

To determine the level of expression, the differences (Δ) between the threshold cycle (C_t) or crossing points (C_p) are measured. Thus the mentioned methods can be summarized as the ΔC_p methods (Morse et al., 2005; Livak and Schmittgen, 2001). But the complexity of the relative quantification procedure can be increased. In a further step a second relative parameter can be added, e.g. comparing the GOI expression level relative to:

- a nontreated control
- a time point zero
- healthy individuals

These more complex relative quantification methods can be summarized as $\Delta\Delta C_P$ methods (Livak and Schmittgen, 2001).

3.3   Normalization

To achieve optimal relative expression results, appropriate normalization strategies are required to control for experimental error (Vandesompele *et al.*, 2002; Pfaffl *et al.*, 2004), and to ensure identical cycling performance during real-time PCR. These variations are introduced by various processes required to extract and process the RNA, during PCR set-up and by the cycling process. All the relative comparisons should be made on a constant basis of extracted RNA, on analyzed mass of tissue, or an identical amount of selected cells (e.g. microdissection, biopsy, cell culture or blood cells) (Skern *et al.*, 2005). To ensure identical starting conditions, the relative expression data have to be equilibrated or normalized according to at least one of the following variables:

- sample size/mass or tissue volume
- total amount of extracted RNA
- total amount of genomic DNA
- reference ribosomal RNAs (e.g. 18S or 28S rRNA)
- reference messenger RNAs (mRNA)
- total amount of genomic DNA
- artificial RNA or DNA molecules (= standard material)

But the quality of normalized quantitative expression data cannot be better than the quality of the normalizer itself. Any variation in the normalizer will obscure real changes and produce artefactual changes (Bustin, 2002; Bustin *et al.*, 2005).

It cannot be emphasized enough that the choice of housekeeping or lineage specific genes is critical. For a number of commonly used reference genes, processed pseudogenes have been shown to exist, e.g. for β-actin or GAPDH (Dirnhofer *et al.*, 1995; Ercodani *et al.*, 1988). Pseudogenes may be responsible for specific amplification products in a fully mRNA independent fashion and result in specific amplification even in the absence of intact mRNA. It is vital to develop universal, artificial, stable, internal standard materials, that can be added prior to the RNA preparation, to monitor the efficiency of RT as well as the kinetic PCR respectively (Bustin, 2002). Usually more than one reference gene should be tested in a multiple pair-wise correlation analysis and a summary reference gene index be obtained (Pfaffl *et al.*, 2004). This represents a weighted expression of at least three reference genes and a more reliable basis of normalization in relative quantification can be postulated.

There is increasing appreciation of these aspects of qRT-PCR software tools were established for the evaluation of reference gene expression levels. *geNorm* (Vandesompele *et al.*, 2002) and *BestKeeper* (Pfaffl *et al.*, 2004) allows for an accurate normalization of real-time qRT-PCR data by geometric averaging of multiple internal control genes (http://medgen.ugent.be/~jvdesomp/genorm). The *geNorm* Visual Basic applet for Microsoft Excel® determines the most stable reference genes from a set of 10 tested genes in a given cDNA sample panel, and calculates a gene expression normalization factor for each tissue sample based on the geometric mean of a user defined number of reference genes. The normalization strategy used in *geNorm* is a prerequisite for accurate kinetic RT-PCR expression profiling, which opens up the possibility of studying the biological relevance of small expression differences (Vandesompele *et al.*, 2002). These normalizing strategies are summarized and described in detail elsewhere (Huggett *et al.*, 2005; LightCycler® Relative Quantification Software, 2001).

3.4 Mathematical models

The relative expression of a GOI in relation to another gene, mostly to an adequate reference gene, can be calculated on the basis of 'delta C_p' (ΔC_p, 24) or 'delta delta C_t' ($\Delta\Delta C_t$) values (Livak and Schmittgen, 2001). Today various mathematical models are established to calculate the relative expression ratio (R), based on the comparison of the distinct cycle differences. The C_p value can be determined by various algorithms, e.g. C_p at a constant level of fluorescence or C_p acquisition according to the established mathematic algorithm (see Section 3.6).

Three general procedures of calculation of the relative quantification ratio are established:

1. The so-called 'delta C_t' (eqs. 1–2 using ΔC_p) or 'delta-delta C_t' method (eqs. 3–4 using $\Delta\Delta C_p$) without efficiency correction. Here an optimal doubling of the target DNA during each performed real-time PCR cycle is assumed (Livak, 1997, 2001; Livak and Schmittgen, 2001). Such expression differences on basis of ΔC_p values are shown in *Figure 3.1*.

$$R = 2^{[C_p \text{ sample} - C_p \text{ control}]} \qquad \text{(eq. 1)}$$

$$R = 2^{\Delta C_p} \qquad \text{(eq. 2)}$$

$$R = 2^{-[\Delta C_p \text{ sample} - \Delta C_p \text{ control}]} \qquad \text{(eq. 3)}$$

$$R = 2^{-\Delta\Delta C_p} \qquad \text{(eq. 4)}$$

2. The efficiency corrected calculation models, based on ONE sample (eqs. 5–6) (Souaze *et al.*, 1996; LightCycler® Relative Quantification Software, 2001) and the efficiency corrected calculation models, based on MULTIPLE samples (eqs. 7) (Pfaffl, 2004).

$$\text{ratio} = \frac{(E_{\text{target}})^{\Delta C_p \text{ target (control − sample)}}}{(E_{\text{Ref}})^{\Delta C_p \text{ Ref (control − sample)}}} \qquad \text{(eq. 5)}$$

$$\text{ratio} = \frac{(E_{\text{Ref}})^{C_p \text{ sample}}}{(E_{\text{target}})^{C_p \text{ sample}}} \div \frac{(E_{\text{Ref}})^{C_p \text{ calibrator}}}{(E_{\text{target}})^{C_p \text{ calibrator}}} \qquad \text{(eq. 6)}$$

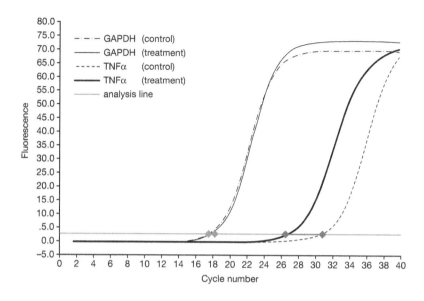

Figure 3.1

Effect of LPS treatment of TNFα target gene expression and on GAPDH reference gene expression in bovine white blood cells. Expression differences are shown by ΔC$_p$ values.

$$\text{ratio} = \frac{(E_{target})^{\Delta C_{P\ target}\ (MEAN\ control\ -\ MEAN\ sample)}}{(E_{Ref})^{\Delta C_{P\ Ref}\ (MEAN\ control\ -\ MEAN\ sample)}} \qquad \text{(eq. 7)}$$

3. An efficiency corrected calculation models, based on MULTIPLE sample and on MULTIPLE reference genes, so-called REF index, consisting at least of three reference genes (eq. 8) (Pfaffl, 2004).

$$R = \frac{(E_{target})^{\Delta C_{P\ target}\ (MEAN\ control\ -\ MEAN\ sample)}}{(E_{Ref\ index})^{\Delta C_{P\ Ref\ index}\ (MEAN\ control\ -\ MEAN\ sample)}} \qquad \text{(eq. 8)}$$

In these models, the target-gene expression is normalized by one or more non-regulated reference gene (REF) expression, e.g., derived from classical and frequently described reference genes (Bustin, 2000; Vandesompele *et al.*, 2002; Pfaffl *et al.*, 2005). The crucial problem in this approach is that the most common reference-gene transcripts from so-called stable expressed housekeeping gene are influenced by the applied treatment. The detected mRNA expressions can be regulated and these levels vary significantly during treatment, between tissues and/or individuals (Pfaffl, 2004; Schmittgen and Zakrajsek, 2000).

Thus always one question appears: which is the right reference to normalize with and which one(s) is (are) the best housekeeping- or reference gene(s) for my mRNA quantification assay? Up to now no general answer can be given. Each researcher has to search and validate each tissue and treatment analyzed for its own stable expressed reference genes. Further, each primer and probe combination, detection chemistry, tubes and the real-time cycler platform interfere with the test performance. However, qRT-PCR is influenced by numerous variables and appears as a multifactorial reaction. Thus, relative quantification must be highly validated to generate useful and biologically relevant information.

The main disadvantage of using reference genes as external standards is the lack of internal control for RT and PCR inhibitors. All quantitative PCR methods assume that the target and the sample amplify with similar efficiency (Wittwer *et al.*, 2001; Livak and Schmittgen, 2001). The risk with external references is that some analyzed samples may contain substances that significantly influence the real-time PCR amplification efficiency of the PCR reaction. As discussed earlier (Pfaffl, 2004), sporadic RT and PCR inhibitors or enhancers can occur.

3.5 Real-time qPCR amplification efficiency

Each analyzed sample generates an individual amplification history during real-time fluorescence analysis. As we know from laboratory practice, biological replicates, even technical replicates, result in significantly different fluorescence curves as a result of sample-to-sample variations (*Figure 3.2*). Changing PCR efficiencies are caused by RT and PCR inhibitors or enhancers, and by variations in the RNA pattern extracted. Thus the shapes of fluorescence amplification curves differ in the background level (noisy, constant or increasing), the take-off point (early or late), the steepness (good

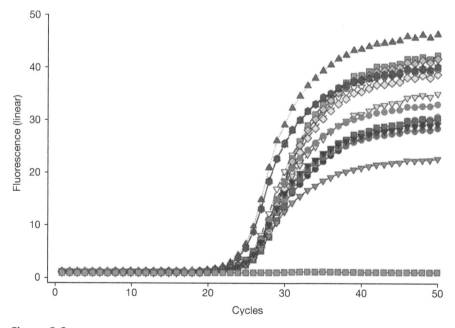

Figure 3.2

Variation of fluorescence amplification plot of three different genes run in quadruplicates.

or bad efficiency), the change-over to the plateau phase (quick or steady), and in the appearance of the PCR plateau (constant, in or decreasing trend) (Tichopad *et al.*, 2003; Tichopad *et al.*, 2004). The PCR amplification efficiency bears the biggest impact on amplification kinetics and is critically influenced by PCR reaction components. Therefore C_P determination of the threshold level and in consequence the accuracy of the quantification results are influenced by the amplification efficiency. The efficiency evaluation is an essential marker and the correction is necessary in real-time gene quantification (Rasmussen, 2001; Liu and Saint, 2002a; Liu and Saint, 2002b; Tichopad *et al.*, 2003; Meijerink *et al.*, 2001).

A constant amplification efficiency in all compared samples is one important criterion for reliable comparison between samples. This becomes crucially important when analyzing the relationship between an unknown and a reference sequence, which is performed in all relative quantification models. In experimental designs employing standardization with reference genes, the demand for invariable amplification efficiency between target and standard is often ignored, despite the fact that corrections have been suggested in the recent literature (Pfaffl, 2001; Pfaffl *et al.*, 2002; Liu and Saint, 2002a; Liu and Saint, 2002b; Soong *et al.*, 2000; Wilhelm *et al.*, 2003). A correction for efficiency, as performed in efficiency corrected mathematical models (eqs. 5–8), is strongly recommended and results in a more reliable estimation of the 'real' expression changes compared with NO efficiency correction. Even small efficiency differences between target and reference generate false expression ratio, and the researcher over- or under-estimates the initial mRNA amount. A theoretic difference in qPCR efficiency (ΔE) of 3% ($\Delta E = 0.03$) between a low copy target gene and medium copy reference gene generate falsely calculated differences in expression ratio of 242% in case of $E_{target} > E_{ref}$ after 30 performed cycles. This gap will increase dramatically by higher efficiency differences $\Delta E = 0.05$ (432%) and $\Delta E = 0.10$ (1,744%). The assessment of the sample specific efficiencies must be carried out before any relative calculation is done. Some tools are available to correct for efficiency differences. The LightCycler® Relative Expression Software (2001), Q-Gene (Muller *et al.*, 2002), qBase (Hellmans *et al.*, 2006), SoFar (Wilhelm *et al.*, 2003), and various REST software applications (LightCycler® Relative Quantification Software, 2001; Pfaffl *et al.*, 2002; Pfaffl and Horgan, 2002; Pfaffl and Horgan, 2005) allow the evaluation of amplification efficiency plots. In most of the applications a triplicate determination of real-time PCR efficiency for every sample is recommended. Therefore efficiency corrections should be included in the relative quantification procedure and the future software applications should calculate automatically the qPCR efficiency (Pfaffl, 2004).

3.6 Determination of the amplification rate

Up to now only one software package can automatically determine the real-time PCR efficiency sample-by-sample. In the Rotor-Gene™ 3000 software package (Corbett Research), it is called the comparative quantification. Amplification rate is calculated on the basis of fluorescence increase in the PCR exponential phase. Further algorithms and methods are described in recent publications to estimate the real-time PCR efficiency. These can be grouped in direct and indirect methods. Direct methods are based on either a dilution method or a measurement of the relative fluorescence increase in the exponential phase. On the other hand, indirect methods are published, doing the efficiency calculation on basis of a fit to a mathematical model, like sigmoidal, logistic models or an exponential curve fitting (for details see http://efficiency.gene-quantification.info).

3.6.1 Dilution method

The amplification rate is calculated on the basis of a linear regression slope of a dilution row (*Figure 3.3*). Efficiency (*E*) can be determined based on eq. 9 (Higuchi *et al.*, 1993; Rasmussen, 2001). But the real-time PCR efficiency should be evaluated sample-by-sample, which is quite laborious and costly, wastes template, and takes time if the dilution method is used. Alternatively, the pool of all sample RNAs can be used to accumulate all possible 'positive and negative impacts' on kinetic PCR efficiency. Applying the dilution method, usually the efficiency varies in a range of *E* = 1.60 to values over 2 (*Figure 3.3*) (Souaze *et al.*, 1996).

$$E = 10^{\,[-1/\text{slope}]}$$ (eq. 9)

Typically, the relationship between C_P and the logarithm of the starting copy number of the target sequence should remain linear for up to five orders of magnitude in the calibration curve as well as in the native sample

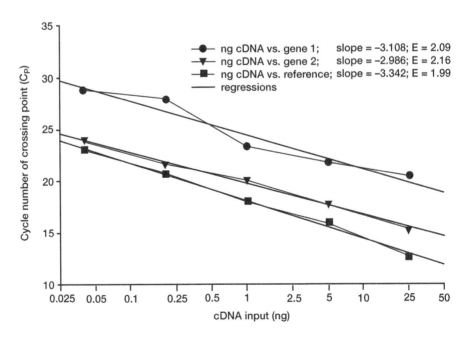

Figure 3.3

On the basis of a dilution row the real-time efficiency is calculated according to eq. 9 (Higuchi *et al.*, 1993; Rasmussen, 2001).

RNA (Muller *et al.*, 2002). The advantage of the dilution method is that it is highly reproducible and constant within one transcript and tissue. The disadvantage of this approach is the high efficiencies, often higher than two ($E > 2.0$), which is practically impossible on the basis of the PCR amplification theory. This indicates that this efficiency estimation is more or less not the best one and it will overestimate the 'real' amplification efficiency.

3.6.2 Fluorescence increase in exponential phase

Efficiency calculation from the fluorescence increases in the exponential phase of fluorescence history plot (in log. scale) (*Figure 3.4*). Fitting can be done by eye, or more reliably by software applications like LinRegPCR (Ramakers *et al.*, 2003) or DART-PCR (Peirson *et al.*, 2003). The investigator has to decide which fluorescence data to include in the analysis and which to omit. A linear regression plot is drawn from at least four data points, where the slope of the regression line represents the PCR efficiency. Therefore this method is more or less arbitrary and dependent on the chosen data points. Resulting efficiencies range between $E = 1.45$, and $E = 1.90$, and seem more realistic than the results mentioned above. This efficiency calculation method might be good estimator for the 'real efficiency,' because data evaluation is made exclusively in exponential phase.

The advantage of both direct methods is the independency of the background fluorescence. We know from several applications that a rising

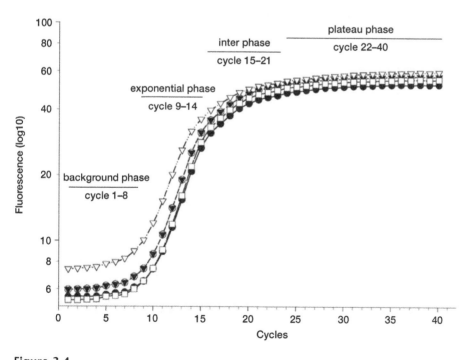

Figure 3.4

Efficiency calculation in the exponential phase.

trend in the background fluorescence will interfere with the indirect curve fit, like sigmoidal, logistic and exponential models. Probe based detection in particular exhibits high and noisy background levels, whereas SYBR® Green I applications show low and constant background fluorescence (*Figure 3.5*).

3.6.3 Sigmoidal or logistic curve fit

A number of publications have suggested an efficiency calculation on the basis of all fluorescence data points (starting at cycle 1 up to the last cycle), according to a sigmoidal or logistic curve fit model (Tichopad *et al.*, 2003; Tichopad *et al.*, 2004; Liu and Saint, 2002a; Liu and Saint, 2002b; Rutledge, 2004). The advantage of such models is that all data points will be included in the calculation process and no background subtraction is necessary. The efficiency will be calculated at the point of inflexion (cycle 27.06 shown in *Figure 3.5*) at absolute maximum fluorescence increase.

$$f(x) = y_0 + \frac{a}{1 + e^{-(\frac{x-x_0}{b})}}$$ (eq. 10)

In the four-parametric sigmoid model (eq. 10), x is the cycle number, $f(x)$ is the computed function of the fluorescence in cycle number x, y_0 is the background fluorescence, a is the difference between maximal fluorescence reached at plateau phase and background fluorescence (i.e. the plateau height), e is the natural logarithm base, x_0 is the co-ordinate of the first derivative maximum of the model or inflexion point of the curve, and b describes the slope at x_0 in the log–linear phase (Tichopad *et al.*, 2004). But

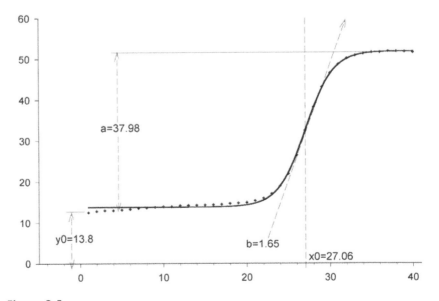

Figure 3.5

Efficiency calculation on the basis of a four-parametric sigmoid model (eq. 10).

the derived slope parameters generated by the sigmoidal or logistic models, e.g. b, can not directly compared with the 'real PCR efficiency.' The advantages of the four-parametric sigmoid model is that it is easy to perform, is a good estimator for the maximum curve slope with high correlation between replicates ($r > 0.99$) and the algorithm can easily implemented in analysis software. The resulting efficiencies are comparable to the latter method and range from 1.35 to 1.65.

3.6.4 Efficiency calculation in the exponential phase using multiple models

Here we describe the efficiency calculation in the exponential phase using multiple models, first, a linear, second, a logistic and third, an exponential model (Tichopad *et al.*, 2003). The background phase is determined with the linear model using studentized residual statistics. The phase until the second derivative maximum (SDM) of the logistic fit exhibits a real exponential amplification behavior (*Figure 3.6*). The phase behind including the first derivative maximum (FDM) shows suboptimal and decreasing amplification efficiencies and therefore has to be excluded from the analysis. Efficiency calculation is only performed between the background and before SDM. Here an exponential model according to a polynomial curve fit is performed, according to eq. 11.

$$Y_n = Y_0 \, (E)^n \qquad \text{(eq. 11)}$$

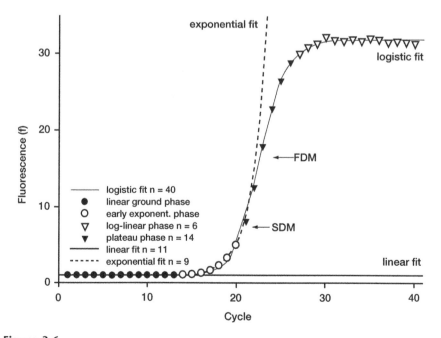

Figure 3.6

Efficiency calculation in the exponential phase using multiple model fitting: linear, logistic and exponential model (Tichopad *et al.*, 2003).

In the polynomial model, Y_n is fluorescence acquired at cycle n, and Y_0 initial fluorescence, and E represents the efficiency. Here in the exponential part of the PCR reaction, kinetic is still under 'full amplification power' with no restrictions. The calculation is performed on each reaction kinetic plot and the amplification efficiency can be determined exactly. They range from $E = 1.75$ to $E = 1.90$, in agreement with the other methods.

A comparable multi-factorial model is used in the SoFAR software application (Wilhelm *et al.*, 2003). Here the background is corrected by a least square fit of the signal curve. Efficiency is determined by an exponential growth function (eq. 11) or a logistic or sigmoidal fit (eq. 10). The sigmoidal exponential function was the most precise one and could increase the amplification efficiency, before and after correction, from around 62% up to 82% (Wilhelm *et al.*, 2003).

All models lead to efficiency estimates, but which model results in the 'right', most accurate and realistic real-time amplification efficiency estimate has to be evaluated in further experiments. From our experiment we know that the detection chemistry, the type of tubes (plastic tubes or glass capillaries), the cycling platform as well the optical system has considerable influence on the estimates of real-time efficiency. Better dyes and much more sensitive optical and detection systems are needed to guarantee a reliable efficiency calculation. In *Table 3.1* an overview of the existing efficiency calculation methods is shown.

Table 3.1 Overview of existing efficiency calculation methods.

Summary	Sample individual determination	Overestimation + Intermediate Ø Underestimation –	Combination of efficiency and C_p determination
Dilution series (fit point or SDM) Rasmussen (2001)	no	+ n = 3–5	
Fluorescence increase Various authors	+	– n = 3–6	
Fluorescence increase Peccoud and Jacob (1996)	+	– n = 3	
Sigmoidal model Lui and Saint (2002a, 2002b) Tichopad *et al.* (2004) Wilhelm *et al.* (2003) Rutledge (2004)	+	– n = 1	+
LinRegPCR Ramakers *et al.* (2003)	+	Ø n = 4–6	
KOD Bar *et al.* (2003)	+	Ø n = 3–5	
Logistic model Tichopad *et al.* (2003) Wilhelm *et al.* (2003)	+	Ø n > 7	+
Rotor-Gene™ 3000 Comparative quantitation analysis	+	Ø n = 4	+

3.7 What is the right crossing point to determine?

The C_p value is the central value in real-time PCR applications. Everything is related to this single point. But not much effort has been put into standardizing and optimizing the determination of this parameter that is so central to quantification. Most software use the so-called 'threshold cycle method' or 'fit point method' and measure the C_p at a constant fluorescence level. But there are other possibilities and options to consider. Let us first think about the background:

- What kind of background fluorescence is evident, a noisy, a constant, a rising or a decreasing background?
- Does the software show me my real raw-fluorescence-data or are the data already manipulated, e.g., additional ROX adjustment?
- What about the curve smoothing of the fluorescence data?
- Which kind of fluorescence background correction and/or subtraction is applied?

Most real-time platforms show pre-adjusted fluorescence data and pre-adjusted C_p. After doing an automatic background correction the C_p value are determined by various methods, e.g., at a constant level of fluorescence. These constant threshold methods assume that all samples have the same DNA concentration at the threshold fluorescence. But measuring the level of background fluorescence can be a challenge. Often real-time PCR reactions with significant background fluorescence variations occur, caused by drift-ups and drift-downs over the course of the reaction. Averaging over a drifting background will give an overestimation of variance and thus increase the threshold level (Livak, 1997, 2001; Rasmussen, 2001; Wilhelm *et al.*, 2003). The threshold level can be calculated by fitting the intersecting line at 10 standard deviations above baseline fluorescence level. This acquisition mode can be easily automated and is very robust (Livak, 1997, 2001). In the fit point method the user has to discard uninformative background points, exclude the plateau values by entering the number of log-linear points, and then fit a log line to the linear portion of the amplification curves. These log lines are extrapolated back to a common threshold line and the intersection of the two lines provides the C_p value. The strength of this method is that it is extremely robust. The weakness is that it is not easily automated and so requires a lot of user interaction, which are more or less arbitrary (Rasmussen, 2001, LightCycler® Software, 2001).

The real problem lies in comparing numerous biological samples. The researcher will have problems in defining a constant background for all samples within one run or between runs. These sample-to-sample differences in variance and absolute fluorescence values lead to the development of a new and user-friendly C_p acquisition model. As discussed in the previous section there are several mathematical models to determine the amplification rate, using a logistic or sigmoidal model. These mathematically fit models can also be used to determine the optimal C_p (*Table 3.1*). They are more or less independent of the background level or calculated on the basis of the background fluorescence and implement the data in the C_p determination model (Tichopad *et al.*, 2004; Wilhelm *et al.*, 2003).

In LightCycler® (Roche Applied Science) and Rotor-Gene™ (Corbett

Research) software packages these approaches are already implemented. In second derivative maximum method the C_p is automatically identified and measured at the maximum acceleration of fluorescence (Ramussen, 2001; LightCycler® Software, 2000). The exact mathematical algorithm applied is still unpublished, but is very comparable to a logistic fit. In the Rotor-Gene family using comparative quantification the 'take of point' is also calculated on basis of a sigmoidal model. Both the sigmoidal and polynomial curve models' work well with high agreement ($P<0.001$; r>0.99) (Tichopad *et al.*, 2004; Liu and Saint, 2002a; Liu and Saint, 2002b; Rutledge, 2004). The sigmoidal exponential function was the more precise and could increase the exactness and precision of the C_p measurement as well as the amplification efficiency rate (Wilhelm *et al.*, 2003). Peirson further discusses the importance of threshold setting in relative quantification in Chapter 6.

3.8 Relative quantification data analysis and software applications

A major challenge is the development of exact and reliable gene expression analysis and quantification software. A 'one-fits-all' detection and application software is the target for future developments and seems the optimal solution. But can we implement various detection chemistries with varying background and fluorescence acquisition modes in one software package? Should we not think about optimized models on each real-time platform and for each applied chemistry? In biological research and in clinical diagnostics, real-time qRT-PCR is the method of choice for expression profiling. On the one hand cycler and chemistry developed much faster than detection and analysis software. However, accurate and straightforward mathematical and statistical analysis of the raw data (cycle threshold/crossing point values or molecules quantified) as well as the management of growing data sets have become the major hurdles in gene expression analyses. Now the 96-well applications are the standard in the research laboratories, but in the near future high throughput 384-well applications will generate huge amounts of data. The data need to be grouped (Hellemans *et al.*, 2006) and standardized by intelligent algorithms. Real-time qPCR data should be analyzed according to automated statistical method, e.g. Kinetic Outlier Detection (KOD), to detect samples with dissimilar efficiencies (Bar *et al.*, 2003). Mostly the statistical data analysis or C_p values is performed on the basis of classical standard parametric tests, such as analysis of variance or t-tests. Parametric tests depend on assumptions, such as normality of distributions, whose validity is unclear (Sheskin, 2000). In absolute or relative quantification analysis, where the quantities of interest are derived from ratios and variances can be high, normal distributions might not be expected, and it is unclear how a parametric test could best be constructed (Pfaffl *et al.*, 2002; Sheskin, 2000). At present, the following relative quantification data analysis and software applications are available.

3.8.1 LightCycler® Relative Quantification Software

The first commercially available software was the LightCycler® Relative Quantification Software (2001). It can be used to calculate and compare relative quantification results of triplicates of a target versus a calibrator gene. Target genes are corrected via a reference-gene expression and calculates on the basis of the median of the performed triplets. Real-time PCR efficiency correction is possible within the software and is calculated from the calibration curve slope, according to the established eq. 9, ranging from $E = 1.0$ (minimum value) to $E = 2.0$ (theoretical maximum and efficiency optimum). A given correction factor and a multiplication factor, which are provided in the product specific applications by Roche Molecular Biochemicals (LightCycler® Relative Quantification Software, 2001), have to be incorporated in eq. 6. Importantly, no statistical comparison of the results by a statistical test is possible.

3.8.2 REST

In 2002, the relative expression software tool (REST, http://rest.gene-quantification.info) was established as a new tool (Pfaffl *et al.*, 2002). The first REST version is Excel®-based and programmed in Visual Basic to compare several gene expressions on C_P level. It compares two treatment groups, with multiple data points in the sample versus control groups, and calculates the relative expression ratio between them. The mathematical model used is published and is based on the mean C_P deviation between sample and control group of target genes, normalized by the mean C_P deviation of one reference gene as shown in eq. 7 (Pfaffl *et al.*, 2002). Further an efficiency correction can be performed, either based on the dilution method (eq. 9) or an optimal efficiency of $E = 2.0$ is assumed. The big advantage of REST is the provision of a subsequent statistical test of the analyzed C_P values by a *Pair-Wise Fixed Reallocation Randomization Test* (Pfaffl *et al.*, 2002). Permutation or randomization tests are a useful alternative to more standard parametric tests for analyzing experimental data (Manly, 1997; Horgan and Rouault, 2000). They have the advantage of making no distributional assumptions about the data, while remaining as powerful as conventional tests. Randomization tests are based on one we know to be true: that treatments were randomly allocated. The randomization test repeatedly and randomly reallocates at least 2000 times the observed C_P values to the two groups and notes the apparent effect each time, here in the expression ratio between sample and control treatment. The REST software package makes full use of the advantages of a randomization test. In the applied two-sided *Pair-Wise Fixed Reallocation Randomization Test* for each sample, the C_P values for reference and target genes are jointly reallocated to control and sample groups (= pair-wise fixed reallocation), and the expression ratios are calculated on the basis of the mean values. In practice, it is impractical to examine all possible allocations of data to treatment groups, and a random sample is drawn. If 2000 or more randomizations are taken, a good estimate of P-value (standard error <0.005 at $P = 0.05$) is obtained. Randomization tests with a pair-wise reallocation are seen as the most appropriate approach for this type of application. In 2005 various new REST versions were developed, calculating with a geometric mean averaged REF index (Vandescompele *et al.*, 2002; Pfaffl *et al.*, 2004), according to the mathematical model described in eq. 8, which can analyze 15 target and reference genes (REST-384) (LightCycler® Relative Quantification Software,

2001). Specialized REST versions can compare six treatment group with one non-treated control (REST-MCS, REST – Multiple Condition Solver) (LightCycler® Relative Quantification Software, 2001), or take individual amplification efficiency into account, exported from the Rotor-Gene (REST-RG). A stand alone application REST-2005 was developed, running independent of Excel® or Visual Basic, comparing 'unlimited' target and reference genes, using newly developed bootstrapping statistical tool, and graphical output showing 95% confidence interval (TUM and Corbett Research, 2005) (Pfaffl and Horgan, 2005).

3.8.3 Q-Gene

Recently a second software tool, Q-Gene, was developed, which is able to perform a statistical test of the real-time data (Muller *et al.*, 2002). Q-Gene manages and expedites the planning, performance and evaluation of quantitative real-time PCR experiments. The expression results were presented by graphical presentation. An efficiency correction according to the dilution method is possible (eq. 9). Q-Gene can cope with complex quantitative real-time PCR experiments at a high-throughput scale (96-well and 384-well format) and considerably expedites and rationalizes the experimental set-up, data analysis, and data management while ensuring highest reproducibility. The *Q-Gene Statistics Add-In* is a collection of several VBA programs for the rapid and menu-guided performance of frequently used parametric and non-parametric statistical tests. To assess the level of significance between any two groups' expression values, it is possible to perform a paired or an unpaired Student's test, a Mann-Whitney U-test, or Wilcoxon signed-rank test. In addition, the Pearson's correlation analysis can be applied between two matched groups of expression values. Furthermore, all statistical programs calculate the mean values of both groups analyzed and their difference in percent (Muller *et al.*, 2002).

3.8.4 qBASE

Comparable software application qBASE was recently developed by colleagues to offer solutions to compare more real-time set-ups (Hellemans *et al.*, 2006). QBASE is an Excel®-based tool for the management and automatic analysis of real-time quantitative PCR data (http://medgen.ugent.be/qbase). The qBASE browser allows data storage and annotation while keeping track of all real-time PCR runs by hierarchically organizing data into projects, experiments, and runs. It is compatible with the export files from many currently available PCR instruments and provides easy access to all the data, both raw and processed. The qBASE analyzer contains an easy plate editor, performs quality control, converts C_P values into normalized and rescaled quantities with proper error propagation, and displays results both tabulated and in graphs. One big advantage of the program is that it does not limit the number of samples, genes and replicates, and allows data from multiple runs to be combined and processed together (Hellemans *et al.*, 2006). The possibility of using up to five reference genes allows reliable and robust normalization of gene expression levels, on the basis of the *geNorm* normalization procedure (Vandescompele *et al.*, 2002). qBASE allows the easy exchange of data between users, and exports tabulated data for further statistical analyses using dedicated software.

3.8.5 SoFAR

The algorithms implemented in SoFAR (distributed by Metralabs) allow fully automatic analysis of real-time PCR data obtained with a Roche LightCycler® (Roche Diagnostics) instrument. The software yields results with considerably increased precision and accuracy of real-time quantification. This is achieved mainly by the correction of amplification independent fluorescence signal trends and a robust fit of the exponential phase of the signal curves. The melting curve data are corrected for signal changes not due to the melting process and are smoothed by fitting cubic splines. Therefore, sensitivity, resolution, and accuracy of melting curve analyses are improved (Wilhelm *et al.*, 2003).

3.8.6 DART-PCR

DART-PCR (Data Analysis for Real-Time PCR) provides a simple means of analyzing real-time PCR data from raw fluorescence data (Peirson *et al.*, 2003) (http://nar.oxfordjournals.org/cgi/content/full/31/14/e73/DC1). This allows an automatic calculation of amplification kinetics, as well as performing the subsequent calculations for the relative quantification and calculation of assay variability. Amplification efficiencies are also tested to detect anomalous samples within groups (outliers) and differences between experimental groups (amplification equivalence). Data handling was simplified by automating all calculations in an Excel® worksheet, and enables the rapid calculation of threshold cycles, amplification rate and resulting starting values, along with the associated error, from raw data. Differences in amplification efficiency are assessed using one-way analysis of variance (ANOVA), based upon the null hypotheses, that amplification rate is comparable within sample groups (outlier detection) and that amplification efficiency is comparable between sample groups (amplification equivalence) (Peirson *et al.*, 2003).

3.9   Conclusion

Facilitating data management and providing tools for automatic data analysis, these software applications address one of the major problems in doing real-time quantitative PCR-based nucleic acid quantification. Nevertheless, successful application of real-time RT-PCR and relative quantification depends on a clear understanding of the practical problems. Therefore a coherent experimental design, application, and validation of the individual real-time RT-PCR assay remains essential for accurate, precise and fully quantitative measurement of mRNA transcripts. An advantage of most described software applications (except SoFAR) is that they are freely available and scientists can use them for their academic research. qBASE intends to be an open source project and interested parties can write their own analysis or visualization plug-ins. All calculation- and statistical-software applications are summarized and described in detail at http://bioinformatics.gene-quantification.info.

References

Bar T, Stahlberg A, Muszta A, Kubista M (2003) Kinetic Outlier Detection (KOD) in real-time PCR. *Nucleic Acids Res* **31**(17): e105.

Burkardt HJ (2000) Standardization and quality control of PCR analyses. *Clin Chem Lab Med* **38**(2): 87–91.

Bustin SA. (2000) Absolute quantification of mRNA using real-time reverse transcription polymerase chain reaction assays. *J Mol Endocrinol* **25**: 169–193.

Bustin SA (2002) Quantification of mRNA using real-time RT-PCR. Trends and problems. *J Mol Endocrinol* **29**(1): 23–39.

Bustin SA (ed) (2004) *A–Z of quantitative PCR*. La Jolla, CA: IUL Biotechnology Series, International University Line.

Bustin SA, Benes V, Nolan T, Pfaffl MW (2005) Quantitative real-time RT-PCR – a perspective. *J Mol Endocrinol* **34**(3): 597–601.

Dirnhofer S, Berger C, Untergasser G, Geley S, Berger P (1995) Human beta-actin retro pseudogenes interfere with RT-PCR. *Trends Genet* **11**(10): 380–381.

Donald CE, Qureshi F, Burns MJ, Holden MJ, Blasic JR Jr, Woolford AJ (2005) An inter-platform repeatability study investigating real-time amplification of plasmid DNA. *BMC Biotechnol* **5**(1): 15.

Ercolani L, Florence B, Denaro M, Alexander M (1988) Isolation and complete sequence of a functional human glyceraldehyde-3-phosphate dehydrogenase gene. *J Biol Chem* **263**(30): 15335–15341.

Freeman TC, Lee K, Richardson PJ (1999) Analysis of gene expression in single cells. *Curr Opin Biotechnol* **10**(6): 579–582.

Fronhoffs S, Totzke G, Stier S, Wernert N, Rothe M, Bruning T, Koch B, Sachinidis A, Vetter H, Ko Y (2002) A method for the rapid construction of cRNA standard curves in quantitative real-time reverse transcription polymerase chain reaction. *Mol Cell Probes* **16**(2): 99–110.

Gibson UE, Heid CA, Williams PM (1996) A novel method for real time quantitative RT-PCR. *Genome Res* **6**: 1095–1101.

Heid CA, Stevens J, Livak KJ, Williams PM (1996) Real time quantitative PCR. *Genome Res* **6**: 986–993.

Hellemans J, Mortier G, Coucke P, De Paepe A, Speleman F, Vandesompele J (2006) qBase: open source relative quantification software for management and automated analysis of real-time quantitative PCR data (submitted to *Biotechniques*).

Higuchi R, Fockler C, Dollinger G, Watson R (1993) Kinetic PCR analysis: Real-time monitoring of DNA amplification reactions. *Biotechnology* **11**(9): 1026–1030.

Horgan GW, Rouault J (2000) *Introduction to Randomization Tests*. Biomathematics and Statistics Scotland.

Huggett J, Dheda K, Bustin S, Zumla A (2005) Real-time RT-PCR normalisation; strategies and considerations. *Genes Immun* **6**(4): 279–284.

Lai KK, Cook L, Krantz EM, Corey L, Jerome KR (2005) Calibration curves for real-time PCR. *Clin Chem* **51**(7): 1132–1136.

LightCycler Relative Quantification Software (2001) Version 1.0, Roche Molecular Biochemicals.

LightCycler Software® (2001) Version 3.5; Roche Molecular Biochemicals.

Liu W, Saint DA (2002a) A new quantitative method of real time reverse transcription polymerase chain reaction assay based on simulation of polymerase chain reaction kinetics. *Anal Biochem* **302**(1): 52–59.

Liu W, Saint DA (2002b) Validation of a quantitative method for real time PCR kinetics. *Biochem Biophys Res Commun* **294**(2): 347–353.

Livak KJ (1997 & 2001) *ABI Prism 7700 Sequence detection System User Bulletin #2 Relative quantification of gene expression*. ABI company publication.

Livak KJ, Schmittgen TD (2001) Analysis of relative gene expression data using real-time quantitative PCR and the $2^{\Delta\Delta}C(T)$ Method. *Methods* **25**(4): 402–408.

Lockey C, Otto E, Long Z (1998) Real-time fluorescence detection of a single DNA molecule. *Biotechniques* **24**: 744–746.

Mackay IM, Arden KE, Nitsche A (2002) Real-time PCR in virology. *Nucleic Acids Res* **30**: 1292–1305.

Manly B (1998) *Randomization, Bootstrap and Monte Carlo Methods in Biology*. London: Chapman & Hall.

Meijerink J, Mandigers C, van de Locht L, Tonnissen E, Goodsaid F, Raemaekers J (2001) A novel method to compensate for different amplification efficiencies between patient DNA samples in quantitative real-time PCR. *J Mol Diagn* **3**(2): 55–61.

Morse DL, Carroll D, Weberg L, Borgstrom MC, Ranger-Moore J, Gillies RJ (2005) Determining suitable internal standards for mRNA quantification of increasing cancer progression in human breast cells by real-time reverse transcriptase polymerase chain reaction. *Anal Biochem* **342**(1): 69–77.

Muller PY, Janovjak H, Miserez AR, Dobbie Z (2002) Processing of gene expression data generated by quantitative real-time RT-PCR. *Biotechniques* **32**(6): 1372–1378.

Orlando C, Pinzani P, Pazzagli M (1998) Developments in quantitative PCR. *Clin Chem Lab Med* **36**(5): 255–269.

Peccoud J, Jacob C (1996) Theoretical uncertainty of measurements using quantitative polymerase chain reaction. *Biophys J* **71**(1): 101–108.

Peirson SN, Butler JN, Foster RG (2003) Experimental validation of novel and conventional approaches to quantitative real-time PCR data analysis. *Nucleic Acids Res* **31**(14): e73.

Pfaffl MW (2001) A new mathematical model for relative quantification in real-time RT-PCR. *Nucleic Acids Res* (2001) **29**(9): e45.

Pfaffl MW (2004) Quantification strategies in real-time PCR. In: Bustin SA (ed), *A–Z of Quantitative PCR*, pp. 87–120. La Jolla, CA: IUL Biotechnology Series, International University Line.

Pfaffl MW, Georgieva TM, Georgiev IP, Ontsouka E, Hageleit M, Blum JW (2002) Real-time RT-PCR quantification of insulin-like growth factor (IGF)-1, IGF-1 receptor, IGF-2, IGF-2 receptor, insulin receptor, growth hormone receptor, IGF-binding proteins 1, 2 and 3 in the bovine species. *Domest Anim Endocrinol* **22**(2): 91–102.

Pfaffl MW, Hageleit M (2001) Validities of mRNA quantification using recombinant RNA and recombinant DNA external calibration curves in real-time RT-PCR. *Biotechn Lett* **23**: 275–282.

Pfaffl MW, Horgan GW (2002) Physiology, (REST©/REST-XL©) Weihenstephan, Technical University of Munich.

Pfaffl MW, Horgan GW (2005) (REST-2005©) Technical University of Munich and Corbett Research.

Pfaffl MW, Horgan GW, Dempfle L (2002) Relative expression software tool (REST) for group-wise comparison and statistical analysis of relative expression results in real-time PCR. *Nucleic Acids Res* **30**(9): e36.

Pfaffl MW, Horgan GW, Vainshtein Y, Avery P (2005) (REST-384©. REST-MCS©, REST-RG©) Physiology, Weihenstephan, Technical University of Munich.

Pfaffl MW, Tichopad A, Prgomet C, Neuvians TP (2004) Determination of stable housekeeping genes, differentially regulated target genes and sample integrity: BestKeeper – Excel-based tool using pair-wise correlations. *Biotechnol Lett* **26**(6): 509–515.

Ramakers C, Ruijter JM, Deprez RH, Moorman AF (2003) Assumption-free analysis of quantitative real-time polymerase chain reaction (PCR) data. *Neurosci Lett* **339**(1): 62–66.

Rasmussen, R (2001) Quantification on the LightCycler. In: Meuer S, Wittwer C, Nakagawara K (eds) *Rapid cycle real-time PCR, methods and applications*. Springer Press, Heidelberg.

Rutledge RG (2004) Sigmoidal curve-fitting redefines quantitative real-time PCR with the prospective of developing automated high-throughput applications. *Nucleic Acids Res* **32**(22): e178.

Schmittgen TD, Zakrajsek BA (2000) Effect of experimental treatment on housekeeping gene expression: validation by real-time, quantitative RT-PCR. *J Biochem Biophys Methods* **46**(1–2): 69–81.

Schmittgen TD, Zakrajsek BA, Mills AG, Gorn V, Singer MJ, Reed MW (2000) Quantitative reverse transcription-polymerase chain reaction to study mRNA decay: comparison of endpoint and real-time methods. *Anal Biochem* **285**(2): 194–204.

Sheskin D (2000) *Handbook of Parametric and Nonparametric Statistical Procedures*. CRC Press LLC, Boca Raton, FL.

Skern R, Frost P, Nilsen F (2005) Relative transcript quantification by quantitative PCR: roughly right or precisely wrong? *BMC Mol Biol* **6**(1): 10.

Soong R, Ruschoff J, Tabiti, K (2000) Detection of colorectal micrometastasis by quantitative RT-PCR of cytokeratin 20 mRNA. Roche Molecular Biochemicals internal publication.

Souaze F, Ntodou-Thome A, Tran CY, Rostene W, Forgez P (1996) Quantitative RT-PCR: limits and accuracy. *Biotechniques* **21**(2): 280–285.

Steuerwald N, Cohen J, Herrera RJ, Brenner CA (1999) Analysis of gene expression in single oocytes and embryos by real-time rapid cycle fluorescence monitored RT-PCR. *Mol Hum Reprod* **5**: 1034–1039.

Tichopad A, Didier A, Pfaffl MW (2004) Inhibition of real-time RT-PCR quantification due to tissue specific contaminants. *Molec Cellular Probes* **18**: 45–50.

Tichopad A, Dilger M, Schwarz G, Pfaffl MW (2003) Standardised determination of real-time PCR efficiency from a single reaction setup. *Nucl Acids Res* **31**(20): e122.

Tichopad A, Dzidic A, Pfaffl MW (2003) Improving quantitative real-time RT-PCR reproducibility by boosting primer-linked amplification efficiency. *Biotechnol Lett* **24**: 2053–2056.

Vandesompele J, De Preter K, Pattyn F, Poppe B, Van Roy N, De Paepe A, Speleman F (2002) Accurate normalization of real-time quantitative RT-PCR data by geometric averaging of multiple internal control genes. *Genome Biol* **3**(7): 0034.1–0034.11

Wilhelm J, Pingoud A, Hahn M (2003) *SoFAR* – Validation of an algorithm for automatic quantification of nucleic acid copy numbers by real-time polymerase chain reaction. *Anal Biochem* **317**(2): 218–225.

Wittwer CT, Garling DJ (1991) Rapid cycle DNA amplification: Time and temperature optimization. *BioTechniques* **10**: 76–83.

Wittwer CT, Herrmann MG, Gundry CN, Elenitoba-Johnson KS (2001) Real-time multiplex PCR assays. *Methods* **25**(4): 430–442.

Wong ML, Medrano JF (2005) Real-time PCR for mRNA quantitation. *Biotechniques* **39**(1): 75–85.

Normalization

Jim Huggett, Keertan Dheda and Stephen A Bustin

4

4.1 Introduction

The specificity, wide dynamic range and ease-of-use of the real-time quantitative reverse transcription polymerase chain reaction (qRT-PCR) has made it the method of choice for quantitating RNA levels. However, as with any scientific measurement, qRT-PCR suffers from problems caused by error (Huggett *et al.*, 2005). Consequently, data normalization is an indispensable component of the qRT-PCR analysis process, and is essential for accurate comparison of qRT-PCR measurements between different samples. Normalization controls for variation in the total mass of RNA analyzed and, ideally, corrects for any biological and technical variability. Both of these pose serious problems for gene expression analyses and manifest themselves as noise that can either mask or overstate true expression levels and patterns. It is vexing that whilst the requirement to incorporate some normalization method to control for error is obvious, in practice it remains one of the most discussed, but unsolved problems of qRT-PCR analysis (Huggett *et al.*, 2005). This chapter will outline normalization, address the associated problems when using real-time PCR and discuss data interpretation once normalization has been performed.

4.2 General error and directional shift

There are two types of error that must be controlled for in a qRT-PCR assay. One is general error (noise) and directional shift: *Figure 4.1* shows the many stages at which general error can be introduced into a qRT-PCR assay. Directional shift occurs when poor normalization either creates an artifactual difference or obscures a real difference between experimental

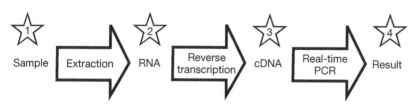

Sample — Extraction — RNA — Reverse transcription — cDNA — Real-time PCR — Result

Figure 4.1

Multistage process required to obtain a real-time RT-PCR result. Stars indicate the points at which normalization can be performed: 1: sampling, 2: RNA, 3: cDNA and 4: PCR dependent standard.

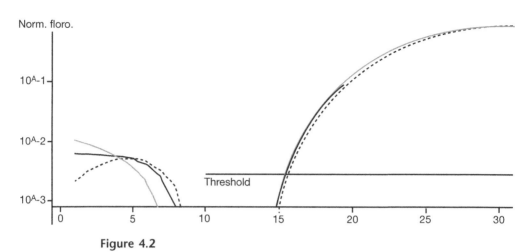

Figure 4.2

Example of a triplicate real-time PCR reaction (using the same template). This demonstrates that real-time PCR can be highly reproducible with coefficients of variation of <5%).

groups. Real-time PCR has been reported to measure two-fold differences in nucleic acid amounts (Bubner *et al.*, 2004) and is certainly capable of coefficient of variation (CV) of <1 % (*Figure 4.2*). However, achieving consistently low CVs in qRT-PCR assays is not as simple, with even the best CVs significantly higher than 1% (Melo *et al.*, 2004). The main reason for this is that it is not feasible to determine directly the target RNA. Instead, RNA quantification requires the inclusion of several steps that eventually measure cDNA copy numbers, but that also serve as sources of systematic variation, increasing CV and limiting the resolution of the assay. Variability is introduced during sampling, RNA extraction, cDNA synthesis and comparison with PCR dependent controls, and appropriate normalization of each step is essential (*Figure 4.1*).

4.3 Methods of normalization

4.3.1 Sample size

The most basic method of normalization ensures that an experiment compares similar sample sizes and this is achieved by measuring tissue weight, volume or cell number. If no attempt to use similar sample sizes is made, unnecessary error may be introduced at the first stage of the assay. Furthermore, different extraction protocols may add to variability, especially as the extraction of RNA from differing amounts of tissue is usually subject to differing efficiencies. This is a particular problem when the extraction method becomes saturated and results in inefficient extraction of RNA from large samples. Unfortunately sample sizing is often not as simple as might be assumed. The examples below demonstrate that use of a similar sample size may still mean extracting RNA from highly dissimilar samples.

1. Sampling blood from HIV patients (Huggett *et al.*, 2005) allows their stratification into groups based on their CD4+ T-cell numbers. Patients with >200 cells/ml may be otherwise healthy whereas those with counts of <200/ml are considered to have acquired immune deficiency syndrome (AIDS). Clearly, since AIDS patients have fewer CD4+ cells per volume of blood, normalizing cell mRNA copy numbers against blood volume alone would result in inaccurate and biologically meaningless quantification.

2. *In vivo* biopsies contain numerous cell types from different lineages and may contain different types of tissue, e.g. adjacent normal and cancer (*Figure 4.3*, color plate, between pages 38 and 39). This is resulting in a more widespread use of laser microdisssection and makes it feasible to normalize against the dissected area, allowing the reporting of copy numbers per area excised (*Figure 4.4*, color plate, between pages 38 and 39). Of course, one has to assume that the extraction efficiency and integrity of the RNAs extracted from different areas are similar, which may not be the case. It is acceptable to do this when comparing expression patterns from the same slide, as the tissue will have been subject to identical treatment. It is probably also acceptable to use this method for normalization when comparing different sections from the same tissue block. It remains to be seen whether it is acceptable to use this method for comparison between samples obtained from different tissue blocks.

3. Another problem when using similar sample sizes may occur when comparing samples that differ histologically. In situations where fibrosis has occurred or if young tissue is being compared with older tissue the efficiency by which nucleic acid is extracted from the sample may be very different. If different nucleic acids amounts are extracted from the same amount of histologically different samples then this will lead to a directional shift and any measurement will be influenced by the decrease in extraction efficiency from one of the groups. Another problem relates to samples containing cell types that differ significantly in their respective sizes; again, this can interfere with accurate quantification (*Figure 4.5*, color plate, between pages 38 and 39).

4. Many of the problems associated with *in vivo* samples might be expected to be overcome in *in vitro* systems. However, culture can provide its own problems when sampling (*Figure 4.6*). Non-adherent cells that do not aggregate represent the ideal starting material however this scenario is the exception rather than the norm. Furthermore the cells may become phenotypically very different as a result of the variable being tested by the experiment. Cells can be treated with buffers and/or enzymes to cause them to detach or segregate facilitating counting, however these treatments are almost certainly going to effect gene expression.

Cleary, whilst it is essential to compare samples of similar size and composition, ensuring that this is actually the case is often more complicated than it may at first appear. Consequently, we stress the importance of careful consideration of sample selection and emphasize that additional methods of normalization must be used.

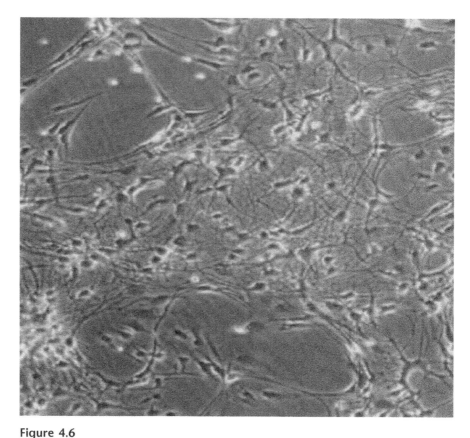

Figure 4.6

Normalization against cell number. It may be impossible to normalize against total cell number even when using tissue culture cells, as demonstrated by this neuronal cell line.

4.3.2 Total RNA amount

Once the RNA has been extracted, the next stage of normalization can be performed by ensuring that the same amount is used for each reverse transcription reaction. Not only does this facilitate normalization but circumvents problems associated with the linearity of the reverse transcriptase step. However, normalizing against total RNA requires accurate measurement of RNA quantity and again this is not straightforward. A number of different methods are available; most commonly, optical methods using spectrophotometry or fluorimetry with fluorescent dyes (Ribogreen) are used and provide varying degrees of accuracy. What is also important, but often overlooked, is the need to assess the quality of RNA because degraded RNAs may adversely affect results (Bustin and Nolan, 2004). Most commonly, gel electrophoresis or analysis using the Agilent Bioanalyser/Biorad Experion is used to determine the integrity of the ribosomal RNA (rRNA) band.

There are several drawbacks to normalization against total RNA: it is not easy to take into consideration differences in mRNA integrity, it does not allow for differences caused by the presence of inhibitors and ignores the efficiency of converting RNA into cDNA. Even if RNA integrity is assessed, any quality statement refers to rRNA, not mRNA. Total RNA comprises ~80% rRNA (with the protein coding mRNA comprising ~2–5% depending cell type); consequently, degradation of rRNA may not directly reflect degradation of mRNA. In addition, the relative amounts of rRNA and mRNA often differ between samples, especially when the mRNA originates from highly proliferating tissue, e.g., cancers. Of course, it is possible to extract mRNA, but this involves additional steps thereby potentially increasing experimental error, which can significantly reduce yield and may result in long-term stability problems. Furthermore, mRNA extraction is currently limited to the extraction of poly-adenylated mRNAs so will not purify non-adenylated mRNAs, for example those specifying histones.

4.3.3 Quantification of cDNA

Following the reverse transcription a method for measuring the amount of cDNA that has been reverse transcribed would be desirable; the only reliable method to do this is to quantify the synthesis of radio labeled cDNA. This strategy was more commonly used 10 years ago but the requirement for using radio isotopes coupled with the extensive increase in the use of qRT-PCR has resulted in this method rarely being used today.

Alternative methods have been proposed using spectrophotometry or DNA specific binding dyes but the major problem is that there is usually much more RNA than cDNA present after a reverse transcription. However, these are not satisfactory, as the background noise levels generate additional errors that tend to mask our ability to quantitate accurately the amount of synthesized cDNA. However, new methods for more accurate, non-radioactive quantification of cDNA synthesis efficiency are being developed and it is likely that normalization against cDNA will become feasible in the not too distant future.

4.3.4 PCR dependent reference

The use of a PCR dependent reference is by far the most attractive normalization option. In theory it is also the most straightforward method, as both the target gene and PCR dependent reference are measured by real-time PCR and all that is required for their measurement are additional PCR reactions. There are four types of PCR dependent references available:

Genomic DNA

Genomic DNA (gDNA) measures the amount of gDNA co-extracted with the RNA. This method does not require reverse transcription and so does not control for error introduced by this method. However, the problem with measuring gDNA is that most RNA extraction techniques try and eliminate co-extracted DNA as much as possible since DNA can inhibit the RT-PCR assay. Nevertheless, gDNA PCR dependent referencing has played an important role in research into early stage embryonic development as it can provide very accurate information about which division stage is being measured (Hartshorn *et al.*, 2003).

Spike

Spiking the sample with a distinctive (alien) nucleic acid allows an assessment of extraction efficiency and can be used for normalization. The spike, as long as it is not found in the sample of interest, can be synthetic or from another organism. This method benefits from the fact that accurately defined amounts of the spike can be introduced prior to extraction allowing good estimation of error introduced through most of the stages of processing. If RNA spikes are used the reverse transcription reaction can also be controlled for which is essential for fine measurements. The main criticism of using spikes is that, while they can be introduced prior to extraction, unlike the cellular RNAs they are not extracted from within the tissue. Consequently, there may be situations (e.g. if the samples differ histologically) when the spike may not be a good control for the extraction procedure. The concept of the RNA spike has been taken to its maximum level with RNA viral titer measurements from human plasma; here similar whole viruses (e.g., brome mosaic virus when measuring HIV) can be used as spikes providing detailed information on extraction efficiency, reverse transcriptase efficiency, can be used to control for PCR inhibition and defining the dynamic range linear limit of the multistaged process outlined in *Figure 1*.

Reference genes

Reference genes represent the by far most common method for normalizing qRT-PCR data. This strategy targets RNAs that are, it is hoped/assumed/ demonstrated, universally and constitutively expressed, and whose expression does not differ between the experimental and control groups, to report any variation that occurs due to experimental error. Theoretically, reference genes are ideal as they are subject to all the variation that affects the gene of interest. The problem occurs when a reference gene is used without validation.

Reference genes (previously termed housekeeping genes) such as GAPDH (glyceraldehyde-3-phosphate dehydrogenase) are historical carry-overs from RNA measurement techniques that generate more qualitative results (northern blotting, RNase protection assay). These genes were found to be essential for cell metabolism and, importantly, always switched on. Since northern blot analysis is mainly concerned with gross up-or-down regulation, they were perfectly adequate as loading controls.

The development of qRT-PCR transformed RNA analysis from a qualitative and, at best semi-quantitative assay, to a quantitative high-throughput one. Suddenly it became possible to measure far more than crude alterations in RNA expression; it was possible to detect reproducibly small changes in mRNA levels and provide numerical values that could be subject to statistical analysis.

The increasing emphasis on quantification meant that the requirement for ubiquitous expression was no longer sufficient. Now its expression also had to be stable and not be affected by experimental design. At best a poorly chosen reference gene would reduce the resolution of the assay by introducing additional noise and, at worst, the reference gene would be directly affected by the experimental system; directional shift would either hide a true change in the gene of interest or present a completely false result.

What has become apparent over recent years is that there is no single reference gene for all experimental systems (at least none has been discovered). Vandesompele *et al.*, (2002) quantified errors related to the use of a single reference gene as more than three-fold in 25% and more than six-fold in 10% of samples. Tricarico *et al.*, (2002) showed that VEGF mRNA levels could be made to appear increased, decreased, or unchanged between paired normal and cancer samples, depending only on the normaliser chosen. Today it is clear that reference genes must be carefully validated for each experimental situation and that new experimental conditions or different tissue samples require re-validation of the chosen reference genes (Schmittgen, 2000). There are two strategies that can be used to validate a reference gene:

(a) the first method is the whole error approach; this strategy directly measures the error of the raw data of candidate reference genes within the experimental system as illustrated by Dheda *et al.* (2004). The measured error comprises both the experimental error and the biological fluctuation of the chosen reference gene. This method is dependent on good laboratory technique not introducing technical error that will reduce the resolution, may require one of the other normalization methods discussed above to increase this resolution, and is dependent on choosing the correct reference. This approach scores over strategies that normalize against total RNA in that it defines the resolution of the individual assay and so provides a measurable parameter for assessing the likelihood of a quantitative result being meaningful. This method provides a simple strategy for reference gene normalization and is ideal when resources are limiting and when the measurement of larger differences (>five-fold) are required. However, this strategy is inappropriate when measuring more subtle changes in mRNA levels (<two-fold). This is due to the fact that the initial validation must be normalized to another factor (e.g. total RNA) and so is limited by the resolution of this factor, which may not be accurately measurable.

(b) The second method uses the error induced trends approach. This strategy also measures the error of the raw data of candidate reference genes within the experimental system. However this strategy differs from the single validated reference gene approach by comparing the fluctuations between the respective reference genes. Consequently, should error be introduced by a particular sample that artificially increases the measurement this will be observed within all the measured reference genes and can be compensated. Furthermore, by measuring the geometric mean as opposed to the arithmetic mean the influence of outliers can be greatly reduced (Vandesompele *et al.*, 2002). As with the single validated gene approach, this method defines the assay resolution; however it can give a far better estimation of this resolution as it is independent of technical error so that measurements as small as 0.5 fold have been reported

(Depreter *et al.*, 2002). The drawback to this method is that it requires the measurement of at least three reference genes and, for very fine resolution, may require as many as 10. This is hardly feasible when comparing expression patterns in numerous tissues, or when using several different treatment regimes. Furthermore, it may be necessary to revalidate reference genes when changing extraction procedures, using new enzymes or analysis procedures.

The two strategies constitute two extremes: one is applicable when measuring relatively large changes on a small budget (whole error approach), the other when the emphasis is more on measuring subtle changes in mRNA levels (error induced trends approach).

When choosing potential candidate reference genes the resolution can be increased if candidate genes are chosen based on their function not being involved in the experimental question. For example, there is no point in using GAPDH as a reference gene when analyzing glycolytic pathways as GAPDH is directly involved in this and so would be likely to be influenced by studies that looked at sugar metabolism, for example.

Data can also be supported by measuring a number of different reference genes which can strengthen conclusions as the experiment can be validated using the error induced trends approach on an individual experimental basis; this is of particular value when small measurements need to be made. Finally, it is important that intergroup comparisons must take into account any variability of the reference genes, as this defines the overall error and provides an estimate of the reliability of measurements of mRNA levels.

4.4 Conclusion

qRT-PCR has revolutionized RNA expression measurement that led to an explosion of publications over the last 8 years. Ironically, acceptable normalization methods, particularly with reference genes, have remained a contentious issue and it has only been over the last 4 years that this has been addressed. Despite this, even today RNA expression data are published with no mention of whether the methods used for normalization have been validated. However, without this validation the assay may not have the resolution to detect small changes or worse produce artefactual results. Worryingly, papers are still being published even though the authors have not demonstrated any validation of their normalization strategies.

References

Bubner B, Gase K, Baldwin IT (2004). 'Two-fold differences are the detection limit for determining transgene copy numbers in plants by real-time PCR.' *BMC Biotechnol* **4**(1): 14.

Bustin SA, Nolan T (2004). Pitfalls of quantitative real-time reverse-transcription polymerase chain reaction. *J Biomol Tech* **15**(3): 155–166.

Depreter M, Vandesompele J, Espeel M, Speleman F (2002) Modulation of the peroxisomal gene expression pattern by dehydroepiandrosterone and vitamin D: therapeutic implications. *J Endocrinol* **175**: 779–792.

Dheda KJ, Huggett F, Bustin SA, Johnson MA, Rook G, Zumla A (2004) Validation of housekeeping genes for normalizing RNA expression in real-time PCR. *Biotechniques* **37**(1): 112–114, 116, 118–119.

Hartshorn C, Rice JE, Wangh LJ (2003) Differential pattern of Xist RNA accumulation in single blastomeres isolated from 8-cell stage mouse embryos following laser zona drilling. *Mol Reprod Dev* **64**(1): 41–51.

Huggett J, Dheda K, Bustin S, Zumla A (2005) Real-time RT-PCR normalization; strategies and considerations. *Genes Immun* **6**(4): 279–284.

Melo MR, Faria CDC, Melo KC, Reboucas NA, Longui CA (2004) Real-time PCR quantitation of glucocorticoid receptor alpha isoform. *BMC Molecular Biology* **5**: 19.

Schmittgen TD, Zakrajsek BA (2000) Effect of experimental treatment on housekeeping gene expression: validation by real-time, quantitative RT-PCR. *J Biochem Biophys Methods* **46**(1–2): 69–81.

Tricarico C, Pinzani P, Bianchi S, Paglierani M, Distante V, Pazzagli M, Bustin SA, Orlando C (2002). Quantitative real-time reverse transcription polymerase chain reaction: normalization to rRNA or single housekeeping genes is inappropriate for human tissue biopsies. *Anal Biochem* **309**(2): 293–300.

Vandesompele J, De Preter K, Pattyn F, Poppe B, Van Roy N, De Paepe A, Speleman F (2002). Accurate normalization of real-time quantitative RT-PCR data by geometric averaging of multiple internal control genes. *Genome Biol* **3**(7): 1–0034.11.

High-throughput primer and probe design

5

Xiaowei Wang and Brian Seed

The design of appropriate primers for the specific quantitative generation of DNA amplicons from diverse transcripts is an important requirement for most applications of real-time PCR. For the simultaneous measurement of multiple transcripts in a single experiment, an additional constraint, that all reactions proceed efficiently under the same conditions, must be imposed. In the first section of this chapter, we describe general primer and probe design guidelines for real-time PCR. In the second part, we present an online real-time PCR primer database encompassing most human and mouse genes identified to date. The primer database contains primers that perform well under a single set of conditions, allowing many simultaneous determinations of mRNA abundance to be carried out.

5.1 Primer and probe design guidelines

5.1.1 Primer specificity

Non-specific amplification is one of the greatest challenges for the successful deployment of real-time PCR methods intended to be used for the validation or discovery of transcript abundance variation. In addition to the sequence of interest, many thousands of other sequences can be expected to be present in such applications. The design of PCR primers for this purpose should therefore take into account the potential contribution of all possible off-target template sequences, in order to prevent mispriming. This is usually achieved by comparing sequence similarity between the primers and all other template sequences in the design space.

5.1.2 Primer length

PCR primers are typically 16–28 nucleotides long. If the length is too short, it is difficult to design gene-specific primers and choose optimal annealing temperature. On the other hand, very long oligos unnecessarily increase oligo synthesis cost. In addition, longer primers are more likely to form secondary structures that result in decreased PCR efficiency or promote primer dimer formation, since the primers constitute the nucleic acid sequences at highest concentration in the reaction.

5.1.3 Primer GC content

In most PCR applications the primer GC content lies between 35% and 65%. If the GC content is too high, mispriming frequently results, because even a short stretch of oligo sequence may form a stably annealed duplex with non-target templates. On the other hand, very low GC content may result in poor primer binding, leading to decreased PCR efficiency.

5.1.4 Primer 3′ end stability

The 3′ end residues contribute strongly to non-specific primer extension by Taq DNA polymerase, especially if the binding of these residues is relatively tight to the non-target template. Therefore, primers with very high 3′ terminal stability should be rejected. The binding stability can be calculated from the free energy profile (ΔG). Typically the more computationally demanding aspects of calculating free energy, such as loop entropy, can be ignored because of the unfavorable energetics of opening small loops (internal denaturation) compared to end melting.

5.1.5 Primer sequence complexity

Low-complexity sequences should be identified and discarded during primer design, due to the likelihood of mispriming with such primers. In addition, some types of low sequence complexity pose a challenge for oligonucleotide synthesis. For example, a large number of contiguous guanosine residues in a primer can lead to poor synthesis yield due to decreased chemical coupling efficiency. In addition the resulting primers can exhibit poor solubility in aqueous media.

5.1.6 Primer melting temperature

The melting temperature (Tm) is the most important factor in determining the optimal PCR annealing temperature. An ideal PCR reaction should have forward and reverse primers with similar Tm values. Tm is not only determined by primer sequence, but also by other parameters, such as salt concentration and primer concentration. In recent years, extensive thermodynamic studies have been carried out to accurately determine oligo Tm values.

Currently, the following methods for Tm calculation are adopted by most primer design programs.

The '4 + 2' rule

$$Tm = 4 * (G + C) + 2 * (A + T).$$

This is a simple equation solely based on primer GC content. Tm is calculated by counting the total number of G/C and A/T. Each G/C contributes 4°C and each A/T contributes 2°C to Tm. This method is sometimes used to quickly approximate Tm values for very short oligos (<10 residues). However PCR primers are usually much longer and this Tm calculation method is not generally recommended.

Simple equation based on GC content and salt concentration

$$Tm = 81.5 + 16.6 * Log_{10}[Na^+_{eq}] + 0.41 * (\%GC) - 600/L$$

where [Na^+_{eq}] is the equivalent sodium molar concentration, (%GC) is the percent of G/C in the primer, and L is the primer length (Sambrook *et al.*, 1989). This method is widely used in many computer programs, partially because it is relatively easy to implement. Despite its simplicity, Tm values calculated in this way usually have reasonable accuracy and are only a few degrees different from empirically determined Tm values. In addition, this method is considered to be the most accurate way to calculate Tm for very long DNA duplex, such as PCR amplicon.

The nearest neighbor method

$$Tm = \Delta H°/(\Delta S° - R \ln (C_T/4))$$

where R is the gas constant (1.987 cal/Kmol), C_T is the primer concentration, $\Delta H°$ is the enthalpy change, and $\Delta S°$ is the entropy change (Breslauer *et al.*, 1986). This is considered to be the most accurate method to calculate oligo thermodynamic stability, and thus is the recommended method for primer Tm determination. $\Delta H°$ and $\Delta S°$ are calculated using the empirically determined thermodynamic parameters for neighboring bases in a primer (SantaLucia, Jr., 1998).

The entropy change is significantly affected by salt concentration. Thus, an entropy correction is required: $\Delta S° = \Delta S°$ (1 M Na^+) + 0.368 (L − 1) $\ln[Na^+_{eq}]$, where L is the primer length and [Na^+_{eq}] is the equivalent sodium molar concentration from all salts in a PCR reaction. According to a recent study (von Ahsen *et al.*, 2001), [Na^+_{eq}] can be determined by the following equation:

$$[Na^+_{eq}] = [\text{Monovalent cations}] + 120 \sqrt{[Mg^{2+}] - [dNTPs]},$$

where monovalent cations are typically present as Na^+, K^+ and $Tris^+$ in PCR buffer.

5.1.7 Primer location in the sequence

Oligo dT and random primers are commonly used in reverse transcription (RT) to produce cDNA template for real-time PCR. Oligo dT primers anneal specifically to mRNA poly(A) tails, thus minimizing non-coding cDNA products. However, this priming strategy introduces 3′ bias in cDNA synthesis because it is difficult to produce full-length cDNAs due to limited RT extension capability. If oligo dT primers are used in RT, the real-time PCR primers should be picked from the 3′ region of a gene sequence to gain maximum assay sensitivity.

In contrast, random primers are often used in RT for full transcript coverage. Random primers can potentially bind to any site in a transcript sequence. Because all primers are extended toward the 5′ end of the transcript, there will be a linear gradient of sequence representation, with highest cDNA abundance in the 5′ regions. If random priming strategy is adopted, real-time PCR primers should be picked close to the 5′ end of the target sequence for maximum sensitivity in real-time PCR.

5.1.8 Amplicon size

PCR efficiency can be affected by amplicon size. Very long amplicons leads to decreased PCR efficiency. Since PCR efficiency is one of the most important factors for accurate expression quantification, the amplicon should be smaller than 250 bp. Typically the size range is 100–250 bp.

5.1.9 Cross-exon boundary

To minimize the effect of DNA contamination in RNA template, the forward and reverse primers can be designed from different exons and to span exon–intron boundaries. In this way one can reduce the genomic DNA contribution to expression quantitation. This design strategy is most important for accurate quantitation of low-expressing genes or genes with many loci in the genome. However, in general, DNA contamination has minimal effect on real-time PCR because a typical transcript copy number is much higher than the number of gene loci and standard RNA preparation procedures remove genomic DNA efficiently. In any case, this strategy may fail if pseudogenes of the target gene are present in the genome.

5.1.10 Primer and template sequence secondary structures

Primer or target template secondary structures can retard primer annealing, leading to decreased amplification efficiency. The likelihood of secondary structure is greatest in regions rich in complementary base pairing, such as the stem of a stem-loop structure (Mount, 2001). If part of a primer or target sequence is inaccessible due to secondary-structure formation, the primer annealing efficiency may decrease dramatically.

In addition, primer secondary-structure may lead to primer dimer formation, which is one of the biggest challenges for accurate quantification by real-time PCR, especially when DNA intercalating dyes (e.g., SYBR® Green I) are used. Primer dimers can be produced by primer self-annealing, or by annealing between the forward and reverse primers. Therefore, the forward and reverse primers should be evaluated together during design to avoid potential primer dimer formation.

5.1.11 TaqMan® probe design

Many of the guidelines for primer design are also applicable to TaqMan® probe design. It is recommended to use Primer Express® software (Applied Biosystems) for TaqMan® assay design.

- The probe melting temperature in general should be ~10°C higher than the forward or reverse primer.
- Do not put G at the 5' end of the probe as this will quench reporter fluorescence.
- In general the GC content should be 35–65%.
- The probe should not self-anneal to form secondary structure and should not be picked from a gene region with high likelihood of secondary structure. Secondary structure formation may reduce hybridization efficiency, leading to reduced assay sensitivity.

- The probe should be selected from gene-specific regions with reasonable sequence complexity to avoid cross-hybridization.
- The probe should be as close to the forward and reverse primers as possible, without overlapping the primer sequences. The amplicon size is usually in the range of 60–150.

5.1.12 Molecular beacon probe design

Molecular beacon probes form stem-loop hairpin structures at low temperature. In this closed state, the fluorophore and the quencher are held in close vicinity, and thus no fluorescence signals are detected (Tyagi and Kramer, 1996). The hairpin loop contains gene specific sequence for hybridization in real-time PCR. Thus, the rules for TaqMan® probe design also apply to molecular beacon loop design. The hairpin structure is disrupted by hybridization to the amplicon, which separates the fluorophore from the quencher for fluorescence detection. Because of this unique structural requirement, one major task for molecular beacon probe design is to identify a suitable hairpin structure that melts 7–10°C higher than the PCR primers. The Tm of the hairpin stem cannot be calculated using the Tm formulas for PCR primers because the stem is formed by intramolecular folding. In general, programs for secondary structure prediction, such as Mfold (Zuker, 2003), can be used to predict hairpin stem melting temperature. The stem usually consists of 5–7 base pairs with 75–100% GC content. A G residue should not be placed at the end of the stem because it will quench the fluorophore.

5.2   PrimerBank – an online real-time PCR primer database

5.2.1 Primer design algorithm

We have developed a real-time PCR primer design algorithm based on the general guidelines described in Section 1 (Wang and Seed, 2003a). An outline of the algorithm is presented in *Figure 5.1*. The algorithm is implemented by a design tool called uPrimer, with which we have designed more than 300,000 primers encompassing most human and mouse genes. The following comprise detailed descriptions of the properties of these primers.

Primer specificity

Although primer specificity is one of the most important requirements in real-time PCR, most primer design programs take only one target sequence without considering mispriming to off-target templates. At the present time, to address the mispriming problem, one can design a number of primer pairs and then individually check cross-matches of each primer with BLAST (Altschul *et al.*, 1990). However this screening step is incomplete and does not consider some important design criteria. For example, cross-matches at the 3' end of a primer are more likely to produce non-specific amplicons. In our experience, only about two thirds of the primers designed in the conventional way can be used in real-time PCR experiments.

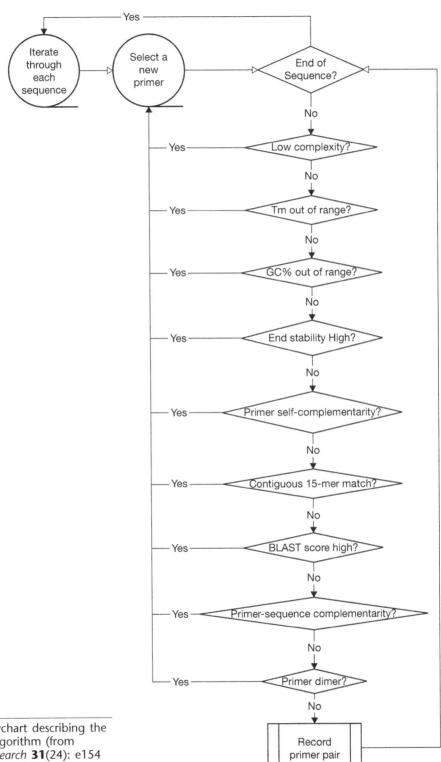

Figure 5.1

A simplified flowchart describing the primer design algorithm (from *Nucleic Acids Research* **31**(24): e154 with permission).

Mismatches are known to significantly affect binding stability, and sometimes even a single mismatch may destabilize a DNA duplex. Therefore contiguous base pairing is expected to be one of the most important contributing factors for duplex stability. Our primary filter for primer cross-reactivity is to reject a stretch of contiguous residues that is also found in off-target sequences. Every possible 15-mer in a primer sequence is compared to all known gene sequences in a genome. Primers with repetitive 15-mers are quickly identified and rejected using a computational hashing technique (Wang and Seed, 2003b). To further reduce cross-reactivity, BLAST searches for primer sequence similarity are carried out and all qualified primers have BLAST scores of less than 30.

The primer 3′ end is essential to prevent non-specific amplification because Taq polymerase extension can be greatly retarded by terminal mismatches (Huang *et al.*, 1992). Therefore a more stringent filter is applied to cross-matches at the 3′ ends. In our algorithm the cross-hybridizing Tm for the 3′ end perfect match does not exceed 46 degrees.

Random priming in RT reactions leads to significant amount of non-coding cDNA products, which are a potential source for mispriming. Because of their abundance, more stringent primer specificity filters are applied against non-coding RNA sequences.

Primer and template sequence secondary structures

Secondary structure is most likely to form by Watson–Crick base-pairing. To reduce self-complementarity, a primer is rejected if it has five contiguous base-pairs with the primer complementary sequence. To prevent primer dimer formation, the 4 residues at the 3′ end of a primer should not pair with its complementary sequence. Similarly, complementarity between the forward and reverse primers in a primer pair is also evaluated to prevent detrimental inter-primer dimer formation.

To avoid selecting primers from a template sequence region with potential secondary structure, a primer is rejected if it has 9 or more contiguous base-pairs with the target complementary strand. As one additional check, BLAST similarity search is also carried out to compare the primer and the complementary sequence of the target and the score is required to be less than 18.

Primer melting temperature

Similar Tm values are one important requirement for uniform PCR reactions. Primer Tm is calculated using the nearest neighbor method with the latest thermodynamic parameters. The formula for Tm calculation is described in details in Section 5.1.6. 250 nM primer and 0.15 M Na_{eq}^{+} concentrations are used as the default parameters. The default primer and salt concentrations are commonly used in real-time PCR. Variations in other typical real-time PCR setups do not significantly affect the Tm values.

Other primer properties

The forward and reverse primers should have similar biochemical properties to ensure a balanced PCR reaction. If several PCR reactions are carried out

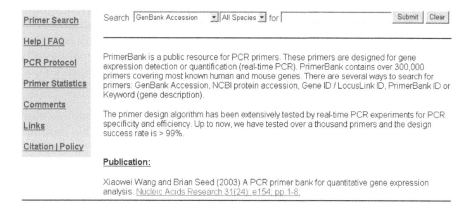

Figure 5.2

A screenshot of the web interface for PrimerBank. There are several ways to search for primers: GenBank Accession, NCBI protein accession, Gene ID, PrimerBank ID or Keyword. PrimerBank currently contains over 300,000 primers designed for most known human and mouse genes.

simultaneously, it is important that all primers in the reactions be similar. To provide flexibility in PCR setup, all of our *in silico* designed primers have similar properties.

- The primer length is 18–24 residues
- The primer GC content is 35%–65%.
- 3' end stability. The ΔG value for the five 3' end bases is at least 9 kcal/mol.
- The PCR product length is 100–250 bp. Occasionally if this requirement cannot be satisfied, alternative length ranges are used.
- No primer is designed from low-complexity regions. The low-complexity regions are identified and excluded from primer design by the NCBI DUST program (Hancock and Armstrong, 1994).
- A primer does not contain 6 or more contiguous same nucleotides.
- A primer does not contain any ambiguous nucleotide.

5.2.2 PrimerBank

To provide easy access to our computationally designed primers, we have constructed an online primer database, PrimerBank with a freely accessible web interface (http://pga.mgh.harvard.edu/primerbank). *Figure 5.2* is a screen shot of the primer search page. There are several ways to search for primers: GenBank Accession, NCBI protein accession, gene ID, PrimerBank ID or Keyword. PrimerBank currently contains over 300,000 primers designed for most known human and mouse genes. This web site also presents a detailed experimental protocol for SYBR® Green real-time PCR and common issues concerning real-time PCR. A short version of the protocol is given later in this chapter.

5.2.3 Experimental validation of the primer design

We have tested our *in silico* designed primers for over 1,000 mouse genes. These genes are known to be expressed in mouse livers by EST library analysis. Therefore we have used mouse liver total RNA in the validation experiments. Real-time PCR melting curve analysis and gel electrophoresis are performed to evaluate real-time PCR success. In most cases, we observe a single peak in the melting curve and a single band on gel for one PCR reaction. Occasionally, the gel and melting curve analyses do not agree with each other. In these cases, we have also performed DNA sequencing to make sure there is only a single PCR product. The design success rate, defined as a single PCR product with reasonable amplification efficiency, was 99% using these validation methods in an initial screen. A somewhat lower rate (approximately 93%) has been observed for a primer collection spanning nearly all of the mouse genome (A. Spandidos, personal communication).

Figure 5.3 shows a validation experiment for 16 cytochrome P-450 genes. Cytochrome P-450 is a large gene family with ~90% sequence similarity between some family members. Despite this very high sequence homology, all 16 P-450 primer pairs lead to single PCR products, validated by both gel electrophoresis and DNA sequencing. The melting curve analyses also indicate single PCR products for most of the PCRs (6 examples shown in *Figure 5.3(B)*).

5.3 Experimental protocol using PrimerBank primers

Although all PrimerBank primers are designed to be gene specific, the choice of PCR conditions is important and has a major impact on the success rate. Non-specific primer extension of only a few bases at room temperature by Taq polymerase can lead to non-specific PCR amplifications. Therefore, some variety of hot-start PCR is strongly recommended. AmpliTaq Gold® DNA polymerase has been our preferred enzyme as we found it to be more specific than other hot-start DNA polymerases we tested.

The annealing temperature may affect PCR specificity. To avoid non-specific PCR products, a high annealing temperature is recommended. In general, the annealing temperature should be at least 55°C or higher. Here we present a standard SYBR® Green real-time PCR protocol using PrimerBank primers. A more detailed and complete protocol is also presented in Chapter 7.

5.3.1 Reverse transcription (RT)

We have tested Invitrogen's SuperScript™ First-Strand Synthesis System and Ambion's RetroScript™ RT kit for cDNA synthesis from total RNA. Both kits produce high-quality cDNA templates for real-time PCR. Random primers are recommended in RT reactions because they provide full transcript

(A)

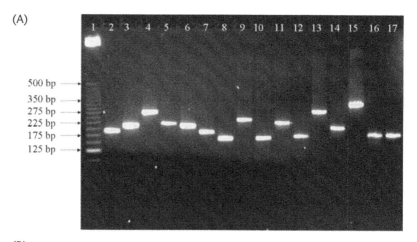

(B)

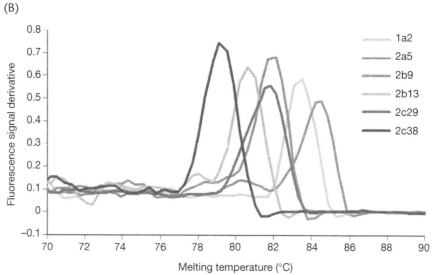

Figure 5.3

Real-time PCR primer validations for 16 cytochrome P450 genes.
A. PCR amplifications of 16 cytochrome P450 genes. Lane 1: 25 bp DNA
ladder; lane 2–17: 10 μl PCR products of P450 1a2, 2a5, 2b9, 2b13, 2c29,
2c38, 2c40, 2d26, 2d1, 2j5, 3a16, 3a25, 4a10, 4a12, 4a14, and 7a1.
B. Melting curves of 6 genes from cytochrome P450 family 1 and 2 (plotted as
the first derivative of the absorbance with respect to temperature). Adapted
from *Nucleic Acids Research* **31**(24): e154 with permission.

coverage. The random priming approach gives highest detection sensitivity
toward the 5′ end of the sequence.

5.3.2 Real-time PCR

We recommend using SYBR® Green PCR Master Mix from Applied
Biosystems.

1. Mix the gene-specific forward and reverse primer pair. Each primer (forward or reverse) concentration in the mixture is 5 pmol/μL.
2. A real-time PCR reaction volume can be either 50 μL or 25 μL. Prepare the following mixture in each optical tube. Do not use more than 1 μL cDNA template.
   - 25 μL SYBR® Green Mix (2x)
   - 0.5 μL cDNA template
   - 2 μL primer pair mix (5 pmol/μL each primer)
   - 22.5 μL H_2O

   or
   - 12.5 μL SYBR® Green Mix (2x)
   - 0.2 μL cDNA template
   - 1 μL primer pair mix (5 pmol/μL each primer)
   - 11.3 μL H_2O
3. Set up the experiment on a real-time PCR machine with the following PCR program.
   1) 50°C 2 minutes, 1 cycle
   2) 95°C 10 minutes, 1 cycle
   3) 95°C 15 seconds → 60°C 30 seconds → 72°C 30 seconds, 40 cycles
   4) 72°C 10 minute, 1 cycle

5.3.3 Troubleshooting

Below are a few commonly encountered problems in the use of real-time PCR using PrimerBank primers.

5.3.4 Little or no PCR product

Poor quality of PCR templates, primers, or reagents may lead to PCR failure. It is important to always include appropriate PCR controls to identify these possibilities. Some transcripts are expressed transiently or show tight lineage-specific regulation. In this regard an important caveat is that microarrays sometimes produce misleading information about gene expression levels, particularly when the inferred transcript abundance is low. False positives are relatively common when low microarray hybridization signal is observed. After switching to appropriate templates, we have obtained positive results following what initially appeared to be primer failures (Wang and Seed, 2003a).

5.3.5 Poor PCR amplification efficiency

The accuracy of real-time PCR is highly dependent on PCR efficiency. The amplification efficiency should be at least 80%. Following an inefficient PCR reaction the asymptotic value (plateau) for the final fluorescence intensity in the amplification curve is significantly lower than that seen with other reactions. If this is the case, the primers should be re-synthesized.

5.3.6 Primer dimer

Primer dimer may be observed if the gene expression level is very low. In this case, increasing the template amount may help in eliminating primer dimer formation. Primer dimer formation is a greater problem if the PCR is not conducted under hot-start conditions and the annealing temperature is less than 60°C.

5.3.7 Multiple bands on gel or multiple peaks in the melting curve

Agarose gel electrophoresis or melting curve analysis may not always reliably measure PCR specificity. Based on our experience, bimodal melting curves are sometimes observed for long amplicons (>200 bp) even when the PCRs are specific. The observed multiple peaks may be the result of amplicon sequence heterogeneity rather than non-specificity. On the other hand, for short amplicons (<150 bp) very weak and fussy bands migrating ahead of the major specific bands are sometimes observed on agarose gel. These weak bands are super-structured or single-stranded version of the specific amplicons in equilibrium state and therefore should be considered specific. Although gel electrophoresis or melting curve analysis alone may not be 100% reliable, the combination of both can always reveal PCR specificity in our experience.

5.3.8 Non-specific amplification

Non-specific amplicons, identified by both gel electrophoresis and melting curve analysis, lead to inaccurate real-time PCR results. To avoid this problem, please make sure to perform hot-start PCR and use 60°C annealing temperature. We have noticed that not all hot-start Taq polymerases are equally efficient at suppressing polymerase activity during sample set-up. AmpliTaq Gold® DNA polymerase is highly recommended. If the non-specific amplicon is persistent, you have to choose a different primer pair for the gene of interest. This primer design failure is usually the result of mispriming to a previously unidentified splice isoform (Wang and Seed, 2003a).

5.4   Web resources about primer and probe design

5.4.1 Real-time PCR primer and probe databases

- PrimerBank (http://pga.mgh.harvard.edu/primerbank). PrimerBank is a public database of real-time PCR primers. PrimerBank contains over 300,000 primers covering most known human and mouse genes.
- Other primer and probe databases. These databases are collections of validated real-time PCR primers and probes.
  RTPrimerDB (http://medgen.ugent.be/rtprimerdb)
  Real-Time PCR Primer Set (http://www.realtimeprimers.org)
  QPPD (http://web.ncifcrf.gov/rtp/gel/primerdb/default.asp)

5.4.2 Primer and probe design tools

- *Primer3* (http://www.genome.wi.mit.edu/cgi-bin/primer/primer3_www.cgi). Primer3 is a popular free design tool for PCR primers and probes. Users can customize many PCR parameters.

- *Primer Express®* (http://www.appliedbiosystems.com). This is a commercial desktop program from Applied Biosystems. It is widely used for design of TaqMan primers and probes.
- *Premier Biosoft International* (http://www.premierbiosoft.com). This is a commercial web site for real-time PCR primer and probe design.

5.4.3 Other useful web sites

- *Protocol Online* (http://www.protocol-online.org). This web site has a large collection of biological protocol links including the ones for real-time PCR.
- *Real-Time PCR Review* (http://www.dorak.info/genetics/realtime.html). This is a review article with many links about real-time PCR.
- *Bioinformatics.net* (http://www.bioinformatics.vg). This is a useful gateway to many bioinformatics tools. It has a collection of program links for PCR primer design.
- *Gene-Quantification Info* (http://www.gene-quantification.de/primers.html). This is a web site dedicated to real-time PCR technology. The primer page contains a list of design programs for real-time PCR.
- *Public Health Research Institute* (http://www.molecular-beacons.org). This is a comprehensive web site dedicated to the Molecular Beacon technology.
- *QPCR Listserver* (http://groups.yahoo.com/group/qpcrlistserver). This is a popular discussion group about practical issues related to real-time PCR.

References

Altschul SF, Gish W, Miller W, Myers EW, Lipman DJ (1990) Basic local alignment search tool. *J Mol Biol* **215**: 403–410.

Breslauer KJ, Frank R, Blocker H, Marky LA (1986) Predicting DNA duplex stability from the base sequence. *Proc Natl Acad Sci USA* **83**: 3746–3750.

Hancock JM, Armstrong JS (1994) SIMPLE34: an improved and enhanced implementation for VAX and Sun computers of the SIMPLE algorithm for analysis of clustered repetitive motifs in nucleotide sequences. *Comput Appl Biosci* **10**: 67–70.

Huang MM, Arnheim N, Goodman MF (1992) Extension of base mispairs by Taq DNA polymerase: implications for single nucleotide discrimination in PCR. *Nucleic Acids Res* **20**: 4567–4573.

Mount DW (2001) *Bioinformatics: sequence and genome analysis*. Cold Spring Harbor Laboratory Press, New York.

Sambrook J, Fritsch EF, Maniatis T (1989) *Molecular cloning: a laboratory manual*. Cold Spring Harbor Laboratory Press, New York.

SantaLucia J, Jr (1998) A unified view of polymer, dumbbell, and oligonucleotide DNA nearest-neighbor thermodynamics. *Proc Natl Acad Sci USA* **95**: 1460–1465.

Tyagi S, Kramer FR (1996) Molecular beacons: probes that fluoresce upon hybridization. *Nat Biotechnol* **14**: 303–308.

von Ahsen N, Wittwer CT, Schutz E (2001) Oligonucleotide melting temperatures under PCR conditions: nearest-neighbor corrections for Mg(2+), deoxynucleotide triphosphate, and dimethyl sulfoxide concentrations with comparison to alternative empirical formulas. *Clin Chem* **47**: 1956–1961.

Wang X, Seed B (2003a) A PCR primer bank for quantitative gene expression analysis. *Nucleic Acids Res* **31**: e154.

Wang X, Seed B (2003b) Selection of oligonucleotide probes for protein coding sequences. *Bioinformatics* **19**: 796–802.

Zuker M (2003) Mfold web server for nucleic acid folding and hybridization prediction. *Nucleic Acids Res* **31**: 3406–3415.

Quantitative analysis of ocular gene expression

6

Stuart N. Peirson

Summary

This chapter addresses a range of problems that arise with quantifying gene expression with quantitative PCR (qPCR), drawing upon the author's experience using the eye as a model system. The eye (or more specifically the light-sensitive retina), as well as mediating the primary events of vision, provides an accessible model of the nervous system. However, the light-sensitive nature of this tissue makes analysis of ocular tissues difficult as dynamic changes in gene expression may occur during tissue collection. The diverse cellular makeup of the retina also presents a problem in the form of assessing gene expression in a heterogeneous tissue. When dealing with models of retinal degeneration, these problems are compounded by very small tissue samples and, furthermore, when the pathways of interest involve numerous transcripts, high-throughput assays become essential.

To address some of these issues, a kinetic approach to high-throughput relative quantification of gene expression has been successful. This chapter covers topics such as assay design and power analysis, RNA extraction and quality assessment, use of suitable internal controls, approaches to data analysis, and methods of ensuring comparable amplification efficiency across all samples, all of which are vital for successful qPCR. Most of these issues are not unique to the eye, and as such the aim of this chapter is to provide generalized advice for researchers using qPCR, as well as demonstrating how qPCR assays may be optimized for specific research challenges.

6.1 Introduction

The following section provides an introduction to the eye as a model system, as well as some of the problems facing researchers using this model. The aim of this introduction is to highlight how the sample under analysis should be considered prior to the implementation of a quantitative assay, thus minimizing tissue-specific problems that may bias subsequent results.

6.1.1 Gene expression in the eye

The mammalian eye is in many respects similar to the eyes of most other vertebrates, consisting of a focal apparatus, the pupil and lens, that direct light onto the photosensitive retina to produce an image. The retina is a discretely layered structure containing two primary classes of photoreceptive cells, the rods and cones, which appose the retinal pigment epithelium. The additional proximal layers of the retina consist of secondary neurons and their processes, and are involved in the first steps of visual processing. The output from the retina is via the axons of the retinal ganglion cells which form the optic nerve, projecting to the visual centers of the brain. In addition to the primary image-forming role in vision, recent studies have highlighted an additional photoreceptive pathway within the eye, based upon directly light-sensitive retinal ganglion cells expressing the photopigment melanopsin (Provencio *et al.*, 2000; Berson *et al.*, 2002; Sekaran *et al.*, 2003; Foster and Hankins, 2002). This pathway is involved in detecting gross changes in light levels (irradiance) and is involved in the regulation of circadian rhythms as well as regulating other non-image forming tasks such as the pupillary light response and acute suppression of pineal melatonin (Lucas *et al.*, 2001; Lucas *et al.*, 1999).

The eye, and more specifically the retina, forms an ideal substrate for the study of neural physiology, as it forms a discrete outpost of the central nervous system, containing a wide range of neurons specialized to a variety of different tasks (Dowling, 1987). Investigating patterns of gene expression within the eye is therefore an essential aspect of the analysis of retinal physiology. qPCR provides an ideal tool for the investigation of temporal changes in gene expression in response to stimuli, as well as enabling characterization of pathways involved in retinal development and retinal degeneration.

6.1.2 Problems associated with ocular gene expression

Analysis of gene expression in ocular tissues has highlighted a number of potential problems, some of which are specific to the eye, and others that are not unique to this tissue. These may be broadly summarized as:

a) light sensitivity
b) RNA quantity
c) RNA quality
d) cellular heterogeneity
e) multiple target genes
f) biological variance

These problems are expanded upon below, and possible resolutions are discussed in the following sections.

Light sensitivity

The first major obstacle to the study of ocular tissues is their light-sensitivity. The manner in which ocular tissues are collected may alter the expression of the transcripts of interest. This scenario is not unique to the eye. As changes in gene expression are dynamic, the methods used to dissect tissues or dissociate cell types may consequently affect the expression of either target or internal control genes (Dougherty and Geschwind, 2005). This technical consideration is rarely controlled for, and may result in potentially misleading data. These problems are compounded when temporal changes in gene

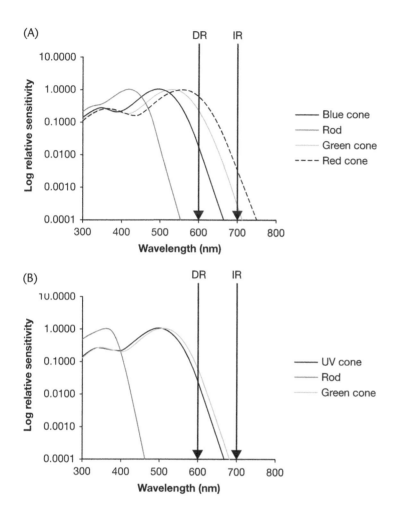

Figure 6.1

Spectral sensitivity profiles for human **A**. and mouse **B**. visual pigments. Data plotted on logarithmic scale. Arrows indicate dim red stimulus, at 600 nm (DR) and longer-wavelength infra-red stimulus at 700 nm (IR), illustrating the increased long-wavelength sensitivity of the human visual system conferred by the presence of the red cone class. Visual pigment profiles based upon Govardoskii template (Govardovskii *et al.*, 2000). λ_{max} values taken from Dartnall *et al.* (1983), Jacobs *et al.* (1991), Bridges (1959) and Sun *et al.* (1997).

expression are of interest, as not only tissues must be collected without light exposure, but they must be collected rapidly to prevent changes in gene expression or RNA degradation (see 'RNA quality' below).

Visual scientists have traditionally collected tissues under dim red light (>600 nm) to minimize the effects of photopigment activation. For many studies, this works extremely well, although it should be noted that the success of this approach is dependent upon the species and its innate photoreceptor complement in comparison to human spectral sensitivity. The visual sensitivity functions of photopigments can extend well into the long-wavelength region of the spectrum and use of even dim red light may result in a significant absorption by the LWS cones of some species. *Figure 6.1* illustrates the photopigment complements of humans (A) and mice (B), illustrating the difference in long wave sensitivity between these species.

RNA quantity

RNA yields may be variable, and for models of retinal degeneration, where the outer retina is lost, tissue yields may be very low. qPCR is ideal for the analysis of such small samples, and due to the sensitivity of the technique, pooling samples to obtain enough RNA should not be necessary. This is an important consideration, as it enables gene expression to be determined on an individual basis, enabling biological variance in gene expression to be assessed.

RNA quality

RNA degradation is a major problem with qPCR, as it introduces additional variability between samples that must be subsequently accounted for. Even if using equal quantities of RNA for each reverse transcription, differences in RNA quality may render samples incomparable without normalization (see Section 6.2 below). RNA quality may be adversely affected by tissue collection protocols as discussed above (see 'Light sensitivity' above), as well as freeze-thawing or improper storage, all of which give endogenous RNases the opportunity to degrade the RNA within the sample. Furthermore, the addition of exogenous RNases from contaminated plasticware and laboratory surfaces is all too easy due to the ubiquitous nature of these enzymes and their resilience to many forms of denaturation (Sambrook and Russell, 2001).

Cellular heterogeneity

Like most tissues, the eye as a structure contains a great diversity of cell types. These range from the cells of the lens and the iris, to the melanin-containing retinal pigment epithelium, the cells of the choroid and retinal vasculature, and of course the numerous photoreceptors, neurons and supportive Muller cells of the retina. Such cellular heterogeneity will restrict the detection of changes in gene expression, as if a transcript is expressed in a rare cell type, even a massive change in expression may be undetectable as it reflects only a small fraction of the RNA population measured. Alternatively, changes in gene expression in one cell type may be masked by compensatory changes in other cell types, for example rhythms of gene expression in anti-phase. Even when changes in expression can be detected in heterogeneous tissues, determining the source of the change may be impossible. Overall changes could be due to a change in all cells within the tissue, a change in only a fraction of the cells, a change in the cellular makeup of the tissue, e.g., neuronal loss, reactive gliosis or invasion of inflammatory cell types (Dougherty and Geschwind, 2005).

The counterpoint to these problems is that dissection or dissociation of specific compartments or cell types may affect the dynamic expression profile within the tissue. Reduction in starting tissue volume will also result in a decreased RNA yield and may even necessitate pooling individual samples, resulting in a decreased sample size or, in the worst case scenario, a single pooled sample. Moreover, the dissociation of cell types within heterogeneous tissues may lead to cell damage and release of endogenous RNases that will compromise RNA quality. Finally, the tissue yield obtained even from similar samples may be more variable following dissection or dissociation. This is certainly the case with retinal dissection, as the whole eye represents a discrete organ, and even given very careful dissection, retinal yields are considerably more variable.

As such, the use of heterogeneous tissues presents a trade-off between the problems of studying mixed cell populations and the problems of introduced variance and potential RNA loss associated with producing cell-specific populations.

Multiple target genes

Due to the ever-increasing knowledge of intracellular signaling pathways, it is often desirable to examine multiple transcripts of interest within the same biological samples. This is taken to extremes in the case of microarrays, where it is now possible to examine the whole transcriptome from a single sample. qPCR retains its essential place in the canon of molecular techniques, as in the majority of studies, candidate genes are already known or implicated. The need to examine the expression of multiple transcripts is evident within several pathways within the eye, including photopigment complements, elements of the phototransduction cascade, inflammatory markers, markers of neuronal activity and clock genes. High-throughput approaches are therefore essential to face the rising number of candidates, making approaches based upon the construction of accurate exogenous controls cumbersome and rate-limiting.

The need to examine multiple transcripts places an effective limit on the number of genes which may be examined from any single sample. 1 µg of total RNA should provide enough cDNA for at least 20 target genes to be investigated, but with extended pathways of interest, the use of multiple internal controls or with low RNA yields, this can quickly become a limiting factor.

Biological variance

A common problem facing researchers investigating gene expression is that of biological variance. Even when factors such as genetic background and environment may be controlled, biological variance will exist between individuals (Pritchard *et al.*, 2001), resulting in noisy data and necessitating large sample sizes. Whilst many qPCR platforms enable 96-sample formats, many researchers utilize sample replicates, typically triplicates, to ensure each data point is well-resolved. The use of large samples may become problematic due to inter-assay variability, as data from samples split across different experimental runs is more difficult to directly compare.

6.2 Relative quantification

Like many basic research applications, analysis of ocular gene expression is particularly amenable to relative quantification using qPCR (Bustin, 2000; Bustin, 2002; Ginzinger, 2002; Klein, 2002; Walker, 2002) (also discussed in Chapters 2 and 3). This involves calculating changes in gene expression relative to a control population, which should be present in the majority of experimental designs. The control population is totally dependent upon the type of study, and may, for example, be an untreated group (in stimulus or drug-treatment regimes), a specific tissue or cell type (in cross tissue/cell studies), time point zero (in time course studies), wildtype (in transgenic studies) or a specific age group (in developmental or aging studies). Changes are then expressed as a fold change relative to this control population.

6.2.1 The R_0 method

Relative quantification may be most easily carried out using the $2^{-\Delta\Delta Ct}$ method (Livak, 1997; Livak and Schmittgen, 2001). The necessary calculations may be simplified by use of the term R_0, the theoretical starting fluorescence. This concept is most easily understood by thinking of the reaction as a simple exponential amplification with a doubling of product every cycle. If both the fluorescence at a common threshold (termed R_n), and the number of cycles required to reach this threshold is known (the threshold cycle, C_t), then it becomes a simple matter to back-calculate to the starting fluorescence, R_0 (*Figure 6.2*). As the fundamental concept of real-time PCR is that fluorescence is proportional to the DNA concentration

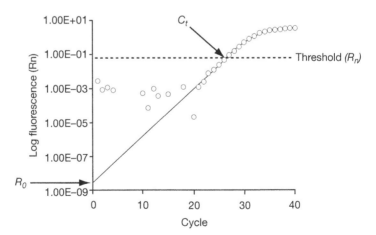

Figure 6.2

The starting fluorescence (R_0) method. If threshold (Rn), threshold cycle (C_t) and amplification rate (E) are known, then the fluorescence at cycle 0 can be determined (R_0), using eq. 2. In this example $R_n = 0.153$, $E = 0.893$ and $C_t = 28.00$, giving a value of $R_0 = 2.67 \times 10^{-9}$. Several methods may be used to determine E, including standard curves, assumed 100% E or kinetic approaches.

within a reaction, R_0 reflects the expression within that sample. This may be expressed as:

$$R_0 = R_n \times 2^{-Ct} \qquad\qquad \text{eq. 1}$$

The main advantage of such an algebra rather than using the $2^{-\Delta\Delta Ct}$ method is that it simplifies subsequent calculations, particularly those involving statistical calculations such as standard deviations. Transformation of expression values into a linear series rather than use of C_t values results in a data set which are more intuitive for subsequent manipulation. Consequently, normalization of data to an internal control (implicit within the $2^{-\Delta\Delta Ct}$ equation) involves simply dividing the individual sample target gene R_0 values by their respective internal control gene R_0 values. This approach facilitates the use of multiple internal controls, as discussed below.

This approach should be unaffected by changes in the threshold (R_n), as a change in R_n will result in a corresponding alteration of C_t, and the R_0 value should remain unchanged; e.g. if $R_n = 0.05$ and $C_t = 21.3$, then $R_0 = 1.94 \times 10^{-8}$. If R_n is increased to 0.10, then $C_t = 22.3$, and $R_0 = 1.94 \times 10^{-8}$. This is of course an over-simplification, as amplification efficiency dynamically changes throughout the course of a PCR. However, this approach offers a simple means of investigating how the placement of the threshold affects the data obtained (see below).

This approach can also simply incorporate efficiency corrections, such as those described by Pfaffl (2001). By conducting a cDNA standard curve, the amplification efficiency (hereafter, termed E) for an individual primer/probe set may be determined, and incorporated as a correction to improve accuracy:

$$R_0 = R_n \times (1+E)^{-Ct} \qquad\qquad \text{eq. 2}$$

Reaction efficiency may be determined by constructing a standard curve based upon the serial dilution of a sample of known concentration. This may be a plasmid, oligonucleotide or a cDNA sample, where C_t is plotted on the y-axis and logarithm (base 10) concentration is plotted on the x-axis. E may be derived from a standard curve using the following formula:

$$E = 10^{(-1/S)} \qquad\qquad \text{eq. 3}$$

Where E equals the amplification efficiency and S is the slope of the standard curve (which should be a negative slope). E may also be derived from the kinetics of the reactions under analysis, as described in Section 6.2.2.

Finally, it should be noted that the R_0 method is mathematically equivalent to the $2^{-\Delta\Delta Ct}$ as well as standard curve methodologies, and will result in identical fold changes when using the same E (Peirson et al., 2003). The main factor that therefore differs between these commonly used approaches to data analysis is the means by which E is derived. Another difference is of course the units, but if normalization to an internal control is conducted then the units will effectively cancel out, resulting in a ratio of target gene to internal control gene expression. This approach represents no great novelty in qPCR data analysis, but enables a simple common methodology for data analysis to be implemented.

6.2.2 Kinetic approaches to qPCR

Whilst the term kinetic PCR was originally used for what is now more commonly referred to as real-time PCR (Higuchi *et al.*, 1993; Higuchi and Watson, 1999), it has come to be used to describe those approaches to qPCR where the reaction efficiency is derived from the kinetics of the samples under analysis. A number of approaches to kinetic PCR have been proposed (Liu and Saint, 2002a; Liu and Saint, 2002b; Tichopad *et al.*, 2002; Ramakers *et al.*, 2003, Peirson *et al.*, 2003), though the main consideration of such approaches here is that they enable E to be derived without the use of additional standards. At their simplest form, kinetic approaches involve calculating the slope of an amplification plot to determine E (*Figure 6.3A*).

The region of the amplification plot used for kinetic analysis presents a similar problem to threshold setting. If the region used is too low then E will be inaccurate due to background noise, and too high, then E will be too low as the reaction will be approaching a plateau (Kainz *et al.*, 2000). This raises the concern of selecting an 'optimum threshold' for both kinetic analysis and threshold setting, which is a trade-off between background noise and plateau effects (Peirson *et al.*, 2003). Changing the placement of a threshold will therefore affect the R_0 value obtained. The nature of these inaccuracies is dependent upon the data under analysis. Analysis of the *c-fos* data presented in Peirson *et al.* (2003) suggests that a two-fold change in the threshold only alters the R_0 value by less than 2% of its midpoint value, a five-fold change in threshold yields around a 6% change, and a ten-fold change in threshold results in a change of around 18%. In all cases, lowering the threshold created a larger deviation in R_0 value than increasing it, illustrating that the placement of a threshold too close to background noise may introduce considerable errors when evaluating expression levels by qPCR. This possible source of assay noise also suggests that platforms with a narrow fluorescent signal range between background noise and reaction plateau may be more prone to errors introduced due to threshold placement.

An advantage of kinetic approaches is that the individual E values may be used as a criterion for outlier detection and sample exclusion. As the raw data from qPCR is in the form of C_t values, even a small amount of noise associated with these measurements may result in the introduction of large systematic errors. Bar *et al.*, (2003) suggested the use of kinetic outlier detection (KOD) to circumvent this problem, whereby aberrant kinetics may be detected based upon deviations of E from a training set (*Figure 6.3B*).

Problems with kinetic approaches include that such approaches may only be possible on certain platforms where there is a wide fluorescent signal range between background noise and reaction plateau. If this signal range is low, limits are imposed on the amount of data available for kinetic analysis, and such analysis may be inaccurate. Similarly, inappropriate background subtraction may affect the slope of amplification plots, with oversubtraction creating artificially high values for E (Bar *et al.*, 2003).

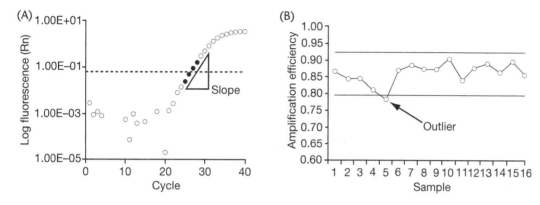

Figure 6.3

Kinetic qPCR. **A.** Use of linear regression to determine slope of amplification plot. Slope was determined in the region indicated by the filled circles, providing a slope of *0.277*. Amplification efficiency (*E*) = $10^{0.277}$ = 0.893. **B.** Amplification efficiency (*E*) determined from the kinetics of 16 ocular samples (circles), with kinetic outlier detection limits indicated by horizontal lines. Sample *5* demonstrated abnormally low *E*, and was therefore excluded from further analysis.

Kinetic approaches appear to offer the promise of correcting each sample for slight differences in their individual amplification rates. However, this is unfortunately a false hope, as any calculation of *E* from raw data, like any measurement, is only an estimate, limited by the precision of the measuring device. Due to the technical noise present within the raw data, data pre-processing as well as the algorithm used, *E* will vary, and based upon the sum of errors from multiple sources, theoretically should conform to a normal distribution. As shown in eq. 2, the efficiency correction is applied as an exponent, and therefore even small differences in *E* will produce enormous differences in R_0. For example, a 1% difference in the efficiency used will produce around a 13% difference in R_0 after 26 cycles, whereas a 5% difference will produce nearly a two-fold change (Freeman *et al.*, 1999). As such, the optimum approach is to use the mean efficiency (ideally after KOD), which if based on repeated measures, should provide an accurate estimate of *E*.

Consideration of this problem also suggests that the main source of inter-assay variability in any form of qPCR is due to the method used to calculate *E* between experiments. A standard curve run on multiple consecutive days will never provide exactly the same *E*, due to the presence of cumulative small errors in, for example, spectrophotometry, pipetting and fluorescent measurement. Once again, barring dramatic outliers, such errors would be expected to result in values of *E* that conform to a normal distribution.

The kinetic approach described above, based on use of the R_0 method is automated in DART-PCR, an Excel®-based worksheet for qPCR data analysis. This is available as supplementary data from Peirson *et al.*, (2003), or freely available on request.

6.2.3 Accurate normalization

Relative quantification of gene expression is dependent upon accurate normalization of data to one or more internal controls (ICs), which are typically housekeeping genes present and unchanged in all samples (Sturzenbaum and Kille, 2001). Issues related to normalization are discussed in greater detail in Chapter 4.

What is the most suitable internal control for qPCR? Unfortunately, no simple answer exists, as every IC has its own benefits and limitations, and what may be ideal in one experimental scenario may be unsuitable in others. Arguably, the most sensible approach to qPCR data normalization was provided by Vandesompele *et al.* (2002), who recommended using multiple ICs, enabling the reliability of these control genes to be assessed against each other. The geometric mean of the most reliable genes is then used in the form of a normalization factor (NF_x, where x is the number of ICs used), which is less susceptible to distortion by outliers than the arithmetic mean.

Comparing internal controls

A quick method of checking the reproducibility of ICs is to plot the R_0 values as a line graph, which given equal amounts of high quality RNA across all samples, should give a straight line. This is of course an ideal, and due to differences in RNA quality, spectrophotometric inaccuracy, differences in RT efficiency and biological variance, such a plot is likely to be far from linear. However, a second IC conducted on the same cDNA should provide a similar profile if IC expression accurately reflects RNA loading, and indeed this is the case (*Figure 6.4A*). As the level of IC expression may differ quite dramatically, samples should be normalized to the mean to transform the data to a common scale. Additional ICs should provide comparable profiles, with deviations due to differential regulation, differences in expression variance and measurement errors. The presence of a common pattern of expression with different ICs should be representative of RNA loading, and differences in the expression profile of ICs may indicate that one of the ICs is affected by the experimental procedure. Such an approach provides additional confidence that the normalization procedure used is in fact reliable.

A quick method of checking the suitability of ICs is to calculate the mean squared error (MSE) of the expression for each sample (mean normalized as noted above). This is simply a measure of the within sample variance, which will be low if the expression patterns of all IC genes are comparable across all samples, and high if the expression profiles are dissimilar. The contribution of each IC and each sample to the total sum of squares (SS_{total}) can be used to determine which samples and/or genes are the greatest source of variance, and may be applied as a basis for sample or IC exclusion (*Figure 6.4B* and *6.4C*, respectively).

Advantages of accurate normalization

What is the advantage of additional ICs? One may consider that the variability between the expression of any two samples is due to three factors: (1) true differences in expression, (2) differences in the RNA content of the sample,

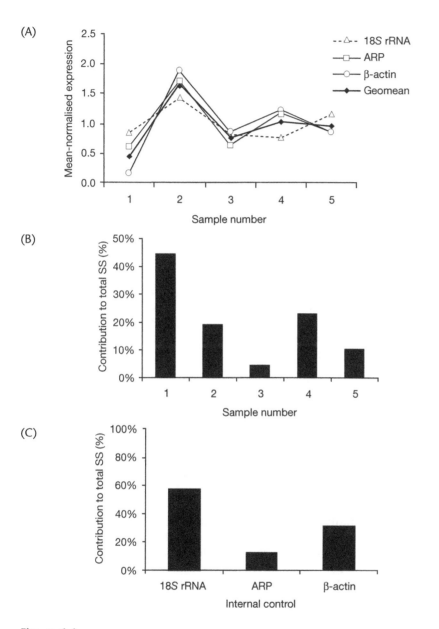

Figure 6.4

Comparison of expression levels of three internal control (ICs) genes across in 5 samples of murine retina. ICs used are: 18S rRNA, acidic ribosomal phosphoprotein (ARP) and β-actin. **A.** IC expression corrected to the mean expression of that IC across all five samples, illustrating similar profiles of expression for all three genes. The geometric mean of all three genes is also shown. **B.** The variance within each sample can be measured, and the contribution to the total sum of squares (SS_{total}) determined. Sample 1 demonstrates the greatest contribution to the contribution to the SS_{total}, whereas Sample 3 demonstrates the highest correlation between expression values. **C.** Contribution to SS_{total} by gene, showing that 18S rRNA demonstrates a greater variance than either ARP or β-actin.

and (3) measurement errors. True differences in expression will reflect basal expression levels as well as changes in response to stimuli, and will include biological variation (6.1.2). It is this level of expression that is the desired endpoint of any experiment. Measurement errors are present in any assay, and should be minimal with most qPCR platforms when analyzing samples in parallel (ideally well below 10%). Differences in RNA load can have a dramatic effect upon the normalized expression. Poor normalization can produce quite misleading data, and for this reason the use of at least two ICs is recommended to ensure that the results obtained are not simply an artifact of normalization. Use of additional ICs also will provide a more accurate measure of RNA loading, and this should ensure that this source of error is minimized. The end result should be a reduction in the overall noise, enabling an improved assay resolution and a better ability to discriminate differences. Or in simple terms, more ICs should result in smaller error bars.

6.3   Assay considerations

The final section of this chapter aims to bring together the problems arising with the analysis of ocular gene expression (6.1) with the approaches to data analysis described above (6.2) to provide potential solutions that may enable researchers to obtain the most reliable data with the minimum expenditure of resources.

Assay design

Most researchers typically think about the application of statistics after an experiment has run and the data have been collected. However, using statistical methodology before an assay is conducted enables the researcher to get some idea of whether what they are attempting to measure is feasible, and that the sample size used will be suitable for the system studied. Both time and resources may be wasted without such consideration.

Power calculations are based upon the statistical values α and β. The α of a test is the probability of a false positive, that is, the test indicates a significant difference where one does not in fact exist. This is typically set to 0.05, and a p-value below this is taken as evidence of a significant difference. β indicates the reverse scenario, where the test indicates no significant difference when a real difference does occur. The power of a statistical test is 1-β, and represents the ability of the test to detect a difference when it really does exist. As such, a high power indicates that the test is likely to detect a real difference if one exists, and a low power suggests that the test will not be capable of detecting a real difference. Unlike α, in experimental biology β is typically set to around 0.20, as false negatives are typically deemed less detrimental than false positives. However, certain assays may require higher β values, so the context of the assay is a critical factor. For example, in clinical detection of disease, a false negative indicates that a patient has a disease but the disease is not diagnosed, and as such a low β and therefore a higher power would be essential (Bland, 2000; Ott and Longnecker, 2001).

The power of a test may be estimated before conducting experiments, and this can provide researchers with an indication as to whether their experimental design is sufficiently robust to detect the differences in expression

Table 6.1 Calculation of the sample size required to detect a given change in gene expression, using power calculations based upon a two-tailed t-distribution for specified degrees of biological variance (expressed as coefficient of variance). Calculations conducted using http://www.univie.ac.at/medstat.

Power 0.80

		Biological variance (CV%)				
		10%	20%	30%	40%	50%
Fold change detectable	1.1	8	32	71	126	197
	1.2	2	8	18	32	50
	1.3	1	4	8	14	22
	1.4	1	2	5	8	13
	1.5	1	2	3	6	8
	1.6	1	1	2	4	6
	1.7	1	1	2	3	5
	1.8	1	1	2	2	4
	1.9	1	1	1	2	3
	2.0	1	1	1	2	2

Power >0.95

		Biological variance (CV%)				
		10%	20%	30%	40%	50%
Fold change detectable	1.1	13	52	117	208	325
	1.2	4	13	30	52	82
	1.3	2	6	13	24	37
	1.4	1	4	8	13	21
	1.5	1	3	5	9	13
	1.6	1	2	4	6	10
	1.7	1	2	3	5	7
	1.8	1	1	2	4	6
	1.9	1	1	2	3	5
	2.0	1	1	2	3	4

they are looking for. The worst case scenario is that experiments are conducted and no significant difference is found, as real differences may be present, but the researcher has insufficient evidence to detect them.

Table 6.1 can be used as a quick lookup chart when designing qPCR experiments. These calculations are based upon an unpaired t-test, although these values will at least provide a guideline for other test procedures, indicating a suitable sample size. An estimation of biological variance must first be made, usually as a coefficient of variance (CV = standard deviation divided by the mean, usually expressed as a percentage). This may be an estimate based upon previous experiments using a similar tissue and/or protocol conducted in house or in existing scientific literature. If such information is unavailable an approximation may be made, with an estimate around 30% not being unreasonable. Using the planned sample size, the table provides the fold change that can be detected with a power of 0.80 (A) or 0.95 (B). This should enable researchers to decide upon a suitable sample size for an assay, and to determine the level of resolution that should be possible with that sample size.

Tissue collection

For light-sensitive tissues, such as the eye, procedures should ideally be conducted under infra-red light (>700 nm). Infra-red optical devices are available from a wide range of suppliers, although complicated dissection is extremely difficult using such devices, and the exact wavelength of emission may vary. Lacking such facilities, dim red light may be used if a suitable cut-off filter is used to prevent stimulation of the photopigments present (see *Figure 6.1*). Based on the known long-wavelength sensitivity of human and mouse visual pigments, use of a light source with a cut-off of 600 nm would result in absorption of around 60% by the human red cone (certainly enough to see by), but only around 5% by the mouse green cone, or around a ten-fold greater sensitivity. Moving up to over 700 nm would result in less than 1% absorption by the human red cone (detectable at very high intensities), but over a hundred-fold less absorption by the mouse visual pigments.

Snap-freezing tissues is ideal for RNA extraction, although all dissections must be conducted before doing so as freeze–thawing can have dramatic effects on RNA quality (Sambrook and Russell, 2001). One solution to this problem is the use of an RNA preservation medium, such as RNAlater™ (Ambion). With ocular tissues, the eye can be enucleated under infra-red light, the eye pierced to facilitate access to the retina, and then placed into RNAlater™ in a light-tight container. The tissue can then be stored overnight at 4°C before the retina is dissected out in light.

To validate this approach, we investigated expression of the immediate-early gene and marker of neuronal activity *c-fos* in response to a 15 min light pulse. *C-fos* expression has been shown to rapidly increase in response to such a stimulus, corresponding to depolarization of retinal ganglion cells (Nir and Agarwal, 1993; Yoshida *et al.*, 1993). Dissection of tissues stored in RNAlater™ followed by qPCR using SYBR® Green I results in a clear induction of *c-fos*, peaking at 30 min, as demonstrated previously (*Figure 6.5*).

RNA extraction protocols

A wide variety of reagents are available for RNA extraction. One of the most commonly used methods is based upon that of Chomczynski and Sacchi (Chomczynski and Sacchi, 1987, Chomczynski, 1993), using a monophasic lysis reagent containing phenol, guanidinium salts and solubulizing agents (available under different trade-names from a variety of suppliers). This approach enables simultaneous extraction of RNA, DNA and protein, and is effective for even small tissue samples.

Use of such monophasic lysis reagents is particularly effective on ocular tissues, including retina, RPE, lens as well as whole eye, and has also been applied to a diverse range of tissues throughout the body. When using a different tissue type for the first time, it is always worth conducting a test extraction, evaluating RNA quantity and quality to ensure that the RNA extracted suitable for subsequent studies. Using a monophasic lysis reagent, it is possible to routinely extract 10–15 µg of high quality total RNA from paired whole eyes, and 1–5 µg of total RNA from paired retinae. When studying models of retinal degeneration, these yields decrease, with 0.5–2 µg total RNA being more typical.

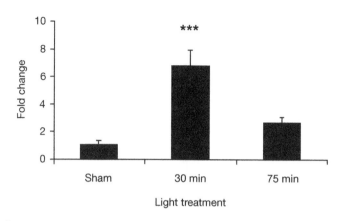

Figure 6.5

C-*fos* induction in wildtype retina, dissected from tissue preserved in RNAlater™ (Ambion). Peak expression of *c-fos* was found 30 min after light pulse administration. Significant differences were assessed using one-way ANOVA, $F_{2,11}$ = 13.56, P<0.01 (n = 4–5 per group).

Analysis of RNA samples using a microfluidic system, such as the Agilent bioanalyzer, enables RNA quality to be assessed with minimum loss of valuable RNA (*Figure 6.6*). Such lab-on-a-chip methods typically utilize just 500 ng of RNA, and enable a quantitative measure of RNA quality as well as quantity.

Reliable normalization

The use of at least two internal controls is recommended, as described above. If these correlate well, then additional ICs may be unnecessary (*Figure 6.4A*). However, if a different profile is found, then it is recommended that additional controls are used. Given that normalization can have dramatic effects on the results of relative quantification of gene expression, accurate normalization ensures that researchers do not report artifacts of normalization which may be misleading. Low expression of an internal control in just one sample can result in the normalized expression appearing erroneously higher. Such outliers can have dramatic effects when small sample sizes are used, and are capable of producing quite misleading data.

The use of additional metrics, such as gene stability index (Vandesompele *et al.*, 2002) or the variance of IC expression within samples and across IC genes as described above (*Figures 6.4B and 6.4C*) enable a quantitative approach to IC selection.

Heterogeneous tissues

The trade-off between using discrete tissue masses containing multiple cell types or attempting to reduce cellular heterogeneity and possibly affecting

(A)

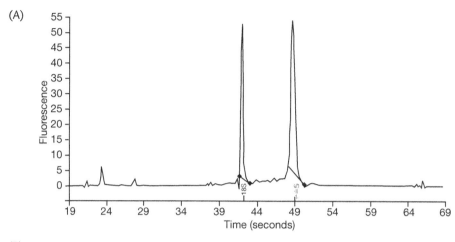

(B)

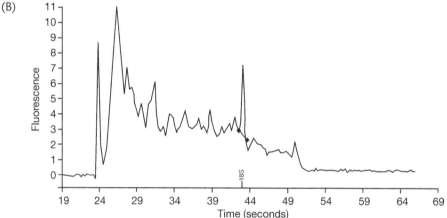

Figure 6.6

Analysis of RNA quality using Agilent Bioanalyser with *RNANano* chip. **A.** 500 ng of total RNA was loaded per well, producing two distinct peaks for the 18S and 28S ribosomal RNA subspecies. **B.** In degraded samples, RNA fragmentation produces a shift to smaller molecular weight products. Data courtesy of Helen Banks.

gene expression or RNA quality presents a major problem with no simple solution. Without extensive validation to establish the suitability of disso-ciation/dissection techniques in a variety of experimental scenarios, the answer remains unclear. Approaches such as fluorescent-assisted cell sorting (FACS) and laser capture microdissection (LCM) offer potential solutions to these problems, but validation for specific research applications would still be required. The best advice would be to ensure that whatever approach is used, when comparisons are made between tissues both should be isolated in a similar manner.

Preliminary analysis of *c-fos* induction within the murine retina using whole eye or just retinae suggests that both approaches can detect the increase in expression (Semo *et al.*, 2003). However, the amplitude of the change is lower in the whole eye (1.6-fold increase relative to sham) compared with just retinae (6.8-fold increase) as would be predicted due to a smaller contribution to the total RNA pool (Dougherty and Geschwind, 2005).

High throughput qPCR

A reverse transcription (RT) reaction containing 1 µg total RNA should yield enough cDNA for at least 20 genes to be examined. Based upon a total RNA yield from whole eye of 10 µg, this should provide enough material for 200 transcripts to be examined. In contrast, a retinal extraction yielding 3 µg will enable 60 transcripts to be profiled.

When approaching the analysis of multiple transcripts in this manner, the inclusion of an IC on every qPCR run is unnecessary, and simply consumes reagents and space on a plate. As comparisons for a given transcript are made relative to each other, normalization to ICs run on different plates on different days is valid. One way of illustrating this point is to divide the expression of a target gene to the mean target gene expression, so that target gene expression is corrected relative to the mean of all samples. Second, carry out the same procedure for an IC measured for the same samples, correcting to the mean IC expression. If the mean-corrected target gene expression is now normalized to the mean-corrected IC expression, the relative fold change and variance observed between any group of samples will be preserved. This offers a great advantage for relative quantification, as it means that multiple ICs can be used based upon different experimental runs, and barring quite dramatic freeze–thaw effects, which may be evaluated by differences in IC profiles, a target gene run one day may be normalized to a IC run days or even weeks earlier (or later). For this reason, any samples between which comparisons are to be made should be included within the same plate to ensure that they are truly comparable.

Once ICs are well-characterized within a group of cDNA samples, and a normalization factor (NF) is calculated for each sample, this value can be used to normalize all subsequent target genes. Given the synthesis of 20 µl of cDNA, and using 1 µl per target gene, a researcher would be wise to first analyze the expression of two or three (or even more) ICs to ensure that an accurate NF can be calculated before moving on to investigate up to 16 or 17 target genes. However, if a second RT is performed on the same RNA, re-deriving the NF is necessary, as differences in the RT efficiency may occur between the first and second batches of cDNA, and the same relationship between target gene and IC expression may not hold in the second batch of cDNA. This raises an unresolved issue with one-step RT-PCR protocols, where a different reverse transcription is conducted for every reaction.

The use of kinetic approaches to qPCR enables researchers to improve assay throughput, removing the need for using space on each plate for standard curves (which may even demonstrate different reaction kinetics to the cDNA samples under analysis). The use of kinetic data already collected during the course of a qPCR to determine E, coupled with a two-step procedure of cDNA synthesis with subsequent NF derivation enables researchers to move from candidate gene to expression profile in a single reaction, greatly improving assay throughput and saving both time and resources.

6.4 Conclusions

qPCR provides an ideal method for the rapid quantification of ocular gene expression. By carefully considering tissue-specific problems, such as light sensitivity, cellular heterogeneity as well as the consideration of suitable sample sizes based upon known or expected biological variation, researchers may save time and resources, enabling a single sample set to be used for quantification of anything up to a hundred or more transcripts.

Working along similar guidelines, qPCR may be successfully adapted for the relative quantification of gene expression in any tissue, and the underlying principles of assay design, data analysis and normalization of gene expression are universally applicable. Consideration of such fundamental aspects of qPCR assays also highlight potential sources of error, such as the effects of using different amplification efficiencies, threshold values and blind reliance on single internal control genes for normalization.

Finally, the use of kinetic analysis of qPCR data enables a high-throughput approach suitable for the rapid quantification of multiple transcripts, enabling additional measures of quality control to be incorporated. This approach is ideal for examining extended signaling pathways and validation of micro-array data, enabling researchers to move from tissue to gene expression data with unparalleled speed.

Acknowledgments

Many thanks to M. Tevfik Dorak for his kind invitation to contribute to this project. Additional thanks to Russell G. Foster for his unstinting support in these studies, as well as to Jason N. Butler for his continued contribution. Data for Figure 6.2 were kindly provided by Helen Banks. Finally, thanks must also be extended to all the researchers with whom I have spoken on the subjects contained herein for providing a constant improvement of my understanding of this rapidly moving field. Support for this work was provided by the British Biotechnology and Biological Sciences Research Council, the Wellcome Trust, and Children with Leukaemia.

References

Bar T, Stahlberg A, Muszta A, Kubista M (2003) Kinetic outlier detection (KOD) in real-time PCR. *Nucleic Acids Res* **31**: E105.

Berson DM, Dunn FA, Takao M (2002) Phototransduction by retinal ganglion cells that set the circadian clock. *Science* **295**: 1070–1073.

Bland M (2000) *An introduction to medical statistics. Third Edition.* Oxford University Press, Oxford.

Bridges C (1959) The visual pigments of some common laboratory animals. *Nature* **184**: 727–728.

Bustin SA (2000) Absolute quantification of mRNA using real-time reverse transcription polymerase chain reaction assays. *J Mol Endocrinol* **25**(2): 169–193.

Bustin SA (2002) Quantification of mRNA using real-time reverse transcription PCR (RT-PCR): trends and problems. *J Mol Endocrinol* **29**(1): 23–39.

Chomczynski P, Sacchi N (1987) Single-step method of RNA isolation by acid guanidinium thiocyanate-phenol-chloroform extraction. *Anal Biochem* **162**(1): 156–159.

Chomczynski P (1993) A reagent for the single-step simultaneous isolation of RNA, DNA and proteins from cell and tissue samples. *Biotechniques* **15**(3): 532–534, 536–537.

Dartnall HJ, Bowmaker JK, Mollon JD (1983) Human visual pigments: microspectrophotometric results from the eyes of seven persons. *Proceedings of the Royal Society London B Biological Science* Nov 22, **220**(1218), 115–130.

Dowling JE (1987) *The retina: an approachable part of the brain.* Harvard University Press, Cambridge, MA.

Dougherty JD, Geschwind DH (2005) Progress in realising the promise of microarrays in systems neurobiology. *Neuron* Jan 20, **45**(2): 183–185.

Foster RG, Hankins MW (2002) Non-rod, non-cone photoreception in the vertebrates. *Prog Retin Eye Res* **21**: 507–527.

Freeman WM, Walker SJ, Vrana KE (1999) Quantitative RT-PCR: pitfalls and potential *Biotechniques* **26**(1): 112–122, 124–125.

Ginzinger DG (2002) Gene quantification using real-time quantitative PCR: an emerging technology hits the mainstream. *Exp Hematol* **30**(6): 503–512.

Govardovskii VI, Fyhrquist N, Reuter T, Kuzmin DG, Donner K (2000) In search of the visual pigment template. *Vis Neurosci* **17**: 509–528.

Higuchi R, Fockler C, Dollinger G, Watson R (1993) Kinetic PCR analysis: real-time monitoring of DNA amplification reactions. *Biotechnology* (NY). **11**(9): 1026–1030.

Higuchi R, Watson R (1999) Kinetic PCR analysis using a CCD camera and without using oligonucleotide probes. PCR Applications: Protocols for Functional Genomics. Innis MA, Gelfand DH, Sninsky JJ (edn). Academic Press, San Diego, CA.

Jacobs GH, Neitz J, Deegan JF (1991) Retinal receptors in rodents maximally sensitive to ultraviolet light. *Nature* **353**: 655–656.

Kainz P (2000) The PCR plateau phase: towards an understanding of its limitations. *Biochimi and Biophys Acta* **1494**(1–2): 23–27.

Klein D (2002) Quantification using real-time PCR technology: applications and limitations. *Trends Mol Med* **8**(6): 257–260.

Liu W, Saint DA (2002a) A new quantitative method of real-time RT-PCR assay based on simulation of PCR kinetics. *Analytical Biochemistry.* **302**(1): 52–59.

Liu W, Saint D (2002b) Validation of a quantitative method for real time PCR kinetics. *Biochemical and Biophysical Research Communications* **294**: 347–353.

Livak KJ (1997) ABI Prism 7700 Sequence Detection System. *User Bulletin* #2. PE Applied Biosystems.

Livak KJ, Schmittgen TW (2001) Analysis of relative gene expression data using real-time quantitative PCR and the $2^{-\Delta\Delta}_{Ct}$ method. *Methods* **25**(4): 402–408.

Lucas RJ, Freedman MS, Munoz M, Garcia-Fernandez JM, Foster RG (1999) Regulation of the mammalian pineal by non-rod, non-cone photoreceptors. *Science* **284**(5413): 505–507.

Lucas RJ, Douglas RH, Foster RG (2001) Characterization of an ocular photopigment capable of driving pupillary constriction in mice. *Nat Neurosci* **4**: 621–626.

Nir I, Agarwal N (1993) Diurnal expression of c-fos in the mouse retina. *Molecular Brain Research* **19**: 47–54.

Ott RL, Longnecker M (2001) *An introduction to statistical methods and data analysis.* Duxbury, Pacific Grove, CA.

Peirson SN, Butler JB, Foster RG (2003) Experimental validation of novel and conventional approaches to quantitative real-time PCR data analysis. *Nucleic Acids Res* **31**(14): e73.

Pfaffl MW (2001) A new mathematical model for relative quantification in real-time RT-PCR. *Nucleic Acids Research* **29**(9): 2002–2007.

Pritchard CC, Hsu L, Delrow J, Nelson PS (2001) Project normal: Defining normal variance in mouse gene expression. *Proc Natal Acad Sci USA* **98**(23): 13266–13271.

Provencio I, Rodriguez IR, Jiang G, Hayes WP, Moreira EF, Rollag MD (2000) A novel human opsin in the inner retina. *J Neurosci* **20**: 600–605.

Ramakers C, Ruijter JM, Lekanne Deprez RH, Moorman AFM (2003) Assumption-free analysis of quantitative real-time polymerase chain reaction (PCR) data. *Neuroscience Letters* **339**: 62–66.

Sambrook J, Russell DW (2001) *Molecular cloning: a laboratory manual. Third Edition.* Cold Spring Harbor Laboratory Press, Cold Spring Harbor, NY.

Sekaran S, Foster RG, Lucas RJ, Hankins MW (2003) Calcium imaging reveals a network of intrinsically light-sensitive inner-retinal neurons. *Curr Biol* **13**: 1290–1298.

Semo M, Lupi D, Peirson SN, Butler JN, Foster RG (2003) Light-induced *c-fos* in melanopsin retinal ganglion cells of young and aged rodless/coneless (*rd/rd cl*) mice. *European Journal of Neuroscience* **18**: 1–11.

Sturzenbaum SR, Kille P (2001) Control genes in quantitative molecular biological techniques: the variability of invariance. *Comparative Biochemistry and Physiology Part B* **130**: 281–289.

Sun H, Macke JP, Nathans J (1997) Mechanisms of spectral tuning in the mouse green cone pigment. *Proc Natal Acad Sci USA* **94**: 8860–8865.

Tichopad A, Dzidic A, Pfaffl MW (2002) Improving quantitative real-time RT-PCR reproducibility by boosting primer-linked amplification efficiency. *Biotechnology Letters* **24**: 2053–2056.

Vandesompele J, De Preter K, Pattyn F, Poppe B, Van Roy N, De Paepe A, Speleman F (2002) Accurate normalization of real-time quantitative RT-PCR data by geometric averaging of multiple internal control genes. *Genome Biol* **3**(7): RESEARCH0034 e-Pub.

Walker NJ (2002) Tech.Sight. A technique whose time has come. *Science* **296**(5567): 557–559.

Yoshida K, Kawamura K, Imaki J (1993) Differential expression of *c-fos* mRNA in rat retinal cells: regulation by light/dark cycle. *Neuron* **10**: 1049–1054.

Quantitative gene expression by real-time PCR: a complete protocol

7

Thomas D. Schmittgen

7.1 Introduction

Real-time quantitative PCR has been applied for many different biological applications including identifying genomic DNA copy number, single nucleotide polymorphism (SNPs), DNA methylation status, viral load and many others. One of the more popular applications of real-time PCR is to quantify levels of RNA (Wong *et al.*, 2005). Using this technique, RNA is first converted to cDNA and then amplified by the PCR. The products of the PCR are detected in real-time using dedicated instrumentation.

Quantifying gene expression using PCR is one of the more intimidating techniques of modern molecular biology because of the perception that the RNA will be degraded and/or inaccurately measured. The advent of real-time quantitative PCR has greatly simplified gene expression analysis. The purpose of this protocol is to describe each of the steps that are involved in measuring gene expression using real-time PCR. These include sample preparation, isolation of RNA, reverse transcription, primer design, plate set-up and data analysis. Variations of the protocol are included for both cell lines and tissues. An emphasis has been placed on various tips and tricks to increase throughput and reduce error. Implementation of a well-organized and methodical protocol will produce accurate, precise and reproducible data no matter what the tissue or gene of interest.

7.2 Materials

7.2.1 Reagents and consumables

Isopropanol
Ethanol
Molecular biology grade water (Sigma Chemical Co.)
Chloroform
Trizol® Reagent (Invitrogen)
Glycogen for molecular biology (Roche)
SYBR® Green PCR master mix (Applied Biosystems)

RNase-free DNase I (10 U/μl) (Roche)
RNA guard, Porcine (Amersham Pharmacia Biotech)
SuperScript II (200 U/μl) (Invitrogen)
100 mM dNTP set (Invitrogen)
BSA (RNase-free) (Amersham Pharmacia Biotech)
Random primers (Invitrogen)
Optical Adhesive Cover (Applied Biosystems)
96-well real-time PCR plate (Applied Biosystems)
Phase lock gel, 2 ml (Eppendorf)
Aerosol-resistant pipette tips (Rainin)

7.2.2 Equipment

DNA Engine® Thermal Cycler PTC-0200 (MJ Research)
ABI Prism® 7900 HT real-time PCR instrument (Applied Biosystems)
ND-1000 micro spectrophotometer (NanoDrop Technologies)
Allegra 25R centrifuge with microplate adapters (Beckman Coulter)
Stainless steel mortar and pestle sets (Fisher)
L-2, L-20, L-200 and L-1000 Pipettes (Rainin)
L8–10 and L8–200 multichannel pipettes (Rainin)
LTS-200 repeating pipette (Rainin)

7.3 Procedure

7.3.1 Sample preparation

1. Cell lines or blood cells – For attached cells, first trypsinize or scrape attached cells from the culture flask. Pellet suspension cells by centrifugation. Determine cell number using a hematocytometer and place ~3 × 10⁶ cells into to a labeled, microcentrifuge tube. Centrifuge for 5 min at 600 × G, aspirate the supernatant and store the tubes at –80°C.
2. Tissues – Place 10 mg of tissue (2 mm³) into a microcentrifuge tube. Flash freeze the tube by dropping it into a dewer containing liquid nitrogen and store the tubes at –80°C.

7.3.2 Isolation of RNA from cultured cells or blood

With the exception of step 7, the protocol for isolating RNA is identical to that described in the Trizol® protocol.

3. Remove cells from the –80°C freezer and place the tubes on dry ice to prevent thawing.
4. Remove one sample from the dry ice and add 1 ml of Trizol®. It is not necessary to thaw the cells prior to adding the Trizol®.
5. Lyse the cells using a L-1000 pipettor by repeated pipetting until the solution is homogeneous. Incubate at room temperature for 5 min.
6. Add 200 μl of chloroform, shake by hand 15 times and incubate at room temperature for 3 min.
7. Transfer the mixture to a 2 ml Eppendorf phase lock tube. The phase lock tube will prevent mixing of the organic and aqueous layers.

8. Centrifuge the phase lock tube in the cold for 15 min at 12,000 × G. Remove the supernatant and place into a 1.5 ml colored, microcentrifuge tubes. Colored tubes will enhance visualizing the RNA pellet.
9. Add 500 µl of isopropanol and precipitate the RNA for 10 min at room temperature. This is a good stopping point. If necessary, the samples may be placed in the −20°C freezer overnight.
10. Place the tubes in a microcentrifuge in the cold (e.g. cold room, refrigerated microcentrifuge or microcentrifuge placed in a refrigerated unit). Orient the caps to the outside of the centrifuge's rotor. Centrifuge for 10 min at 12,000 × G. The RNA pellet should be visible at the bottom of the tube on the side that was oriented to the rotor's outside.

Refer to section on troubleshooting

11. Decant the supernatant into a 2 ml microcentrifuge tube. It is not necessary to remove all of the supernatant. Add 1,000 µl of 75% ethanol. Centrifuge for 5 min at 7,500 × g in the cold.
12. Decant the supernatant into the same 2 ml microcentrifuge tube used for the isopropanol. Briefly spin the tube containing the RNA for several seconds to bring the residual ethanol to the bottom of the tube.
13. Using a pipette, remove most of the residual ethanol, being careful not to disturb the pellet. To remove the remaining ethanol, place the tubes into a dessicator containing a porcelain platform filled with Drierite. Connect to house vacuum for 5 min. It is not necessary to completely dry the sample since the residual water will not affect the reverse transcription step.
14. Dissolve the RNA pellet in 30–50 µl of molecular biology grade water and place the tubes on ice. RNA may be stored at −80° C, however it is recommended that RNA be converted to cDNA on the same day of the isolation.

7.3.3 Isolation of RNA from whole tissue

15. Remove the tissues from the −80°C freezer and place on dry ice.
16. Pre-chill the stainless steel mortar and pestle sets on dry ice. Place the frozen tissue into the cold mortar. Place the pestle onto the mortar and pulverize the tissue into a frozen powder by forcefully striking with a 2-pound hammer.
17. Transfer the frozen, pulverized tissue as quickly as possible into a 1.5 ml microcentrifuge tube that has been pre-chilled on dry ice. Do not let the pulverized tissue thaw. A sterile disposable surgical blade is ideal for transferring the powdered tissue. Use one new blade per tissue sample.
18. Place the tubes containing the pulverized tissue on dry ice until all of the samples are ready for lysis.
19. Proceed with steps 4–14 to isolate the RNA.

Use the mortar and pestle sets only once per tissue. Clean the mortar and pestle sets with soap and water, dry and re-chill on dry ice prior to isolating any additional tissues.

7.3.4 RNA quantification

The RNA must be quantified to ensure that equal amounts are added to each reverse transcription reaction.

20. Place 1 µl of undiluted RNA onto the lower measurement pedestal of the ND-1000 micro spectrophotometer. Lower the sample arm and take the reading.
21. Wipe off the sample from both the upper and lower pedestals using a clean Kimwipe. It is not necessary to rinse the instrument with water.

7.3.5 DNase treatment

Briefly treat the RNA with RNase-free DNase to remove any residual genomic DNA that may be present in the RNA. Prepare the DNase master mix accounting for 1–2 additional reactions.

Ingredient	Per reaction	For x reactions
RNase-free DNase I (10 U/µl)	1.8 µl	$x \times 1.8$ µl
RNA guard	0.3 µl	$x \times 0.3$ µl
25 mM MgCl$_2$	2.4 µl	$x \times 2.4$ µl
Total	4.5 µl	$x \times 4.5$ µl

22. Add 1.2 µg of RNA, 4.5 µl of master mix and water to 30 µl into 200 µl PCR strip tubes.
23. Mix by gentle flicking, and briefly spin on a minicentrifuge that can handle strip tubes. Using a PCR Thermal Cycler, incubate at 37°C for 10 min and then 90°C for 5 min to inactivate the DNase.

7.3.6 cDNA synthesis

Refer to section on troubleshooting
24. Prepare the following master mix, accounting for 1–2 additional reactions per gene.

Ingredient	Per reaction	For x reactions
5X SuperScript II buffer	10 µl	$x \times 10$ µl
10 mM dNTPs	5 µl	$x \times 5$ µl
0.1 M DTT	5 µl	$x \times 5$ µl
BSA (RNase-free, optional)	1.25 µl	$x \times 1.25$ µl
Random primers	0.25 µl	$x \times 0.25$ µl
RNA guard	1.25 µl	$x \times 1.25$ µl
Superscript II reverse transcriptase	1.25 µl	$x \times 1.25$ µl
Molecular biology water	1 µl	$x \times 1$ µl
Total	25 µl	$x \times 25$ µl

25. Using the LTS-200 repeating pipette, add 25 µl of the reverse transcription master mix to 200 µl PCR strip tubes labeled with the date and sample number on the side.
26. Add 25 µl of the DNase-treated RNA to each tube. This will contain 1 µg of RNA.

27. Incubate at the following temperatures using a PCR thermal cycler: 26°C for 10 min (to allow the random hexamers to anneal), 42°C for 45 min (reverse transcription) and 75°C for 10 min (to inactivate the reverse transcriptase).
28. The resulting cDNA may be analyzed immediately by real-time PCR or stored at –20°C or –80°C.

7.3.7 SYBR® Green

SYBR® Green is a minor groove binding dye that binds double-stranded (ds) DNA but not single stranded DNA. A signal from SYBR® Green will be detectable only after significant amplification (>15 cycles for most cDNA). Since SYBR® Green binds to dsDNA, all dsDNA present in the reactions (e.g. primer dimers or non-specific amplification products) will also be detected. Primers should always be validated to insure that a single amplicon is generated by the PCR. Reactions may be run on agarose gels to ensure that only one product is present. The thermal denaturation protocol on the Applied Biosystems and most other real-time PCR instruments should always be run following SYBR® Green reactions. This heats the products after the final reaction from 60°C to 90°C and will produce a characteristic peak at the Tm of each amplicon. If only one amplicon is present, then only one peak will be present. If primer dimers are present along with the product then a second peak (usually with reduced Tm) will be seen.

7.3.8 Primer design

Use a computer program such as Primer Express® (Applied Biosystems) or equivalent to design primers. Alternatively already designed primers in several open access databases may be used (see Chapter 5). For analysis of cDNA from messenger RNA, design PCR primers to span one intron. Should any genomic DNA be present following the DNase treatment, the thermal denaturation protocol will detect two different PCR products, the shorter cDNA and the much longer genomic DNA. A convenient method to determine where the intron/exon junctions exist in a cDNA sequence is to BLAST the cDNA sequence of your gene of interest against the genomic sequence for the same gene. The BLAST report will locate the sequence of the intron/exon junctions. Adhere to the rules of primer design listed in *Table 7.1*. Dissolve primers in molecular biology grade water and store at –20°C.

Table 7.1 Design criteria for real-time PCR primers, SYBR® Green detection

Primer length 18–24 nt
Amplicon length <250 bp (ideally <150 bp)
Both sense and anti-sense primers should have a Tm ≤2°C of each other
Primer Tm 50–60°C (ideal range 55–59°C)
No consecutive runs of the same nucleotide more than six times
No runs of more than three consecutive Gs
% GC content of primers ~50% (no less than 35% and no more than 65%)
No 3' GC clamp on primers (i.e. GG, CC, CG or GC)
≤2 GC in the last five nucleotides of the 3' end of the primer

See also Chapter 5 for more considerations on primer design.

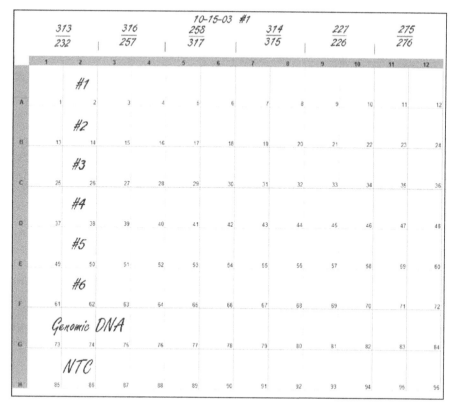

Figure 7.1

An example of a PCR set-up sheet. The numbers (1 to 6) represent different samples of cDNA. The numbers on the top of the columns (e.g. 313/232) represent the pair of primers.

Working stock solutions of primers are diluted to 50–100 µM and stored at –20°C.

7.3.9 Real-time PCR

Perform triplicate PCRs per gene, per cDNA sample. Most real-time PCR instruments are configured with a 96-well block. Use a 'PCR set-up sheet' (Figure 7.1) to organize the pipetting of the plate as well as to coordinate the data during the analysis.

29. Prepare the following master mix, accounting for 1–2 additional reactions per gene:

Ingredient	Per reaction	For x reactions
2X SYBR® Green reagent	12.5 µl	$x \times 12.5$ µl
Forward/Reverse primer mix (50 µM each)	0.125 µl	$x \times 0.125$ µl
Molecular biology grade water	7.375 µl	$x \times 7.375$ µl

30. Dilute the cDNA 1:50 or 1:100 by first placing molecular biology grade water into a disposable sterile basin. Add 99 µl of the water into the PCR strip tube using a multichannel pipette.
31. Add 1 µl of cDNA to the labeled strip tubes using a multichannel pipette (e.g. Rainin L8–10). Use a multichannel pipette (e.g. Rainin L8–200) to mix the solution 20 times. For efficient mixing, set the pipette at 75% of the solution's volume (75 µl in this example).

Refer to section on critical steps

32. Recap the undiluted cDNA with new strip caps to prevent cross contamination.
33. Use a fine tip marker to mark the plate to the location of the different master mixes similar to the set-up sheet in *Figure 7.1*.
34. Use the repeating pipette (Rainin, E12–20) to add 20 µl of master mix to each sample. Add one row at a time.
35. Add 5 µl of dilute cDNA to each of the wells using a multichannel pipette.
36. Add the optical adhesive cover and seal using the sealing tool. Perform a brief spin (up to 1500 RPM) on a centrifuge equipped with a 96 well plate adapter.
37. Perform PCR using the real-time instrument per the manufacturers' protocol. Typically, 40 cycles of 15 sec at 95°C and 60 sec at 60°C followed by the thermal dissociation protocol.

7.3.10 Data analysis

There are two common methods of analyzing gene expression data generated by real-time PCR; absolute quantification and relative quantification (see also Chapters 2, 3, and 6). For routine gene expression analysis, it is not necessary to know precise copy number and the relative changes in expression will suffice. For example, it is not important that a drug increased the copy of gene x from 10,000 to 50,000 copies per cell, only that the change in expression was increased by five-fold. For this reason, only an example of the relative quantification method will be given here. The most commonly used method for relative quantification is the $2^{-\Delta\Delta Ct}$ method. Derivation and examples of this method have been described elsewhere (Livak *et al.*, 2001).

Calculate the relative difference in gene expression using the $2^{-\Delta\Delta Ct}$ method.

Relative fold change in gene expression = $2^{-\Delta\Delta Ct}$

Where: $\Delta\Delta C_t = \Delta C_{t\ treated} - \Delta C_{t\ untreated}$

and $\Delta C_t = (C_{t\ target\ gene} - C_{t\ reference\ gene})$

Refer to section on critical steps

38. The reference gene should be properly validated. Selection of the reference or internal control gene is critical as the expression of the reference gene will often fluctuate with treatment or disease. While the user is welcome to try commonly used internal controls (e.g. β-Actin, GAPDH, or β-2 microglobulin), in our experience, highly expressed, non-protein coding genes, such as 18S rRNA or U6 RNA, have performed the most consistently for most applications. Primer sequences have been

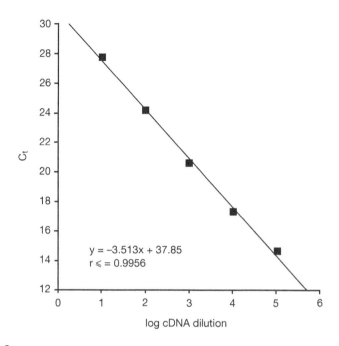

Figure 7.2

An example of the plot used in amplification efficiency calculation.

published for 18S rRNA (Schmittgen *et al.*, 2000) and U6 RNA (Schmittgen *et al.*, 2004). It may be the best strategy to use multiple reference genes as described elsewhere in this book (Chapters 4, 6, and 8).

Refer to section on critical steps

39. Determine the amplification efficiency of the reaction for both the target and reference genes (*Figure 7.2* and see below). The $2^{-\Delta\Delta Ct}$ method assumes that the amplification efficiency for both the target and reference genes is similar (Livak *et al.*, 2001).

Calculation of fold-change in gene expression

40. Export the raw C_t values from the real-time PCR analysis into Microsoft Excel®. Assign sample number, gene name and treatment to each data set.
41. Sort the data such that it groups the sample number (in triplicate), treatment and gene. Calculate the mean C_t value for the internal control gene.
42. Enter the mean C_t for the internal control into one column of the spreadsheet and write a macro to calculate the $2^{-\Delta\Delta Ct}$. A sample spreadsheet is presented (*Figure 7.3*).
43. Calculate the mean fold change, standard deviation and coefficient of variation for the triplicate PCRs for each sample. Acceptable coefficient of variation for triplicate PCRs from the identical sample of dilute cDNA is 2–30%.

Refer to section on troubleshooting

Sample #	Well	Type	Gene	Time (h)	Primer/Probe	C_T Time X	Mean C_T Time 0	$2^{\Delta\Delta CT}$	Mean Fold Change in gene expression	S.D.	C.V.
1	A1	UNKN	fos-glo-myc	0	PR1	22.3	21.9	1.023			
2	B1	UNKN	fos-glo-myc	0	PR1	22.0	21.9	0.786	1.02	0.228	22.4
3	C1	UNKN	fos-glo-myc	0	PR1	21.5	21.9	1.243			
4	D1	UNKN	fos-glo-myc	0.5	PR1	19.8	21.9	1.845			
5	E1	UNKN	fos-glo-myc	0.5	PR1	20.2	21.9	2.019	2.00	0.152	7.61
6	F1	UNKN	fos-glo-myc	0.5	PR1	20.0	21.9	2.149	2^-((20.0-21.7)-(21.9-22.5))		
7	G1	UNKN	fos-glo-myc	1	PR1	19.5	21.9	2.271			
8	H1	UNKN	fos-glo-myc	1	PR1	18.9	21.9	3.466	2.86	0.598	20.9
9	A2	UNKN	fos-glo-myc	1	PR1	19.2	21.9	2.855			
.	.	.	.		.	.	.	.	.	.	.
1	A5	UNKN	Beta Actin	0	PR2	22.9	22.5				
2	B5	UNKN	Beta Actin	0	PR2	22.3	22.5				
3	C5	UNKN	Beta Actin	0	PR2	22.4	22.5				
4	D5	UNKN	Beta Actin	0.5	PR2	21.2	22.5				
5	E5	UNKN	Beta Actin	0.5	PR2	21.7	22.5				
6	F5	UNKN	Beta Actin	0.5	PR2	21.7	22.5				
7	G5	UNKN	Beta Actin	1	PR2	21.2	22.5				
8	H5	UNKN	Beta Actin	1	PR2	21.2	22.5				
9	A6	UNKN	Beta Actin	1	PR2	21.3	22.5				

Figure 7.3

A sample spreadsheet showing the calculation of relative quantification by the $2^{-\Delta\Delta Ct}$ method. (Reproduced by permission of Elsevier.)

7.4 Troubleshooting

Problem	Solution
RNA pellet is not visible after initial precipitation step.	Add 2 µl of glycogen (1 mg/ml) to the isopropanol/RNA mixture and precipitate overnight at −20°C.
RNA pellet is still not visible following the addition of glycogen and overnight precipitation.	Inefficient lysis of sample (step 5). Lyse cells using a 1 cc disposable syringe and 23 gauge needle. Also possible that cells/tissue have low amount of RNA. Increase the amount of cells or tissue (steps 1 and 2).
RNA is very GC-rich or contains high amount of secondary structure.	Synthesize cDNA using a thermostable reverse transcriptase (e.g. Thermoscript, Invitrogen) and increased reaction temperature. Use gene specific primers instead of random primers since random primers will not anneal at elevated temperatures.
Primer pair produces PCR efficiency <1.85.	Optimize the $MgCl_2$ and/or primer concentration in the PCR to increase the efficiency.
High variation (>60%) among triplicate PCRs from the identical sample of diluted cDNA.	Check pipetting technique and pipettes. Use newer and recently calibrated/serviced pipettes. Check the wells in the 96-well plate to see if evaporation has occurred following the PCR.

7.5   Critical steps

Step 31 Dilution and pipetting of cDNA are the most critical steps in the analysis and will introduce the greatest amount of error if done incorrectly. Always make fresh dilutions of cDNA on the day of the analysis and discard any diluted cDNA that remains. Do not freeze and reanalyze the diluted cDNA. Always dilute 50% more cDNA than you think you will need for that day. The original, undiluted cDNA is very stable and may be re-used for many years if stored properly.

Step 35 Apply the technique of 'pipette overloading' to deliver the correct amount of cDNA and reduce the aspiration and adhering of sample to the pipette tip. Depress the plunger of the multichannel pipette set for 5 µl until it stops, then depress the plunger a little bit further. Place the pipette into the PCR strip tubes containing the dilute cDNA and pick up the sample. Add the 5 µl sample to the side of the PCR plate by depressing the plunger until it stops. This will leave some residual solution in each tip. Discard the tips and repeat using new tips.

Step 38 Proper validation of the internal control gene is critical. The internal control gene should not change under experimental conditions (e.g. treated versus untreated or normal versus diseased tissue). Validation of the internal control should be done prior to quantifying the gene(s) of interest. Methodology to validate internal control genes has been described previously (Schmittgen *et al.*, 2000). In brief, triplicate PCRs should be performed on each sample of cDNA (both treated and untreated samples) using primers for the internal control gene. Calculate the mean C_t for each cDNA and present the expression of the internal control as 2^{-Ct}. Perform statistical analyses as described (Schmittgen *et al.*, 2000) or use the students t-test to determine if the internal control gene has fluctuated under the conditions of the experiment.

Step 39 PCR efficiency is determined from the equation $N = N_0 \times E^n$, where N is the number of amplified molecules, N_0 is the initial number of molecules, n is the number of PCR cycles and E is the efficiency which is ideally 2 (Mygind *et al.*, 2002). When the equation is of the form $n = -(1/\log E) \times \log N_0 + (\log N/\log E)$, a plot of log copy number versus C_t yields a straight line with a slope $= -(1/\log E)$ (*Figure 2*). Acceptable amplification efficiencies using this method are 1.85 to 2.05. Determination of amplification efficiency using this method has been described previously (Schmittgen *et al.*, 2004).

7.6   Comments

It is important to remember that RNA will be degraded only in the presence of ribonuclease. Ribonuclease may be present in the lab from previous experiments (e.g. RNAse protection assays), from glassware that once contained serum or from human skin. The likelihood of ribonuclease contamination may be alleviated by performing experiments with RNase in dedicated spaces or in separate laboratories, by using disposable plasticware rather than reusable glassware and by always wearing clean gloves when working with RNA. The two most important aspects of this procedure are to have a visible RNA pellet after the initial precipitation step and accurate pipetting. Due to the sensitivity of the PCR, it is possible to obtain PCR signals by amplifying cDNA synthesized from <1 ng of total RNA. However, we have observed the most consistent results when an RNA pellet is visible following the precipitation. Accurate and conscientious pipetting is critical to reducing error and preventing cross contamination.

References

Livak KJ, Schmittgen TD. (2001) Analysis of relative gene expression data using real-time quantitative PCR and the 2(-Delta Delta C(T)) Method. *Methods* **25**: 402–408.

Mygind T, Birkelund S, Birkebaek N, Oestergaard L, Jensen J, Christiansen G (2002) Determination of PCR efficiency in chelex-100 purified clinical samples and comparison of real-time quantitative PCR and conventional PCR for detection of Chlamydia pneumoniae. *BMC Microbiol* **2**: 17.

Schmittgen TD, Jiang J, Liu Q, Yang L (2004) A high-throughput method to monitor the expression of microRNA precursors. *Nucleic Acids Res* **32**: E43.

Schmittgen TD, Zakrajsek BA (2000) Effect of experimental treatment on housekeeping gene expression: validation by real-time, quantitative RT-PCR. *J Biochem Biophys Methods* **46**: 69–81.

Wong ML, Medrano JF (2005) Real-time PCR for mRNA quantitation. *Biotechniques* **39**: 75–85.

Real-time PCR using SYBR® Green

8

Frederique Ponchel

8.1 Introduction

Real-time PCR has become a well-established procedure for quantifying levels of gene expression, as well as gene rearrangements, amplifications, deletions or point mutations. Its power resides in the ability to detect the amount of PCR product (amplicon) at every cycle of the PCR, using fluorescence. Several approaches have been employed to detect PCR products. The most popular ones are based on specific binding of hydrolysis of hybridization probes to the target sequence. Another approach uses fluorescent dyes, which bind to the double-stranded PCR product non-specifically. Although both assays are potentially rapid and sensitive, their principles of detection and optimization are different, as is the resulting cost per assay. Here we will explore the development and rigorous testing of a real-time PCR assay using the inexpensive SYBR® Green I technology.

8.2 SYBR® Green chemistry

SYBR® Green is a fluorescent dye that binds only to double-stranded DNA. Fluorescence is emitted proportionally to the amount of double-stranded DNA. In a PCR reaction, the input DNA or cDNA is minimal and, therefore, the only double stranded DNA present in sufficient amounts to be detected is the PCR product itself. As for other real-time PCR chemistries, read-outs are given as the number of PCR cycles ('cycle threshold' C_t) necessary to achieve a given level of fluorescence. During the initial PCR cycles, the fluorescence signal emitted by SYBR® Green I bound to the PCR products is usually too weak to register above background. During the exponential phase of the PCR, the fluorescence doubles at each cycle. A precise fluorescence doubling at each cycle is an important indicator of a well optimized assay. After 30 to 35 cycles, the intensity of the fluorescent signal usually begins to plateau, indicating that the PCR has reached saturation. As C_t correlates to the initial amount of target in a sample, the relative concentration of one target with respect to another is reflected in the difference in cycle number ($\Delta C_t = C_t^{sample} - C_t^{reference}$) necessary to achieve the same level of fluorescence.

There are now many commercially available SYBR® Green chemistry kits (SYBR® Green core reagents master mix by ABI; QuantiTech SYBR® Green

PCR master mix by Qiagen; DNA master SYBR® Green by Roche; IQ™ SYBR® Green supermix by Biorad; Superarray by BioScience Corp.; DyNAsin™ II DNA polymerase with SYBR® Green by MJ Research and probably others). There is no clear indication for choosing one over the others, however, once an assay has been optimized with one chemistry, changing kit requires re-optimization. Prices differ and may guide towards one kit or another depending on amount of usage. To minimize variability, assays using master mixes may be recommended, although multi-user facilities may prefer individual reagents. It is also possible to use homemade SYBR® Green master mixes to reduce cost even further (Karsai *et al.*, 2002).

8.3 Primer design

Optimal design of the PCR primers is essential for accurate and specific quantification using real-time PCR. For the assay described herein, detection is based on the binding of the SYBR® Green I dye into double-stranded PCR products, which is a sequence-independent process. While this assay is cheaper than the specific probe-based assays, it loses the additional level of specificity introduced by the hybridization of a specific fluorescent TaqMan® probe to the PCR product. The sensitivity of detection with SYBR® Green may therefore be compromised by the lack of specificity of the primers, primer concentration (which can be limiting) and the formation of secondary structures in the PCR product. The formation of primer-dimers may register false positive fluorescence, however, this can now easily be overcome by running a PCR melting curve analysis. The 5' nuclease assay using TaqMan® probes would also be compromised by the lack of primer specificity and limiting primer concentration, and although these are not detected by the TaqMan® probe, they alter the amplification efficiency of the PCR reaction.

Primer design for SYBR® Green based assays needs to be more stringent than for a classic TaqMan® assay. It largely depends on the sequence targeted, and varies with the type of assay required: gene amplification, rearrangement, deletion, cDNA quantification, splicing variant and others. Primer design is still very restrictive: the annealing temperature is limited to between 58 and 60°C, which correspond to the optimal working conditions for the Taq DNA polymerase enzyme, and the length of the PCR product has to be set between 80 and 150 bp. The use of software dedicated to real-time PCR primer design, such as Primer Express® (Applied Biosystems), is highly recommended. However, it has rarely proven efficient using the TaqMan® primer/probe design option and is rather more successful using the simple DNA PCR option.

The first step in primer design is to determine the sequence that needs to be targeted by the PCR. This is highly dependent on the type of assay (gene amplification, rearrangement, deletion, gene expression, splicing variant, promoter switch, chromatin immunoprecipitation or others), which restricts the sequence available to design primers. The second step is to identify related genes and eliminate regions of high homology, which may further restrict the length of the sequence available. A third step in primer design for cDNA quantification requires the design to overlap exon–exon boundaries.

8.3.1 Step by step primer design: β-actin for a cDNA quantification assay

Identify the target sequence

Browse the human genome project database with β-actin as the keyword to identify the sequence, chromosome location (7p22.1), mRNA accession number (X00351), mRNA length (1761 bp) and gene structure (5 exons).

To choose a suitable accession number to base primer design on, use the graphic representation to select an mRNA that includes all the exons and has 5' and 3' sequence (the database is full of all sorts of partial cDNA sequences (e.g. CR685511) and mRNA with 'alternative' or 'inconsistent' exons (e.g. AK097861)).

Restrict this sequence with regards to regions of homology with related genes (α-actin, γ1 and γ2-actin)

Align these sequences to determine domain of homology. Use the alignment tools on the NCBI website (BLAST 2 sequences) to find homologies.

Determine exon-exon boundaries

Use the human genome browser, click on the accession number for the human mRNA corresponding to the full-length transcript (e.g. X00351). Click on the accession number at the top to display the sequence database information. Click on the accession number under 'query' in the mRNA/Genomic alignment section to obtain 5' and 3' sequences (red), coding region (blue), ATG and STOP codons (underlined), exon–exon boundaries (bold, light blue or orange). Avoid inconsistent splicing sites and polyadenylation signals (underlined but not always).

Design primers

As a result of this preliminary work, the sequence available for designing primer to quantify the expression of β-actin is restricted to two regions, between nucleotides 1 and 90 or between nucleotides 1208 and 1763. There is no exon boundary in the second region. The design should, therefore, be limited to the region between the first and second exons of the gene. Transfer this sequence to the primer design software (e.g. Primer Express® from ABI) and proceed to design.

Copy the mRNA sequence into a simple PCR document. Restrict the target sequence accordingly. Select the option for primer Tm (59°C) and amplicon length (80 to 120). Keep the CG content at default settings and primer length below 24 bp.

See if the software comes up with a possible primer pair. If not, use the primer test document and try to select sections of sequence manually. This sequence is very CG rich, therefore, select a region with A or T as 3' for each primer. Elongate the primer in 5' until reaching a Tm between 58° and 60°C.

Lastly, BLAST the primers (limit to the human species and mRNA bank). If the restriction in the target position has been done thoroughly, the BLAST search should give no matches. Quite a lot of hits are nevertheless reported for these primers. Most of them are on anti-sense strands of cDNAs. If none of them comes in both the forward and the reverse primer BLAST searches, no PCR product can be generated from these hits.

These primers are located in the 5' of the mRNA. This is an issue if the quality of the RNA is not sufficient to ensure reverse transcription up to the 5' end of the mRNA when an oligo-dT cDNA synthesis method is used. Therefore, a random hexamer cDNA synthesis is recommended when using these primers for β-actin as a housekeeping gene.

8.4 Primer optimization

The optimization of the primer concentration is essential. Each set of primers works best at a different concentration. Primer concentration is usually determined to be optimal when the specific amplification relative to primer-dimers is maximal, in a positive versus negative control experiment. However, the recent development of melting curve analysis for PCR products has made the dimers issue obsolete (see below and also the next chapter). One major limitation in primer optimization is, however, the availability of a good positive template for optimization. Finding a tissue that definitely expresses the gene of interest, cloning a rearrangement or a translocation, etc, remain essential to the optimization process. Again primer optimization differs from one type of assay to another.

8.4.1 Absolute quantification of gene expression

This is an easy procedure of primer concentration optimization, where one is looking for the best possible conditions of amplification (lowest C_t). The procedure consists of varying the amounts of primer added to the reaction, in order to find the most efficient concentration, bearing in mind that 'more' does not always mean 'better'. There should be no difference in the other reagents used per reaction and the same input target cDNA should be used in all PCR reactions. For easy pipetting (for a 25 µl reaction), concentrations of 50, 100, 300, 500, and 1,000 nM can be used. An optimization matrix can be set up as shown below. Routinely, a simpler matrix (grey boxes) can be used to find good conditions and then further refined if necessary. An excess of primer, such as 1,000 F/1,000 R nM (as in the above examples) is often limiting and decreases the efficiency of the PCR reaction. An example of this procedure is illustrated in *Figures 8.1A* and *8.1B*.

Following primer concentration optimization, a standard curve should be done, to perform an absolute quantification, but also to find the limits of sensitivity of the assay. A serial dilution of a factor 10 is easy to use and provides an obvious difference of C_t between points. A positive target sample should be used, which either contains a known amount of the target or is used as a default value of 1 (*Figures 8.2* and *8.3*).

8.4.2 Relative quantification of gene expression

This is also an easy optimization procedure, but one is looking for the best possible conditions of amplification (lowest C_t) with a similar PCR efficiency for both the target and the endogenous control(s) as reference

Reverse primer (nM)	Forward primer (nM)				
	50	100	300	500	1000
50	50/50	50/100	50/300	50/500	50/1000
100	100/50	100/100	100/300	100/500	100/1000
300	300/50	300/100	300/300	300/500	300/1000
500	500/50	500/100	500/300	500/500	500/1000
1000	1000/50	1000/100	1000/300	1000/500	1000/1000

Figure 8.1A

A standard protocol for optimizing primer concentrations.

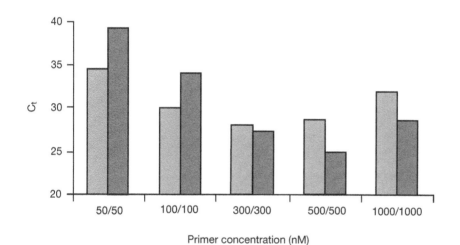

Figure 8.1B

Two sets of primers were designed to amplify genes *grey* and *black*. Primers for gene *grey* worked better at 300/300 nM (C_t 28), but those for gene *black* were better at 500/500 nM (C_t 25).

(usually several housekeeping genes). The procedure consists of finding the most efficient primer concentration for both genes (see above), followed by a comparison of efficiency between target and reference. The same optimization matrix can be set up to find appropriate primer concentration conditions (both for the target and the reference genes). Then, a serial dilution of the input cDNA is used to compare the efficiency of both PCR reactions in parallel. C_t values will be different between the target and the reference but one is looking for parallelism between the two standard curves. PCR efficiencies can be compared by relating the slope of the linear curve of C_t values plotted against the log-dilution. A perfect reaction has a slope of 3.3, which corresponds to an exact two-fold amplification at every cycle of the PCR between dilutions of a factor 10. An example of efficiency determination is shown in *Figure 8.4*.

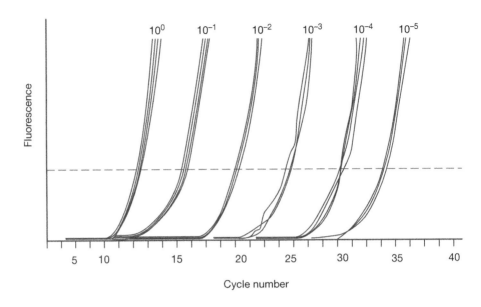

Figure 8.2

An example of amplification plot.

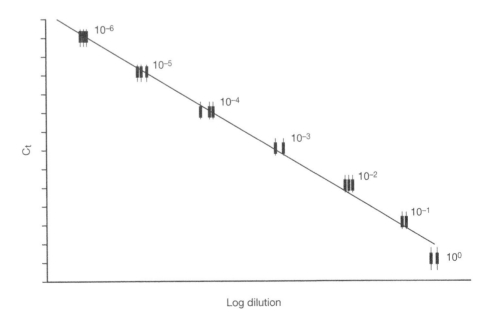

Figure 8.3

An example of standard curve.

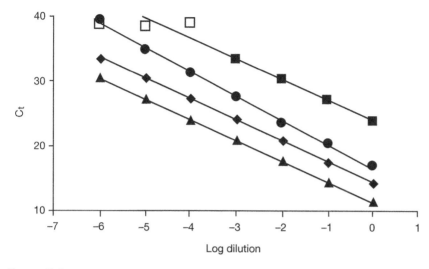

Figure 8.4

Standard curves for different target sequences for comparative efficiency determination. Ideally they should be parallel (same slope) to be assayed together. The same serial dilution (from pure cDNA to 1/1,000,000) was used to analyze the efficiency of PCR reaction for genes *square*, *diamond* and *circle* against the housekeeping gene *GAPDH* (triangle). The efficiency for *GAPDH* was 3.20. It was equivalent to the efficiency of the PCR reaction for gene *circle*, 3.21. However, the PCR efficiency for gene *square* was good (3.19 for the first 4 serial dilutions); the PCR reaction reached its limit after the fourth dilution in a 40 cycles reaction. Finally, the efficiency of the PCR reaction for gene *diamond* was 3.78. This is too far apart for use in a relative quantification. A new set of primers should be designed for gene *diamond*.

8.4.3 Relative quantification of different gene modifications (amplification, deletion, rearrangement, translocation)

The choice of the most appropriate conditions for these types of assay corresponds to the combination of primer concentration allowing a ratio of 1 between the target gene and the reference gene in a DNA sample with no genetic abnormality. This ratio is calculated using the ΔC_t method ($\Delta C_t = C_t$ of target – C_t of reference = 0), a ΔC_t of 0 indicating a ratio of 1. A number of factors can influence this assay, and the optimization consists of compensating these factors to obtain a ratio of 1 in a DNA sample with no genetic abnormality. The inclusion of normal DNA sample in each experiment also provides a reference sample for a $\Delta\Delta C_t$ method of calculation. As the normalization will relate to a value close to 0 for the second ΔC_t, both methods essentially give the same result. For an accurate quantification it is also necessary to verify that PCR efficiency is independent of the initial amount of target DNA. Each pair of primers therefore needs to be tested across several log dilution of a control DNA sample as described above.

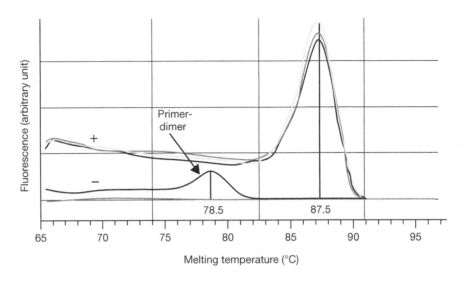

Figure 8.5

An illustration of the use of melting curve analysis to distinguish between specific and non-specific PCR products (see text).

8.5 Melting curve analysis

Heat dissociation (PCR product melting curve) analysis at the end of the PCR confirms whether or not a single product is amplified and that no dimers interfere with the reaction. A melting curve analysis begins with heating the PCR product at the end of the PCR reaction. As the PCR product melts and the SYBR® Green is released into the solution, its fluorescence intensity decreases. A negative first derivation curve of the fluorescence intensity curve over temperature produced by the instrument's software clearly indicates the Tm of the PCR product (peak of the –dF/dT curve) and should be quite close to the predicted Tm of the PCR product.

In the example illustrated in *Figure 8.5*, the PCR product Tm is 87.5°C (curves indicated by +). Complete absence of primer-dimer is rarely achieved in the PCR negative control (curve indicated by –). As seen in this example, 1 out of 3 negative triplicates shows dimers (with a Tm of 78.5°C). The two sets of curves are usually clearly separated with a 10°C shift between Tm of primer-dimers and the specific PCR product. In an experimental negative including a template but no target, dimmers are not usually observed. The degree of an eventual primer-dimer contribution to the overall fluorescent signal of the PCR negative control can also be detected in such a dissociation analysis. Further details of the use of melting curve analysis for allelic discrimination are given in Chapter 9.

8.6 Quantification of gene modification

The detection and quantification of gene rearrangement, amplification, translocation or deletion is a significant challenge, both in research and in a clinical diagnostic setting. Real-time PCR has become a well-established procedure for quantifying these gene modifications. For example, the major and minor break-points of the chromosome 14 to 18 translocation t(14:18)(q32;q21) have been analyzed in non-Hodgkin's lymphoma using a 5' nuclease assay with TaqMan® probes (Estalilla *et al.*, 2000). The diagnosis and treatment of a number of severe conditions, such as the multi-drug resistance phenotype in tumors, could benefit from the development of a quick assay. Compared to Southern blotting or fluorescence in-situ hybridization (FISH) used for the detection of gene amplification, a SYBR® Green real-time PCR assay provides a rapid and accurate alternative. Furthermore, a FISH-based assay would require biologic material, which may not be available or easy to obtain. Genetic counseling could also benefit from a quick and reliable assay to determine carriers of tumor suppressor gene deletions involved in inherited susceptibility to cancer (e.g. P53, RB, WT1, APC, VHL or BRCA1). The general applicability of a SYBR® Green real-time PCR assay quantifying such genetic events (deletion, amplification and rearrangement) has been demonstrated (Ponchel *et al.*, 2003) and requires inclusion of a number of controls.

8.6.1 DNA quantification

A major aim in using real-time PCR is to allow accurate quantification of gene modifications in DNA extracted from relatively few cells. For this purpose, the number of cells present in a given sample can be estimated using real-time PCR quantification of the copy number of a 'reference' gene, each cell having two allelic copies of this gene. In principle, any gene is suitable but examination of deletions, amplifications and rearrangements requires this 'reference' gene not to be involved in any of these processes. The genetic abnormality is then quantified relative to this reference.

It is necessary to establish the validity of the relationship between the amount of DNA in the sample and the 'reference' C_t reading. The relationship between the number of cells in a sample and the DNA concentration of the sample also needs to be linear, indicating a similar efficiency of the DNA extraction method irrespective of the initial number of cells in the sample.

Archived material such as paraffin embedded tissue sections can be used. DNA can be extracted from such material using a simple proteinase K and phenol extraction methodology, with no need for de-waxing, and used in a SYBR® Green I real-time DNA quantification assay.

8.6.2 Gene amplification

Detection of gene amplifications, usually in tumor samples, can also be achieved by SYBR® Green method. In our own experiments, we were able to obtain results that were in agreement with FISH result (*Figure 8.6*).

8.6.3 Gene deletion

Our laboratory has successfully used SYBR® Green detection method to diagnose gene deletions. Point mutations and small insertions or deletions in the *OPA1* gene have been shown to underlie more than half of all cases

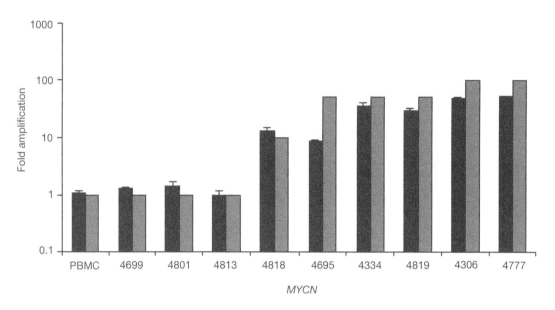

Figure 8.6

MCYN gene amplification in neuroblastoma tumors. We applied a SYBR® Green 1 gene amplification assay to tumor samples for the detection of *MYCN*. These samples had been tested previously by standard procedures for the diagnosis of neuroblastoma using FISH (De Preter *et al.*, 2002). Our results (black bars) compared well with the established amplification factor recorded for these tumors (grey bars) and confirmed the presence of *MYCN* amplification (Ponchel *et al.*, 2003).

of dominant optic atrophy. In addition, two larger deletions in *OPA1* have been detected, each in a single family. One of these deletes exon 20 and the surrounding intronic DNA (Alexander *et al.*, 2000), while the other deletes the entire gene and over half a megabase of surrounding DNA (Marchbank *et al.*, 2002). We designed a SYBR® Green I assay to quantify the number of copies of *OPA1* exon 20 present in any given sample, which should therefore detect both of these deletions. We tested this on genomic DNA from patients and healthy members of the family and were able to confirm a reduction in relative copy number in these patients with the deletion of the entire gene, compared to healthy members of the family (Ponchel *et al.*, 2003).

8.6.4 Gene rearrangement

A variety of gene arrangements can be detected using dsDNA binding dyes. Our laboratory has some experience in the detection of T-cell receptor (TCR) genes. During their passage through the thymus, T-cell precursors rearrange their TCR genes. This step requires the excision of segments of DNA, the ends of which are subsequently ligated to form small circles of episomal DNA (signal and coding) referred to as T-cell receptor excision circles (TREC). The sequences remaining after recombination provide the target for the PCR detection of TREC. The point of recombination in TREC dictates the position of the primers on either side. The sequence around these allows the primers to be designed with more or less flexibility within the Tm and amplicon length limits (Ponchel *et al.*, 2003). Similar examples include Immunoglobulin gene rearrangements, translocations (chromosome 14/18 translocation t (14:18) (q32;q21)).

8.6.5 Gene copy number

A SYBR® Green DNA quantification assay could be used for the quantification of gene copy number in a transgenic mouse model. The gene *Hop* was 5′ FLAG-tagged. The endogenous Hop sequences were used as reference gene for the quantification and the FLAG sequence used to determine the number of gene copies in individual transgenic mice. We were able to identify founders carrying either 1 or 2 copies of the transgene (Ponchel *et al.*, 2003).

8.7   RNA quantification

8.7.1 RNA extraction

There are many methods of RNA extraction available, both manual and commercial. The choice of method needs to be guided by the nature of the initial material, biopsy, tissue culture sample, high or low cell number yield, etc. Commercially available kits perform well enough, have the advantage of not requiring phenol extraction (with the exception of Trizol®, Invitrogen Life Technologies) but are costly. They do provide an advantage when low RNA yield is expected. However, when samples coming from tissue culture, where experiments are repeatable, manual extractions are better justified.

An important point is to verify the absence of genomic DNA contamination. A DNase treatment step may be included to follow RNA extraction as gDNA contamination may interfere with the assay. Different methods and kits are available. None is particularly better, but a column-based technique may be easier for inexperienced users.

8.7.2 cDNA preparation

Synthesis of cDNA from mRNA is the first step of RT-PCR gene expression quantification. The quality of the cDNA is directly related to the quality of the RNA template. If possible, determine the RNA concentration of samples before using them in an RT reaction. Again, there are many methods for cDNA synthesis. Different RT enzyme kits are available (MultyScribe from ABI, SuperScript II from Invitrogen, AMVRT from Promega, Expand RT from Roche, and many others) using initiation with either oligo-dT or random hexamer priming. Random hexamers are preferred for analysis of non polyadenylated RNA. Once prepared, first-strand cDNA should not be thawed and refrozen more than once. During PCR set-up, it has to be kept on ice and used immediately.

Once the cDNA has been synthesized, it remains to be verified that no DNA contaminates the samples. As the assay to determine mRNA quantity is based on PCR, the assay to detect genomic DNA contamination also needs to be performed using PCR (an agarose gel would not be sensitive enough). A PCR across an exon–exon boundary for an abundant mRNA should be used. A genomic DNA target may be used as control for the size of a potential genomic DNA contamination band. The primers ATG GCA AAT TCC ATG GCA and TGA TGA TCT TGA GGC TGT TGT C for the *GAPDH* gene generate an amplicon of 285 bp from cDNA and an amplicon of 453 bp from genomic DNA.

8.7.3 Reference gene validation

A number of reference (housekeeping) genes have been described in the literature and are used at different frequencies. Suzuki and colleagues reviewed 452 papers published in 1999, in six major journals, which used real-time PCR for mRNA quantification (Suzuki *et al.*, 2000) and found that the most frequently used reference gene was *GAPDH* (33%), followed by β-actin (32%) and 18S rRNA (14%). However, the value of some reference genes has been questioned. For example, *GAPDH*, a commonly used reference gene, cannot be used in the analysis of breast tissue as its expression is modulated by hormones (Laborda, 1991). The importance of assay- and tissue-specific reference gene validation has also been illustrated by others (Schmittgen and Zakrajsek, 2000, Sabek *et al.*, 2002, Huggett *et al.*, 2005). The choice of a suitable reference gene is dependent the on individual model and requires investigation of the appropriate literature. The choice of reference genes is the main theme of Chapter 4.

8.7.4 Splice variants and splicing machinery

A number of diseases have recently been linked to mutations in genes coding for proteins involved in mRNA splicing (Khan *et al.*, 2002, Clavero *et al.*, 2004). Real-time PCR can be used to determine whether certain mRNA but not others, are affected in these diseases. Primers are designed to measure the relative expression of mRNA splice variants or the ratio of sliced versus un-spliced mRNA.

8.7.5 Promoter switch

Sometimes the same gene can be expressed from different promoters. A real-time PCR assay can be designed to address this issue. Primer design is very restrictive under these circumstances as it is limited to the very 5′ end of the RNA but, in principle, both TaqMan® and SYBR® Green assays can be used. An example of promoter switch concerns the PEG1/MEST gene causing biallelic expression in invasive breast cancer (Pedersen *et al.*, 2002). Two alternative transcripts have been described for this gene. A SYBR® Green assay was useful to analyze the relative expression of each isoform in breast cancers. The analysis of promoter switch being based on sequences located in the very 5′ of an mRNA an oligo-dT cDNA synthesis may not be recommended.

```
        A*--------------------                    24bp
primer A

Allele A: AGTCGTACGTGTACGGTAGCTGACGTGACGTACGTGTACGTCCATTCAGTC

Allele C: AGTCGTCCGTGTACGGTAGCTGACGTGACGTACGTTACGTCCATTGTCAGTC

        C*--------------------**------------      36bp
primer C
```

Figure 8.7A

Two allele-specific reverse primers (stars indicate mismatch) paired with a common forward primer (not shown).

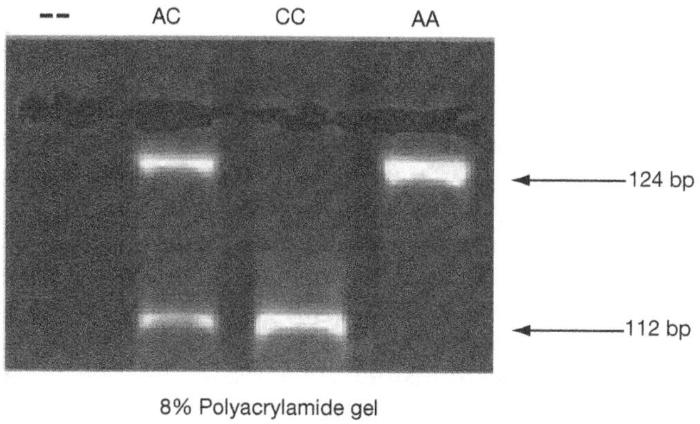

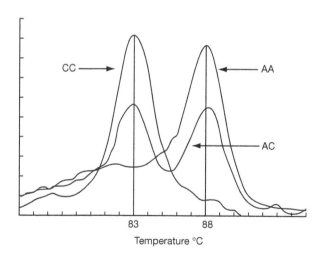

8% Polyacrylamide gel

Figure 8.7B

Joint use of size difference and melting curve analysis in allelic discrimination. The top figure shows the gel electrophoresis for genotyping by size allelic difference and the bottom figure shows the use of (negative derivative of) melting curve to distinguish the three genotypes in a diallelic locus.

8.8 Allelic discrimination

The recent development of the SYBR® Green melting curve analysis assay has allowed allelic discrimination assays to be performed much more easily and less expensively than other methods. Both alleles of the gene of interest can be co-amplified in a single PCR reaction based on one primer discriminating the polymorphic site (allele specific PCR). The discrimination between the two alleles is based on one of these primers also being much longer (additional 10 bp at least) resulting in a size difference for the amplicons. The introduction of a SYBR® Green melting curve analysis assay has helped considerably in resolving the difference in size between the allele specific amplicons by providing a rapid mean of analysis length using Tm. There are numerous examples in the literature (Donohoe *et al.*, 2000, Beuret, 2004) and a theoretical illustration of a basic allelic discrimination assay using SYBR® Green detection method is given in *Figures 8.7A* and *8.7B*.

8.9 Chromatin immunoprecipitation

Chromatin immunoprecipitation (ChIP) was recently developed and is a very powerful tool to study *in vivo* interaction between proteins and DNA. ChIP can be extremely useful in the study of transcription, as it enables the identification of DNA regions that contain a transcription factor binding site or that are susceptible to histone modifications such as methylation or acetylation. It is, however, technically highly demanding and requires efficient antibodies, particularly when analyzing transcription factors as opposed to more abundant histone proteins.

The technique relies on fixing a protein of interest on its DNA target in living cells, in order to isolate this specific complex, followed by PCR amplification to detect the target DNA. Intact cells are fixed using formaldehyde, which cross-links protein to the DNA and therefore preserves specific protein-DNA complexes. The DNA is then sheared into small pieces using ultrasound. The complex between the protein of interest and its target DNA is immunoprecipitated with an antibody specific to the protein, using protein G agarose or magnetic beads. The DNA is extracted using a proteinase K and phenol method, in order to reverse the cross-linking and remove the proteins and antibodies. A PCR based method is then used to detect the presence or absence of the target DNA in the immunoprecipitate. Due to the very limited amount of DNA precipitated, particularly when looking for transcription factor binding sites, a real-time method is highly recommended and SYBR® Green assays perform very well (Wang *et al.*, 2004, Potratz *et al.*, 2005).

Conclusion

This chapter outlines the many uses of SYBR® Green I-based real-time PCR. Experimental protocols using SYBR® Green in gene quantification are presented in Chapters 5 and 7. It appears that this inexpensive alternative to specific probe-based chemistries has overcome the early skepticism about its use. No doubt, developments in melting curve analysis have contributed to its popularity and even allowed multiplexing of allelic discrimination assays. SYBR® Green I or other recently discovered dsDNA binding dye-based real-time PCR techniques can be used in allelic discrimination or quantification assays and will most probably continue to gain popularity.

References

Alexander C, Votruba M, Pesch UE, Thiselton DL, Mayer S, Moore A, Rodriguez M, Kellner U, Leo-Kottler B, Auburger G, Bhattacharya SS, Wissinger B (2000) OPA1, encoding a dynamin-related GTPase, is mutated in autosomal dominant optic atrophy linked to chromosome 3q28. *Nat Genet* 26(2): 211–215.

Beuret C (2004) Simultaneous detection of enteric viruses by multiplex real-time RT-PCR. *J Virol Methods* 115(1): 1–8.

Clavero S, Perez B, Rincon A, Ugarte M, Desviat LR (2004) Qualitative and quantitative analysis of the effect of splicing mutations in propionic acidemia underlying non-severe phenotypes. *Hum Genet* 115(3): 239–247.

De Preter K, Speleman F, Combaret V, Lunec J, Laureys G, Eussen BH, Francotte N, Board J, Pearson AD, De Paepe A, Van Roy N, Vandesompele J (2002) Quantification of MYCN, DDX1, and NAG gene copy number in neuroblastoma using a real-time quantitative PCR assay. *Mod Pathol* 15(2): 159–166.

Donohoe GG, Laaksonen M, Pulkki K, Ronnemaa T, Kairisto V (2000) Rapid single-tube screening of the C282Y hemochromatosis mutation by real-time multiplex allele-specific PCR without fluorescent probes. *Clin Chem* 46(10): 1540–1547.

Estalilla OC, Medeiros LJ, Manning JT, Jr, Luthra R (2000) 5'–>3' exonuclease-based real-time PCR assays for detecting the t(14;18)(q32;21): a survey of 162 malignant lymphomas and reactive specimens. *Mod Pathol* 13(6): 661–666.

Huggett J, Dheda K, Bustin S, Zumla A (2005) Real-time RT-PCR normalisation: strategies and considerations. *Genes Immun* 6(4): 279–284.

Karsai A, Muller S, Platz S, Hauser MT (2002) Evaluation of a homemade SYBR Green I reaction mixture for real-time PCR quantification of gene expression. *Biotechniques* 32(4): 790–792, 794–796.

Khan SG, Muniz-Medina V, Shahlavi T, Baker CC, Inui H, Ueda T, Emmert S, Schneider TD, Kraemer KH (2002) The human XPC DNA repair gene: arrangement, splice site information content and influence of a single nucleotide polymorphism in a splice acceptor site on alternative splicing and function. *Nucleic Acids Res* 30(16): 3624–3631.

Laborda J (1991) 36B4 cDNA used as an estradiol-independent mRNA control is the cDNA for human acidic ribosomal phosphoprotein PO. *Nucleic Acids Res* 19(14): 3998.

Marchbank NJ, Craig JE, Leek JP, Toohey M, Churchill AJ, Markham AF, Mackey DA, Toomes C, Inglehearn CF (2002) Deletion of the OPA1 gene in a dominant optic atrophy family: evidence that haploinsufficiency is the cause of disease. *J Med Genet* 39(8): e47.

Pedersen IS, Dervan P, McGoldrick A, Harrison M, Ponchel F, Speirs V, Isaacs JD, Gorey T, McCann A (2002) Promoter switch: a novel mechanism causing biallelic PEG1/MEST expression in invasive breast cancer. *Hum Mol Genet* 11(12): 1449–1453.

Ponchel F, Toomes C, Bransfield K, Leong FT, Douglas SH, Field SL, Bell SM, Combaret V, Puisieux A, Mighell AJ, Robinson PA, Inglehearn CF, Isaacs JD, Markham AF (2003) Real-time PCR based on SYBR-Green I fluorescence: an alternative to the TaqMan assay for a relative quantification of gene rearrangements, gene amplifications and micro gene deletions. *BMC Biotechnol* 3(1): 18.

Potratz JC, Mlody B, Berdel WE, Serve H, Muller-Tidow C (2005) In vivo analyses of UV-irradiation-induced p53 promoter binding using a novel quantitative real-time PCR assay. *Int J Oncol* 26(2): 493–498.

Sabek O, Dorak MT, Kotb M, Gaber AO, Gaber L (2002) Quantitative detection of T-cell activation markers by real-time PCR in renal transplant rejection and correlation with histopathologic evaluation. *Transplantation* **74**(5): 701–707.

Schmittgen TD, Zakrajsek BA (2000) Effect of experimental treatment on housekeeping gene expression: validation by real-time, quantitative RT-PCR. *J Biochem Biophys Methods* **46**(1–2): 69–81.

Suzuki T, Higgins PJ, Crawford DR (2000) Control selection for RNA quantitation. *Biotechniques* **29**(2): 332–337.

Wang JC, Derynck MK, Nonaka DF, Khodabakhsh DB, Haqq C, Yamamoto KR (2004) Chromatin immunoprecipitation (ChIP) scanning identifies primary glucocorticoid receptor target genes. *Proc Natl Acad Sci USA* **101**(44): 15603–15608.

High-resolution melting analysis for scanning and genotyping

9

Virginie Dujols, Noriko Kusukawa, Jason T. McKinney, Steven F. Dobrowolsky and Carl T. Wittwer

9.1 Introduction

Ever since the introduction of the LightCycler® in 1996 (Wittwer *et al.*, 1997b), melting analysis has been an integral part of real-time techniques. SYBR® Green I allows both quantification without probes (Wittwer *et al.*, 1997a) and verification of product identity by melting analysis (Ririe *et al.*, 1997). However, fine sequence discrimination (e.g. SNP genotyping) has usually required labeled probes. Melting analysis for SNP typing was first achieved with one labeled primer and one labeled probe (Lay and Wittwer, 1997) and later with two labeled probes, each with a single label (Bernard *et al.*, 1998). These techniques are widely used today with conventional real-time instrumentation (Wittwer and Kusukawa, 2004, Gingeras *et al.*, 2005).

New instruments and saturating double-stranded DNA binding dyes for high-resolution melting enable mutation scanning and genotyping without probes. High-resolution melting was first reported with labeled primers (Gundry *et al.*, 2003) for distinguishing different heterozygotes and homozygotes. This was rapidly followed by the introduction of saturating dyes, eliminating any need for fluorescently labeled oligonucleotides (Wittwer *et al.*, 2003). Because no separation steps are required, mutation scanning by melting is inherently simple and inexpensive. The sensitivity and specificity for detecting heterozygous SNPs is at least as good as alternative, more complicated techniques (Reed and Wittwer, 2004, Chou *et al.*, 2005). High-resolution melting for mutation scanning has been reported for c-kit (Willmore *et al.*, 2004), medium chain acyl-CoA dehydrogenase (McKinney *et al.*, 2004), SLC22A5 (Dobrowolski *et al.*, 2005), and BRAF (Willmore-Payne *et al.*, 2005) genes. The method can be used for transplantation matching by establishing sequence identity among related donors within HLA genes (Zhou *et al.*, 2004b). Complete genotyping is possible for most SNPs (Liew *et al.*, 2004, Graham *et al.*, 2005) and bacterial speciation has been reported (Odell *et al.*, 2005). Unlabeled probes can be added when regions of high sequence complexity are analyzed (Zhou *et al.*, 2004a).

Because of its simplicity and power, we suspect high-resolution melting techniques will become integrated into real-time platforms, perhaps even by the publication of this book. In what follows, we briefly describe the instrumentation and saturation dyes that make high-resolution melting analysis possible. Next we present experimental details for mutation scanning, followed by amplicon genotyping and unlabeled probe genotyping. Finally, we compare high-resolution melting analysis to other options for scanning and homogeneous genotyping.

9.2 High-resolution instrumentation

9.2.1 The HR-1™ instrument

High-resolution melting was first developed on an instrument that analyzes individual 10 μl samples in Roche LightCycler® capillaries one at a time (Gundry *et al.*, 2003). Our goal was to investigate the potential power of melting analysis by using the highest resolution possible. The thermal advantages of symmetric capillaries with a high-surface area-to-volume ratio were combined with the stability of a surrounding metal ingot for heating. A single LED with feedback control of intensity for excitation and a photodiode for detection were configured for epi-illumination of the capillary tip. High-resolution 24-bit A-to-D converters were used for digital processing of both temperature and fluorescence. Melting curves with 50–100 points/°C were obtained. The increase in resolution is about 1–2 orders of magnitude greater than in conventional real-time instruments (Zhou *et al.*, 2005). The instrument has been commercialized by Idaho Technology as the HR-1™ high-resolution melting instrument.

Compared to real-time instruments, the HR-1™ is inexpensive. However, it is not designed for thermal cycling and only one sample can be analyzed at a time. Nevertheless, with a 1–2 min turn-around time, throughput is reasonable at about 45 samples per hour. It is most convenient to amplify the samples in a LightCycler® as the real time PCR data is valuable for optimization, and then transfer the capillaries to the HR-1™. However, real-time PCR is not requisite for high-resolution melting analysis, and amplification can occur on any PCR instrument. Transfer of the sample into a capillary is necessary before melting unless a capillary PCR instrument, such as the RapidCycler® 2 (Idaho Technology), is used.

9.2.2 The LightScanner™ instrument

Even though the throughput of the HR-1™ is reasonable, the manual handling of single samples is tedious for large numbers of samples and the possibility of manipulation errors is real. The LightScanner™ (Idaho Technology) is a 96- or 384-well microtiter format high-resolution melting instrument designed for high-throughput use. PCR is first performed on any plate thermocycler. The plate is then transferred to the LightScanner™ for melting analysis. Since the cycle time of the LightScanner™ is 5–10 min, many thermocyclers can feed a single LightScanner™. Sample volumes of 10 μl are standard, although 5 μl or less can be used.

Various commercial PCR plates each have unique optical properties. The best plates have opaque white wells in a black shell (BioRad/MJ Research or Idaho Technology) and are available in 96- and 384-well format. However, not all 96-well PCR instruments are physically compatible with these plates.

An oil overlay (15–20 µl per well for 96-well plates or 10–15 µl per well for 384-well plates) is necessary to prevent sample evaporation during melting analysis in the LightScanner™. Sample evaporation increases the ionic strength and changes the characteristics of the melting profile. The oil is applied before PCR immediately after the samples are placed in the wells (row by row or column by column). Alternatively the samples are underlayed beneath pre-aliquoted oil. In either case, it is wise to spin the plates in a centrifuge (although the adept can 'flick' the plates by hand) to ensure that the oil and sample are properly layered.

9.2.3 The LightCycler® 480 instrument

Real-time instruments not specifically designed for high-resolution melting analysis perform poorly in comparison to the HR-1™ and the LightScanner™ (Herrmann *et al.*, 2006). A possible exception may prove to be the LightCycler® 480 (Roche), the latest in the LightCycler® line of instruments that is a 96/384-well real-time instrument. We were able to test a prototype version of the LightCycler® 480 for mutation scanning and obtained comparable results to the HR-1™ and the LightScanner™ using previously described methods (Reed and Wittwer, 2004). No oil overlay is required, but the recommended sample volumes are larger (20 µl). All three instruments (HR-1™, LightScanner™, LightCycler® 480) perform well for mutation scanning and unlabeled probe genotyping. However, the plate-based instruments are not recommended for amplicon genotyping because even small well-to-well temperature variations may be greater than the absolute Tm difference between some homozygotes (Liew *et al.*, 2004). Characteristics of the instruments are summarized in *Table 9.1*. We suspect that additional instruments will incorporate high-resolution capabilities in the future.

Table 9.1 Characteristics of high-resolution melting instruments

Characteristic	HR-1™	LightScanner™	LightCycler® 480
Number of samples per run	1	96, 384	96, 384
Turn-around time (min)	1–2	10–15	20
Sample volume (µl)	5–10	5–10	10–20
Capabilities			
Mutation scanning	+	+	+
Unlabeled probe genotyping	+	+	+
Amplicon genotyping	+	–	–
Integrated PCR	No	No	Yes
Oil overlay	No	Yes	No
Instrument cost	low	moderate	high (projected)
Running costs (reagents/tubes)	moderate	low	unknown

9.3 Saturating dyes

Using dyes instead of probes for genotyping has been a challenge. Early reports of using SYBR® Green I for genotyping (Marziliano *et al.*, 2000, Pirulli *et al.*, 2000) have been questioned (von Ahsen *et al.*, 2001). Most subsequent work with SYBR® Green I processed the samples after PCR, including product purification and dye addition (Lipsky *et al.*, 2001) or the addition of urea (Elenitoba-Johnson and Bohling, 2001). In our hands, it is not possible to reliably detect heteroduplexes with SYBR® Green I in a closed-tube system. When multiple duplexes are present, SYBR® Green I adequately detects high Tm products during melting, but not lower Tm products (Wittwer *et al.*, 2003). Various reasons for this have been proposed, including dye redistribution during melting (Wittwer *et al.*, 2003), strand reassociation during melting (Gundry *et al.*, 2003), and G:C base pair specificity (Giglio *et al.*, 2003). At least with small amplicons, faster melting rates (0.1–0.3°C/s) can minimize any effect of strand reassociation during melting (Gundry *et al.*, 2003).

Dye redistribution during melting appears to be the major limitation of SYBR® Green I for heteroduplex detection. Heteroduplexes may be detected with SYBR® Green I if high enough concentrations are used (Lipsky *et al.*, 2001). However, at these high concentrations PCR is inhibited (Wittwer *et al.*, 1997a). The ideal dye would saturate all double-stranded DNA produced during PCR (eliminating dye redistribution) and not inhibit or otherwise adversely affect PCR.

9.3.1 LCGreen™ dyes

We have synthesized a number of asymmetric cyanine dyes whose basic structure is shown in *Figure 9.1*. Many of these dyes, generically called 'LCGreen™' dyes, can be used in PCR at saturating concentrations without inhibition. They differ in structure from the more common asymmetric cyanines (Zipper *et al.*, 2004) by the presence of two nitrogens in the right ring, hence they are pyrimidiniums as opposed to pyridiniums. Different properties are obtained by replacing X, R_1, R_2, and R_3 with different moieties. Two of the most useful dyes are LCGreen™ I (introduced with the HR-1™ instrument) and LCGreen™ PLUS (introduced with the LightScanner™) both available from Idaho Technology. LCGreen™ PLUS is much brighter than LCGreen™ I and can be used on most real-time PCR machines, within the resolution constraints of each instrument. The exact

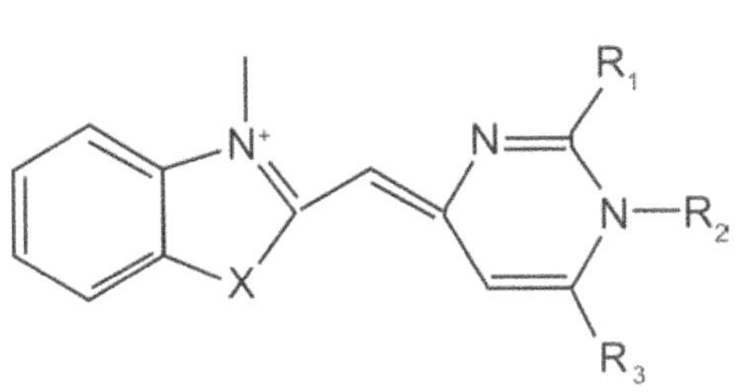

Figure 9.1

The basic structure of the LCGreen™ dye family. LCGreen™ dyes are asymmetric cyanines, different from SYBR® Green I by having two nitrogens in the right ring. Various properties are obtained by modifying the substituents X, R_1, R_2, and R_3. X is usually S or O. R_1, R_2 and R_3 are hydrogens, aliphatic or aromatic moieties that may be branched or linear, charged or uncharged. LCGreen™ I and LCGreen™ PLUS are two commercial versions of the LCGreen™ dye family.

structure of LCGreen™ I and LCGreen™ PLUS is kept a trade secret, as is the usual practice for commercial dyes in this field. The binding mode of the LCGreen™ dyes has not been investigated, but by analogy to SYBR® Green I (Zipper *et al.*, 2004) it is likely that both intercalation and surface binding occur, with fluorescence dependent on the later.

9.4 Mutation scanning

Closed-tube mutation scanning is perhaps the most important application of high-resolution melting analysis. Mutation scanning techniques detect any heterozygous sequence change between the PCR primers. Conventional methods of mutation scanning require downstream processing, a separation step after PCR, or both. In contrast, mutation scanning by melting is performed in the same tube as PCR without any modification to the amplification except the addition of a saturating dye to the reaction prior to PCR. Like SYBR Green® I, LCGreen™ dyes raise the Tm of primers and the PCR product by a few °C, so some re-optimization of PCR temperatures may be necessary. If high melting temperatures are a problem, DMSO (5–10%) can be added without adverse affect to the melting analysis. This is often useful in regions of high GC content.

Primers are chosen to bracket the analysis region, often entire exons including adjacent intron sequence critical to mRNA processing. The sensitivity and specificity of detection depend somewhat on amplicon size. SNP detection approaches 100% for amplicons <400 bp in length, while the sensitivity and specificity for 400–1,000 bp products is 96.1 and 99.4% respectively (Reed and Wittwer, 2004). Longer amplicons often have multiple melting transitions, although such 'domains' may be present in any amplicon depending on sequence. Scanning sensitivity does not appear to depend on the position of the variant within the amplicon or the number of melting domains.

9.4.1 PCR protocols for scanning

Any recipe for PCR can be used with the simple addition of an LCGreen™ dye at 1X concentration. If capillaries are used for amplification, 250–500 µg/ml BSA (final concentration) should be added. We routinely use 50 mM Tris, pH 8.3, 500 µg/ml BSA, 200 µM each dNTP, 1–5 mM $MgCl_2$, 0.4 U heat-stable polymerase, 88 ng of TaqStart™ antibody (ClonTech), 0.2–0.5 µM each primer and 10–50 ng of human genomic DNA in 10 µl reactions. One µl of human genomic DNA with an absorbance of 1.0 at 260 nm provides 50 ng or 15,000 copies of DNA. Either an exonuclease-positive (Roche) or exonuclease-negative polymerase (KlenTaq1™, AB Peptides) can be used. Alternatively, dye can be added to LightCycler® FastStart DNA Master HybProbe PCR kit (Roche). It is likely that by the time of publication of this book, complete PCR master mixes including the LCGreen™ dye will be available from Idaho Technology, Roche, or both. Hot-start procedures (physical temperature, wax barrier, anti-Taq antibody, or heat-activated enzymes), although not required, are recommended. All double-stranded DNA species bind LCGreen™ dyes and affect the melting profile, and although it is often possible to 'read through' undesired amplified products, it is much easier to interpret specific, single product amplifications.

Temperature cycling protocols and optimization will not be detailed here. The goal is a pure PCR product, so rapid cycling, hot-start techniques, and a limited number of cycles (30–40) are recommended. Parameters for rapid capillary PCR have been published elsewhere (Wittwer *et al.*, 2001), although denaturation times must be modified if heat-activated Taq is used. Rapid cycling is not required, and standard PCR on 96- or 384-well plates is the best high-throughput solution. Gradient thermal cyclers are convenient for initial optimization. Real-time monitoring is not necessary, but is convenient and aids in troubleshooting. After PCR the samples should be denatured and cooled rapidly. With capillary systems, a brief '0' sec denaturation is adequate. On plate thermocyclers, a 10–30 sec denaturation is recommended. In either case, cool the samples to below 40°C at the fastest ramp possible. The samples can be melted immediately or stored at room temperature before analysis. If plates have been covered with a film for thermal cycling, remove the film before melting. Melting rates for scanning are usually 0.3°C/sec on the HR-1™ and 0.1°C/sec on the LightScanner™.

9.4.2 Principles of scanning by melting

The principle behind heterozygote detection by melting is shown in *Figure 9.2*. PCR amplification of a heterozygous SNP produces four different single-stranded products, indicated as A, C, G, and T in *Figure 9.2*. If these strands are randomly associated after PCR by denaturation and annealing, four unique duplexes are formed. Two fully base paired fragments are identical to the starting DNA and are close (but usually not identical) in melting temperature. In addition, two imperfectly matched heteroduplexes are

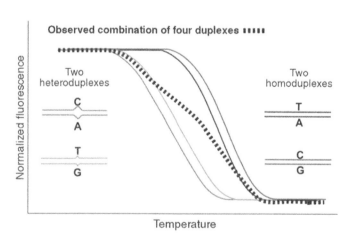

Figure 9.2

Heterozygote detection by melting analysis. When a heterozygous sample is amplified by PCR, denatured, and cooled, the strands re-associate randomly, producing four distinct duplexes. The two homoduplexes and two heteroduplexes that are produced cannot be observed individually during melting. Rather, the composite melting curve of all four duplexes results in a skewed melting curve with a broad melting transition. The altered shape of the melting curve is used to identify heterozygotes.

formed that melt at lower temperatures than the fully base paired fragments. All four duplexes are present when a heterozygous sample is melted, so the observed melting curve is a superposition of all four melting profiles. The result is usually a skewed melting curve with a broad transition. The best indicator for the presence of a heterozygous sample is the altered shape of the melting profile. The melting temperature, or Tm, is less sensitive and not as relevant as melting curve shape.

9.4.3 Software tools for heterozygote identification

Sequence variant identification by visual inspection of the melting curves is surprisingly easy with the appropriate software tools. The process is demonstrated in *Figure 9.3* (A–D) for a 179 bp product of the cholesterol esterase transfer protein gene. The original high-resolution data is shown in (A). The wild type sequence is shown in black with different sequence variants in various colors. At low temperature before any DNA melting transition, fluorescence decreases with increasing temperature. This linear decrease in fluorescence with increasing temperature is a property of fluorescence unrelated to DNA melting.

To normalize the data, two linear regions are selected, one before and one after the major transition. These regions define two lines for each curve, an upper 100% fluorescence line and a lower, 0% baseline. The percent fluorescence within the transition (between the two regions) is calculated at each temperature as the distance to the experimental data compared to the distance between the extrapolated upper and lower lines. The normalized result is shown in (B).

The shape difference between genotypes can be made clearer by eliminating the temperature offsets between samples by shifting the temperature axis of each curve so that the curves are partly superimposed. This is usually done at the high temperature homoduplex region (low percent fluorescence) of the curves so that heteroduplexes can be identified by their early drop in fluorescence at lower temperatures, as shown in (C).

Different genotypes are most easily distinguished by plotting the fluorescence difference between normalized and temperature-shifted melting curves. A reference genotype is selected and the difference between all other curves and the reference is plotted against temperature, as shown in (D). Genotypes other than the reference trace unique paths that can be easily identified visually. Difference plots should not be confused with derivative plots often used to visualize probe Tm values for genotyping (Lay and Wittwer, 1997, Bernard *et al.*, 1998). Although there is some visual similarity, the plots and their applications are very different.

Visual inspection is convenient when there are only a few samples to study. However, when there are many curves to compare, intuitive visual clustering becomes less attractive than automatic clustering. Classical hierarchical clustering can identify different genotypes after high-resolution melting analysis (Zhou *et al.*, 2005). Although automatic clustering algorithms can be used, it is wise to always visually look at the data, especially when multiple domains are present. Commercial software available with the HR-1™ and LightScanner™ instruments incorporates these basic analysis components.

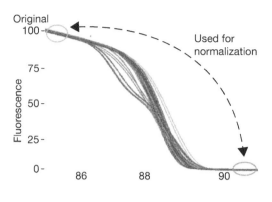

Original

Used for normalization

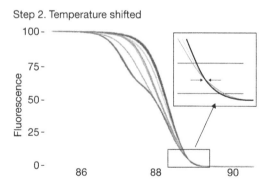

Step 1. Normalized

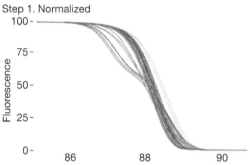

Step 2. Temperature shifted

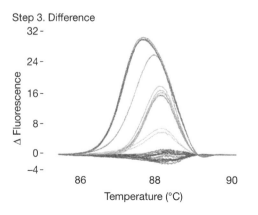

Step 3. Difference

Temperature (°C)

Figure 9.3

Software-aided analysis for mutation scanning. Thirty-two different human genomic DNA samples were amplified in the LightCycler® and analyzed on the HR-1™ instrument. The original melting data (Original) is normalized (Step 1) and temperature-shifted (Step 2) to compare the shape of the melting transitions. Differences in shape are most easily seen on difference plots (Step 3) of the normalized and temperature shifted data. Wild-type samples are in black, with sequence variants in different colors.

9.4.4 Scanning for homozygous variants

Although the most common sequence variants are heterozygous, it is also important to identify homozygous variants. In general, scanning methods do not identify homozygous changes unless the unknown samples are mixed with wild type before analysis. High-resolution melting analysis is an exception to this rule as many homozygous variants can be identified directly (Gundry *et al.*, 2003, Wittwer *et al.*, 2003, Dobrowolski *et al.*, 2005). For example, the A>T base change of HbS can be detected in the homozygous state (sickle cell disease) even though the Tm shift is only 0.1–0.2°C. Sometimes the shape of the melting curve is sufficiently different that homozygous variants can be picked up on standard scanning plots (difference plots of normalized, temperature-shifted data). This occurs, for example, when the melting curve has multiple domains and the sequence change affects only one of them (Wittwer *et al.*, 2003). However, in many cases, the shape of the homozygous variant curve is nearly identical to the wild-type curve. Therefore, as a general rule, temperature shifting is not performed when the objective is to identify homozygous changes and absolute temperatures must be relied upon. This places stringent requirements on the temperature precision of an instrument between samples, either in time (HR-1™) or space (LightScanner™). In practice, the HR-1™ performs better than the LightScanner™ because of temperature variation across sample wells in heating blocks. Fewer homozygous variants will be detected on 96- or 384-well instruments than on the HR-1™.

Systematic studies to estimate the sensitivity and specificity of homozygous variant detection have yet to be performed. Detection of homozygous SNP changes is often easier than small, homozygous insertions or deletions because the later may be very close in Tm to the wild type. For example, the homozygous G542X SNP in cystic fibrosis is detectable by high-resolution melting, while homozygous F508del is not (Chou *et al.*, 2005). Studies with small amplicons have shown that 84% of human SNPs have homozygous Tm differences between 0.8–1.4°C (Liew *et al.*, 2004), suggesting that most SNPs are detectable. Whether they are detected will depend on the amplicon size and the resolution of the instrument used.

9.5 Amplicon genotyping

Because both homozygous and heterozygous variants can be distinguished from wild type, specific loci can be genotyped by amplicon melting. Protocols are identical to mutation scanning except that small amplicons are recommended to increase the melting curve differences between SNP genotypes (Liew *et al.*, 2004). Heterozygotes are always easy to identify because low melting heteroduplexes change the shape of the melting curve. The homozygotes of most SNPs are readily distinguished from wild type with Tm differences of about 1°C. However, about 12% of human SNPs are challenging with differences of about 0.2°C and 4% are impossible because of thermodynamic symmetry. The SNPs that are difficult can be predicted by the type of SNP (Liew *et al.*, 2004). The impossible SNPs can be genotyped after mixing with wild type sample either before or after PCR. When mixed after PCR, equal amounts of wild type and unknown amplicons are mixed, denatured, annealed, and melted again. If the unknown is wild type, no change in the melting curve occurs. If the unknown is homozygous mutant, the new melting curve appears heterozygous. Slightly more elegant is mixing before PCR. In this case, the ideal amount of wild-type DNA is one part added to six parts of unknown DNA. The melting curves of all three genotypes are separated based on the proportion of heteroduplexes formed. When mixing before PCR, only one melting analysis is needed and genotyping remains closed-tube.

Because the Tm differences between homozygous mutant and wild type may be smaller than the temperature variation across 96- or 384-well blocks, only the HR-1™ is recommended for amplicon genotyping (Herrmann *et al.*, 2006). The ability of HR-1™ genotyping to distinguish different heterozygotes was recently demonstrated (Graham *et al.*, 2005). Twenty-one out of 21 heteroduplex pairs tested were distinguishable by high-resolution melting of small amplicons.

9.6 Unlabeled probe genotyping

LCGreen™ dyes are fluorescent in the presence of any double-stranded DNA. For mutation scanning and amplicon genotyping, entire amplicons are labeled with dye. Recently, we found that the melting transition of unlabeled oligonucleotides probes could also be monitored with these dyes (Zhou *et al.*, 2004a). If excess single-stranded product is produced by asymmetric PCR, the Tm of unlabeled probes can be used for genotyping, similar to methods that use covalently-labeled fluorescent oligonucleotides (Lay and Wittwer, 1997, Bernard *et al.*, 1998). Primer asymmetry ratios of 1:5 to 1:10 produce sufficient double-stranded product for amplicon melting and enough single-stranded product for probe annealing. Unlabeled genotyping probes typically incorporate a 3′ phosphate, although any terminator that prevents extension can be used. In contrast to the LCGreen™ dyes, SYBR® Green I does not adequately distinguish heterozygotes queried with unlabeled probes.

Unlabeled probes of 20–30 bases with Tm values from 55–70°C give good peaks on derivative plots. When an exonuclease-negative polymerase is used, the Tm of the most stable hybrid should be below the extension temperature of PCR to avoid possible blockage of the polymerase by the probe. Such blockage can be allele-specific and result in incorrect genotyping. Alternatively, exonuclease-positive polymerases can be used with probe Tm values into the 80s without concern. It is convenient to position the most stable probe transition below that of possible primer dimers to avoid potential confusion. Nearest-neighbor thermodynamic estimates of probe Tm are usually 1–4°C lower than observed Tm values, primarily because of dye stabilization of the hybrid (Zhou *et al.*, 2004a).

9.6.1 PCR protocols for unlabeled probe genotyping

Similar to scanning and amplicon genotyping, any recipe for PCR can be used with the simple addition of LCGreen™ dyes at 1X concentration. We routinely use 50 mM Tris, pH 8.3, 500 µg/ml BSA, 200 µM each dNTP, 1–5 mM $MgCl_2$, 0.4 U heat-stable polymerase, 88 ng of TaqStart™ antibody

(ClonTech), 0.5 μM of the excess primer, 0.05–0.1 uM of the limiting primer, 0.5 uM of the unlabeled probe and 50 ng of human genomic DNA in 10 μl reactions. Make sure that the unlabeled probe is on the same strand as the limiting primer. Polymerase alternatives include regular exonuclease-positive Taq (for a wide range of probe Tm values) and exonuclease-negative KlenTaq1™ (for probe Tm values below the PCR extension temperature). We have also added dye to the LightCycler® FastStart DNA Master HybProbe kit with good results. Hot-start procedures are not as important as in scanning applications, especially if the probes are designed so that alleles melt below potential primer dimers. Asymmetric amplification is usually performed for 45–60 cycles, so rapid cycling techniques are convenient. A final denaturation and cooling step are not required.

9.6.2 Instrumentation for unlabeled probe genotyping

In contrast to mutation scanning and amplicon genotyping, high-resolution melting analysis is not required with unlabeled probes. Results are best on the HR-1™ and LightScanner™, but the LightTyper™ and the LightCycler® produce adequate genotyping results (Zhou et al., 2004a). Although other real-time PCR instruments have not been tested, unlabeled probe geno-typing should be possible with LCGreen™ PLUS. Recommended melting rates for unlabeled probe genotyping are 0.1°C for the LightScanner™ and 0.3°C/sec for the HR-1™. Background fluorescence from single strands (primers and single stranded product) is prominent at low temperatures. However, conversion of the original data directly to derivative curves reveals the melting transitions as peaks on a downward slope (*Figure 9.4*).

9.6.3 Simultaneous genotyping and scanning

Unlabeled probe genotyping can be combined with amplicon scanning in the same reaction (Zhou et al., 2005). Both probe melting (for genotyping) and the amplicon melting (for scanning) are analyzed from the same melting curve. Reaction conditions and design are the same as with unlabeled probes with the exception that denaturation and cooling after PCR is required for heteroduplex formation. High-resolution instrumenta-tion is required. Multiple unlabeled probes can be used in the same reaction, as long as the Tm values of all alleles are unique. The melting temperature of different probe/allele duplexes can be adjusted by probe length, mismatch position, and probe dU vs. dT content. Combined scanning and genotyping with LCGreen™ dyes presents a powerful closed-tube genetic analysis system. The method is only as complex as PCR, yet promises to vastly reduce the need for re-sequencing.

9.7 Simplification of genotyping and mutation scanning

Four alternative designs for genotyping by melting analysis are shown in *Figure 9.4*. Conventional HybProbe® technology requires two probes, each with a single fluorescent label (Bernard et al., 1998). Single HybProbes® (also known as SimpleProbes®) only require one probe with a single label (Crockett and Wittwer, 2001). Unlabeled probe genotyping depends on

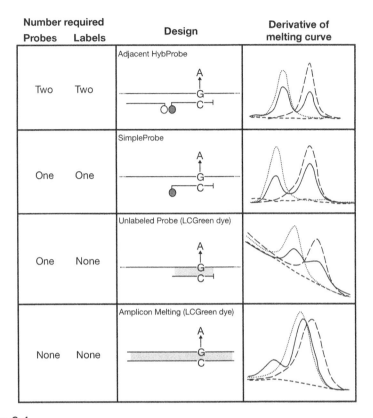

Number required		Design	Derivative of melting curve
Probes	Labels		

Figure 9.4

Four different methods of SNP typing by melting. Shown for each method are the number of probes and fluorescent labels required, the hybridization design, and the negative derivative of the resultant melting curve. The melting curves of each method (dF/dT vs temperature) show plots of wild-type (long dashes), heterozygous (solid lines), homozygous mutant (dots), and no template controls (grey dashes). The top two methods use covalently-attached fluorescent labels (depicted by open and closed circles), whereas the bottom two methods use a saturating DNA dye.

saturation dyes for fluorescence with no covalent fluorescent labels. Finally, genotyping by amplicon melting also relies on saturation dyes and requires no probes.

Closed-tube genotyping methods that use melting analysis are less complex than allele specific alternatives that require one probe for each allele analyzed (*Table 9.2*). Melting techniques also have the capacity to scan for unexpected variants under the probes and/or within amplicons. Allele discrimination by Tm (Wittwer *et al.*, 2001) or curve shape (Wittwer *et al.*, 2003, Graham *et al.*, 2005) are promising alternatives to fluorescent color. Conventional closed-tube genotyping methods require expensive fluorescent labels. Conventional scanning techniques, such as SSCP, DGGE, TGGE, heteroduplex analysis, DHPLC, TGCE, and chemical or enzymatic mismatch detection methods require processing and separation steps after

Table 9.2 Comparison of closed-tube, homogeneous SNP genotyping and scanning methods

Method (references)	Probes/SNP genotype	Modifications/ probe	Scanning region
Amplicon melting (Gundry *et al.*, 2003) (Wittwer *et al.*, 2003) (Liew *et al.*, 2004) (Graham *et al.*, 2005)	0	0	Between primers
Unlabeled probe (Zhou *et al.*, 2004a)	1	0	Between primers and within probes
Simple probe (Crockett and Wittwer, 2001)	1	1	Within probe
Hybridization probes (Bernard *et al.*, 1998)	2	1	Within probes
TaqMan® (Lee *et al.*, 1993)	2	2	None
Beacons (Marras *et al.*, 2003)	2	2	None
Scorpions (Whitcombe *et al.*, 1999)	2	3	None
MGB TaqMan® (de Kok *et al.*, 2002)	2	3	None

PCR (Wittwer and Kusukawa, 2005). Closed-tube genotyping and scanning with LCGreen™ dyes are attractive because: 1) there are no processing or separation steps after the initiation of PCR, 2) only standard PCR reagents, unlabeled oligonucleotides, and a saturating DNA dye are required, and 3) analysis is rapid (1–10 min after PCR).

References

Bernard PS, Ajioka RS, Kushner JP, Wittwer CT (1998) Homogeneous multiplex genotyping of hemochromatosis mutations with fluorescent hybridization probes. *Am J Pathol* **153**: 1055–1061.

Chou L-S, Lyon E, Wittwer CT (2005) A comparison of high-resolution melting analysis to denaturing high performance liquid chromatography for mutation scanning: cystic fibrosis transmembrane conductance regulator gene as a model. *Am J Clin Pathol* **124**: 330–338.

Crockett AO, Wittwer CT (2001) Fluorescein-labeled oligonucleotides for real-time pcr: using the inherent quenching of deoxyguanosine nucleotides. *Anal Biochem* **290**: 89–97.

de Kok JB, Wiegerinck ET, Giesendorf BA, Swinkels DW (2002) Rapid genotyping of single nucleotide polymorphisms using novel minor groove binding DNA oligonucleotides (MGB probes). *Hum Mutat* **19**: 554–559.

Dobrowolski SF, McKinney JT, Amat di San Filippo C, Giak Sim K, Wilcken B, Longo N (2005) Validation of dye-binding/high-resolution thermal denaturation for the identification of mutations in the SLC22A5 gene. *Hum Mutat* **25**: 306–313.

Elenitoba-Johnson KS, Bohling SD (2001) Solution-based scanning for single-base alterations using a double-stranded DNA binding dye and fluorescence-melting profiles. *Am J Pathol* **159**: 845–853.

Giglio S, Monis PT, Saint CP (2003) Demonstration of preferential binding of SYBR Green I to specific DNA fragments in real-time multiplex PCR. *Nucleic Acids Res* **31**: e136.

Gingeras TR, Higuchi R, Kricka LJ, Lo YM, Wittwer CT (2005) Fifty years of molecular (DNA/RNA) diagnostics. *Clin Chem* **51**: 661–671.

Graham R, Liew M, Meadows C, Lyon E, Wittwer CT (2005) Distinguishing different DNA heterozygotes by high-resolution melting. *Clin Chem* **51**: 1295–1298.

Gundry CN, Vandersteen JG, Reed GH, Pryor RJ, Chen J, Wittwer CT (2003) Amplicon melting analysis with labeled primers: a closed-tube method for differentiating homozygotes and heterozygotes. *Clin Chem* **49**: 396–406.

Herrmann MG, Durtschi JD, Bromley LK, Wittwer CT, Voelkerding KV (2006) DNA welting analysis for mutation scanning and genotyping: a cross platform comparison. *Clin Chem* **in press**.

Lay MJ, Wittwer CT (1997) Real-time fluorescence genotyping of factor V Leiden during rapid-cycle PCR. *Clin Chem* **43**: 2262–2267.

Lee LG, Connell CR, Bloch W (1993) Allelic discrimination by nick-translation PCR with fluorogenic probes. *Nucleic Acids Res* **21**: 3761–3766.

Liew M, Pryor R, Palais R, Meadows C, Erali M, Lyon E, Wittwer C (2004) Genotyping of single-nucleotide polymorphisms by high-resolution melting of small amplicons. *Clin Chem* **50**: 1156–1164.

Lipsky RH, Mazzanti CM, Rudolph JG, Xu K, Vyas G, Bozak D, Radel MQ, Goldman D (2001) DNA melting analysis for detection of single nucleotide polymorphisms. *Clin Chem* **47**: 635–644.

Marras SA, Kramer FR, Tyagi S (2003) Genotyping SNPs with molecular beacons. *Methods Mol Biol* **212**: 111–128.

Marziliano N, Pelo E, Minuti B, Passerini I, Torricelli F, Da Prato L (2000) Melting temperature assay for a UGT1A gene variant in Gilbert syndrome. *Clin Chem* **46**: 423–425.

McKinney JT, Longo N, Hahn SH, Matern D, Rinaldo P, Strauss AW, Dobrowolski SF (2004) Rapid, comprehensive screening of the human medium chain acyl-CoA dehydrogenase gene. *Mol Genet Metab* **82**: 112–120.

Odell ID, Cloud JL, Seipp M, Wittwer CT (2005) Rapid species identification within the Mycobacterium chelonae-abscessus group by high-resolution melting analysis of hsp65 PCR products. *Am J Clin Pathol* **123**: 96–101.

Pirulli D, Boniotto M, Puzzer D, Spano A, Amoroso A, Crovella S (2000) Flexibility of melting temperature assay for rapid detection of insertions, deletions, and single-point mutations of the AGXT gene responsible for type 1 primary hyperoxaluria. *Clin Chem* **46**: 1842–1844.

Reed GH, Wittwer CT (2004) Sensitivity and specificity of single-nucleotide polymorphism scanning by high-resolution melting analysis. *Clin Chem* **50**: 1748–1754.

Ririe KM, Rasmussen RP, Wittwer CT (1997) Product differentiation by analysis of DNA melting curves during the polymerase chain reaction. *Anal Biochem* **245**: 154–160.

von Ahsen N, Oellerich M, Schutz E (2001) Limitations of genotyping based on amplicon melting temperature. *Clin Chem* **47**: 1331–1332.

Whitcombe D, Theaker J, Guy SP, Brown T, Little S (1999) Detection of PCR products using self-probing amplicons and fluorescence. *Nat Biotechnol* **17**: 804–807.

Willmore C, Holden JA, Zhou L, Tripp S, Wittwer CT, Layfield LJ (2004) Detection of c-kit-activating mutations in gastrointestinal stromal tumors by high-resolution amplicon melting analysis. *Am J Clin Pathol* **122**: 206–216.

Willmore-Payne C, Holden JA, Tripp S, Layfield LJ (2005) Human malignant melanoma: Detection of BRAF- and c-kit-activating mutations by high-resolution amplicon melting analysis. *Hum Pathol* **36**: 486–493.

Wittwer CT, Herrmann MG, Gundry CN, Elenitoba-Johnson KS (2001) Real-time multiplex PCR assays. *Methods* **25**: 430–442.

Wittwer CT, Herrmann MG, Moss AA, Rasmussen RP (1997a) Continuous fluorescence monitoring of rapid cycle DNA amplification. *Biotechniques* **22**: 130–131, 134–138.

Wittwer CT, Ririe KM, Andrew RV, David DA, Gundry RA, Balis UJ (1997b) The LightCycler: a microvolume multisample fluorimeter with rapid temperature control. *Biotechniques* **22**: 176–181.

Wittwer CT, Kusukawa N (2004) Diagnostic molecular microbiology; principles and applications. In: Persing DH, Tenover FC, Versalovic J, Tang YW, Unger ER, Relman DA, White TJ (eds), Real-time PCR. ASM Press, Washington DC, 71–84.

Wittwer CT, Kusukawa N (2005) Nucleic acid techniques. In: Tietz Textbook of Clinical Chemistry and Molecular Diagnostics, 4th edition Burtis C, Ashwood E, Bruns D (eds), Elsevier, New York, 1407–1449.

Wittwer CT, Reed GH, Gundry CN, Vandersteen JG, Pryor RJ (2003) High-resolution genotyping by amplicon melting analysis using LCGreen. *Clin Chem* **49**: 853–860.

Zhou L, Myers AN, Vandersteen JG, Wang L, Wittwer CT (2004a) Closed-tube genotyping with unlabeled oligonucleotide probes and a saturating DNA dye. *Clin Chem* **50**: 1328–1335.

Zhou L, Vandersteen J, Wang L, Fuller T, Taylor M, Palais B, Wittwer CT (2004b) High-resolution DNA melting curve analysis to establish HLA genotypic identity. *Tissue Antigens* **64**: 156–164.

Zhou L, Wang L, Palais R, Pryor R, Wittwer CT (2005) High-resolution DNA melting analysis for simultaneous mutation scanning and genotyping in solution. *Clin Chem* **51**: 1770–1777.

Zipper H, Brunner H, Bernhagen J, Vitzthum F (2004) Investigations on DNA intercalation and surface binding by SYBR Green I, its structure determination and methodological implications. *Nucleic Acids Res* **32**: e103.

Quantitative analyses of DNA methylation

10

Lin Zhou and James (Jianming) Tang

10.1 Introduction

As an epigenetic and inheritable trait, DNA methylation is well-known for its role in important biological phenomena, including X-chromosome inactivation (in females), genetic imprinting, cell differentiation, and mutagenesis (Robertson and Wolffe, 2000, Feinberg *et al.*, 2002, Fitzpatrick and Wilson, 2003, Muegge *et al.*, 2003). In the human genome, DNA methylation primarily involves about 70% of all cytosines in the context of 5'-CG-3' (CpG) dinucleotides. Fine mapping of CpG methylation has been a major goal of the Human Epigenome Project (HEP), and the first major product derived from this international effort has been published recently (Rakyan *et al.*, 2004).

Methylated cytosine (^mC) differs from cytosine in two biochemical properties. First, treatment of DNA with bisulfite converts normal cytosine to uracil, while ^mC remains intact. Second, many methylation-sensitive restriction endonucleases are unable to cleave target DNA when ^mC is present within the recognition site. These properties form the basis for CpG methylation analyses.

Quantification of CpG methylation can be done with a variety of PCR-based techniques (see *Table 10.1*), each with clear advantages and disadvantages (reviewed by (Geisler *et al.*, 2004)). For example, when a single CpG site is targeted, methylation-specific PCR (MSP) can be used to measure the proportion of ^mC to normal cytosine in bisulfite-treated DNA samples. However, this technique is ineffective for CpG-rich sequences. For detection and quantification of multiple CpG sites in a single assay, the method of choice seems to be pyrosequencing of PCR-amplified, bisulfite-treated DNA (Tost *et al.*, 2003a, Dupont *et al.*, 2004), while MALDI mass spectrometry represents the latest, high-throughput technology for more demanding projects (Tost *et al.*, 2003b).

This chapter will illustrate the technical aspects of quantitative, real-time PCR (qPCR) in studying *de novo* CpG methylation. In our analyses of several genes related to the human immunodeficiency virus type 1 (HIV-1) infection and acquired immune deficiency syndrome (AIDS), the method provides a rapid and cost-effective means of identifying samples with differentially methylated CpG sites.

Table 10.1 Common, PCR-based techniques for quantitative analyses of CpG methylation, in which bisulfite-treated DNA samples serve as the template

Technique	Key features (advantages and disadvantages)	References
MSP: methylation-specific PCR	Simple, sensitive, but only one CpG site at a time	(Herman *et al.*, 1996)
COBRA: combined bisulfite restriction analysis	Easy to use, accurate, but not all CpGs are within restriction sites	(Xiong and Laird, 1997)
BiPS: bisulfite PCR-single-strand conformation polymorphism	Rapid, simple, but sensitivity is limited to small fragments	(Maekawa *et al.*, 1999)
Methylight	High-throughput, homogenous, and expensive	(Eads *et al.*, 2000)
HeavyMethyl	Sensitive, high-throughput, homogenous, and expensive	(Cottrell *et al.*, 2004)
SNuPE: single nucleotide primer extension	Rapid, but only one CpG site at a time	(Gonzalgo and Jones, 1997)
qPCR: real-time, quantitative PCR	Medium- to high-throughput and requires special equipment	(Lo *et al.*, 1999, Zeschnigk *et al.*, 2004)
Direct sequencing of PCR amplicons	Rapid, accurate, but quantification can be a problem for partially methylated sites	(Rakyan *et al.*, 2004)
Sequencing of cloned PCR amplicons	Accurate, time-consuming, and expensive	Too many to list
Pyrosequencing[a] of PCR amplicons	Rapid, accurate, inexpensive, short reads per assay, and platform is not widely available	(Colella *et al.*, 2003, Tost *et al.*, 2003a)
MALDI[b] mass spectrometry	Accurate, expensive, and platform is not widely available	(Tost *et al.*, 2003b, Rakyan *et al.*, 2004)

[a]Detailed information can be found at: http://www.pyrosequencing.com/DynPage.aspx?id=7499.
[b]MALDI, matrix-assisted laser desorption/ionization.

10.2 *MDR1* (*ABCB1*, Gene ID 5243) as a primary target locus

Officially known as ATP-binding cassette, sub-family B member 1 (*ABCB1*), *MDR1* (multidrug resistance protein 1) encodes a P-glycoprotein (GP170) that has been studied extensively for its critical role as a drug transporter. Up-regulation of MDR1 alters the kinetics of many therapeutic agents being used for treating patients with cancer (Rund *et al.*, 1999) and HIV-1 infection (Fellay *et al.*, 2002). In both disease systems, enhanced efflux of intracellular drugs have been associated with germline and somatic mutations at the *MDR1* locus. Uneven distribution of certain *MDR1* variants in human ethnic groups (Rund *et al.*, 1999, Schaeffeler *et al.*, 2001) may be further responsible for the racial disparity in response to certain therapy. In contrast, the influence of *MDR1* promoter methylation on MDR1 function is not so clear.

Table 10.2 Summary of qPCR-based tests of three *MDR1* genomic regions (nucleotide reference positions are based on GenBank sequence X58723).

Specifics	Promoter fragment 1 (MDR1–PCR 1)	Promoter fragment 2 (MDR1–PCR 2)	Intron 1 fragment (MDR1–PCR 3)
Forward primer positions	nt. 768 > 790	nt. 888 > 907	nt. 1316 > 1336
Reverse primer positions	nt. 888 < 907	nt. 1178 < 1193	nt. 1579 < 1598
Amplicon size	140 bp	308 bp	283 bp
GC content	43%	55%	66%
PCR annealing temperature	60°C	60°C	60°C
Amplicon melting temperature	80.5°C	88.5°C	90.0°C
CpG sites within amplicon	14	3	22
EcoR I site (GAATTC)	0	0	0
BstU I site (CGCG)	1	0	6
qPCR standard curve [a] 1	$\log_{10}Y = 7.75-0.29X$	$\log_{10}Y = 7.81-0.27X$	$\log_{10}Y = 7.60-0.26X$
r^2 for curve 1	0.992	0.994	0.994
qPCR standard curve [a] 2	$\log_{10}Y = 7.85-0.30X$	$\log_{10}Y = 7.56-0.26X$	$\log_{10}Y = 7.89-0.28X$
r^2 for curve 2	0.998	0.990	0.998

[a] For all standard curves, Y is the input DNA copy number in each reaction, while X is the threshold cycle (C_t) number determined automatically by the iCycler® (BioRad).

Methylation at various CpG sites within a 1000-bp region (including *MDR1* exon 1 to exon 2) appears to regulate *MDR1* gene expression (Nakayama *et al.*, 1998, Kusaba *et al.*, 1999). Our ongoing studies of several HIV/AIDS cohorts deal with both genetic and epigenetic characteristics of the *MDR1* gene, primarily because *MDR1* is important to the metabolism of HIV-1 protease inhibitors (Kim *et al.*, 1998). The work described here aims to define CpG methylation in the *MDR1* promoter and intron 1 sequences.

10.3 Primer design

The *MDR1* regions (promoter and intron 1) of interest are based on the GenBank sequence X58723 (see *Table 10.2* and *Figure 10.1*). For PCR and related work (e.g. DNA sequencing), the primers are designed to meet six criteria: a) amplicon size <400 bp, b) primer length ≥17 and ≤24 nucleotides, c) primer to target melting temperature (Tm) ≥67°C and ≤71°C (by % GC method), d) Tm for hairpin or homodimer <40°C, e) no apparent heterodimer formation for paired (forward and reverse) primers, and f) no close homology to other known sequences in the human genome. The first four criteria are evaluated using the OligoTech program (Oligos Etc, Wilsonville, Oregon), while the Entrez BLAST search is deemed sufficient in confirming the correct annealing sites for each primer. The oligonucelotides used are purchased from MWG Biotech, Inc. (High Point, North Carolina), where full-length products are recovered following standard HPSF purification, equivalent to reverse phase, high-performance liquid chromatography (HPLC). Each lyophilized oligonucleotide is resuspended in TE80 (10 mM Tris-HCl, pH 8.0, 2 mM EDTA) as 0.5 mM stock. A

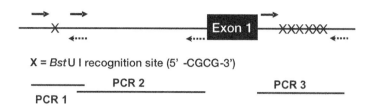

X = *Bst*U I recognition site (5'-CGCG-3')

Figure 10.1

Schematic view (not drawn to exact scale) of qPCR-based analyses of *MDR1* gene, which has several *Bst*U I sites that can be analyzed for methylation status. Solid and broken arrows indicate forward and reverse primers, respectively (see *Table 2*). When *Bst*U I-digested genomic DNA is used as the template, PCR 2 (308 bp) defines the total *MDR1* copy number in each sample, while PCR 1 (140 bp) and PCR 3 (283 bp) qunatifies the methylated DNA copies resistant to digestion by *Bst*U I. The ratio of undigested DNA to total DNA reflects the status of CpG methylation within the *Bst*U I sites.

1:25 dilution in distilled, PCR-grade water makes the 20 µM working solution for PCR.

10.4   Data evaluation: I. assay-to-assay variability

The reliability of qPCR assays was first evaluated by testing the same samples repeatedly, at intervals ranging from one week to three months (see *Figure 10.2*). In 75 genomic DNA samples treated with *EcoR* I and *Bst*U I, correlation analyses were done in a subset of samples ($n = 46$–70) with methylation levels >0.01 (other samples were treated as uninformative). For promoter fragment 1 ($n = 70$ informative observations), *MDR1* sequences with methylated *Bst*U I site accounted for 1% to 40% of the total DNA (see *Figure 10.2A*). The results obtained from two separate experiments were highly consistent, with a correlation coefficient (r) = 0.93 ($P <$0.001), regardless of certain differences (e.g. gender and the amount of input DNA) in the source materials used for qPCR (adjusted $r > 0.90$, $P <$0.001). Results obtained from *MDR1* intron 1 qPCR assays were similar (see *Figure 10.2B*), except that fewer samples (n = 46) could be analyzed for assay-to-assay variability.

10.5   Data evaluation: II. *MDR1* CpG methylation as quantified by qPCR

As expected, CpG methylation levels within the *Bst*U I sites did not fit into a normal distribution ($P \leq$0.01 for all Kolmogorov–Smirnov normality tests). For the 75 samples tested for promoter fragment 1 (MDR1–PCR 1), the ratios of *Bst*U I-resistant DNA to total input DNA ranged from 0.00 to 0.40, with an inter-quartile range of 3% to 9%, implying that the target sequence is hypomethylated. Despite the suspicion that HIV-1 infection can induce DNA methylation (Mikovits *et al.*, 1998, Fang *et al.*, 2001), stratified analyses did not suggest that *MDR1* methylation was elevated in HIV-1 seropositive individuals ($P >$0.20). Results for the intron 1 region

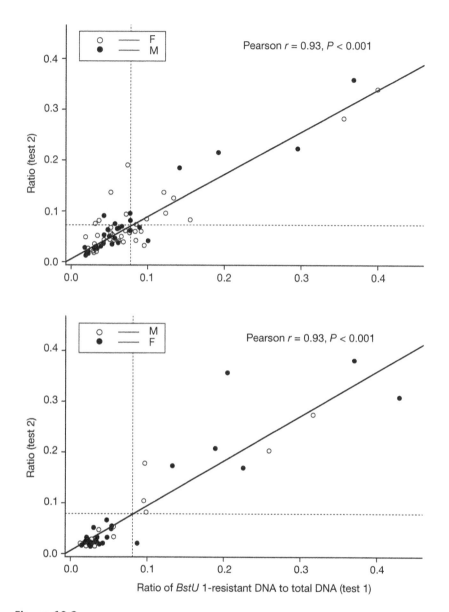

Figure 10.2

Close correlation of qPCR results obtained from different experiments. At intervals ranging from one week to three months, genomic DNA samples (*n* = 75) treated with the methylation-sensitive endonuclease *BstU* I were tested twice for the *MDR1* promoter (panel a) and intron 1 (panel b) sequence, which has one and six *BstU* I (CGCG) sites, respectively (see Table 2 for technical details). Samples from females (F) and males (M) are labeled separately.

(inter-quartile range, 1% to 5%) are similiar, although the assay only measured sequences in which all six *BstU* I sites are resistant to digestion. Additional qPCR assays and pyrosequencing confirmed the hypomethylation status of multiple CpG sites at the *MDR1* locus.

10.6 Expanded analyses

Using the same *BstU* I-digested genomic DNA samples, qPCR-based analyses have been tested for several other genes, including leptin (*LEP*) and tumor necrosis factor alpha (*TNF*). Except for the primers (designed to have similar melting temperatures, with variation <3°C in any direction), qPCR procedures are identical to the ones used for *MDR1*. While the results are still considered preliminary and require further confirmation (e.g. by pyrosequencing and MS-PCR), an important conclusion can be drawn about the comparison of various SYBR® Green qPCR kits: regardless of the gene fragments of interest, the kits from Qiagen consistently produce reliable (clean) amplicons without obvious primer-dimers. Similar kits from three other suppliers either have reduced sensitivity (lower yield) or produce nonspecific amplicons (judged by amplicon size alone) clearly visible on agarose gels stained with ethidium bromide.

For analyses of the *MDR1* locus alone, experiments have been further tested in *EcoR* I-digested gDNA samples treated with other methylation-sensitive DNA endonucleases, including *Aci* I (CCGC), *Hpa*II (CCGG), *Taq* I (ACGT), and *Xho* I (CTCGAG). Other sequences containing any of these sites are also suitable for qPCR-based screening.

Overall, rapid and cost-effective qPCR assays as tested here have at least two major advantages for identifying sequences with differential CpG methylation. First, the tests are quantitative, with highly reproducible results. Second, compared with bisufite-treated, single-stranded DNA samples required for other assays (e.g. MSP and pyrosequencing), the DNA templates treated with methylation-sensitive endonucleases remain double-stranded and can be stored at 4°C for extended and repeated tests. For most conclusive analyses, we recommend qPCR-based screening followed by confirmation with pyrosequencing.

Acknowledgments

This work was supported by a grant (R01 AI51173) from National Institute of Allergy and Infectious Diseases (NIAID). We thank Tarig Hagazi and Thomas R. Unnasch for technical advice; Chengbin Wang and Yufeng Li for statistical analyses; and Jeffery Edberg for supervision of work related to pyrosequencing.

References

Colella S, Shen L, Baggerly KA, Issa JP, Krahe R (2003) Sensitive and quantitative universal Pyrosequencing methylation analysis of CpG sites. *Biotechniques* **35**: 146–150.

Cottrell SE, Distler J, Goodman NS, Mooney SH, Kluth A, Olek A, Schwope I, Tetzner R, Ziebarth, H, Berlin K (2004) A real-time PCR assay for DNA-methylation using methylation-specific blockers. *Nucleic Acids Res* **32**: e10.

Dupont JM, Tost J, Jammes H, Gut IG (2004) De novo quantitative bisulfite sequencing using the pyrosequencing technology. *Anal Biochem* **333**: 119–127.

Eads CA, Danenberg KD, Kawakami K, Saltz LB, Blake C, Shibata D, Danenberg PV, Laird PW (2000) MethyLight: a high-throughput assay to measure DNA methylation. *Nucleic Acids Res* **28**: E32.

Fang JY, Mikovits JA, Bagni R, Petrow-Sadowski CL, Ruscetti FW (2001) Infection of lymphoid cells by integration-defective human immunodeficiency virus type 1 increases *de novo* methylation. *J Virol* **75**: 9753–9761.

Feinberg AP, Cui H, Ohlsson R (2002) DNA methylation and genomic imprinting: insights from cancer into epigenetic mechanisms. *Semin Cancer Biol* **12**: 389–398.

Fellay J, Marzolini C, Meaden ER, Back DJ, Buclin T, Chave J-P, Decosterd LA, Furrer H, Opravil M, Pantaleo G, Retelska D, Ruiz L, Schinkel A, Vernazza P, Eap CB, Telenti A, for the Swiss HIV Cohort Study (2002) Response to antiretroviral treatment in HIV-1-infected individuals with allelic variants of the multidrug resistance transporter 1: a pharmacogenetics study. *Lancet* **359**: 30–36.

Fitzpatrick DR, Wilson CB (2003) Methylation and demethylation in the regulation of genes, cells, and responses in the immune system. *Clin Immunol* **109**: 37–45.

Geisler JP, Manahan KJ, Geisler HE (2004) Evaluation of DNA methylation in the human genome: why examine it and what method to use. *Eur J Gynaecol Oncol* **25**: 19–24.

Gonzalgo ML, Jones PA (1997) Rapid quantitation of methylation differences at specific sites using methylation-sensitive single nucleotide primer extension (Ms-SNuPE). *Nucleic Acids Res* **25**: 2529–2531.

Herman JG, Graff JR, Myohanen S, Nelkin BD, Baylin SB (1996) Methylation-specific PCR: a novel PCR assay for methylation status of CpG islands. *Proc Natl Acad Sci USA* **93**: 9821–9826.

Kim RB, Fromm MF, Wandel C, Leake B, Wood AJ, Roden DM, Wilkinson GR (1998) The drug transporter P-glycoprotein limits oral absorption and brain entry of HIV-1 protease inhibitors. *J Clin Invest* **101**: 289–294.

Kusaba H, Nakayama M, Harada T, Nomoto M, Kohno K, Kuwano M, Wada M (1999) Association of 5′ CpG demethylation and altered chromatin structure in the promoter region with transcriptional activation of the multidrug resistance 1 gene in human cancer cells. *Eur J Biochem* **262**: 924–932.

Lo YM, Wong IH, Zhang J, Tein MS, Ng MH, Hjelm NM (1999) Quantitative analysis of aberrant p16 methylation using real-time quantitative methylation-specific polymerase chain reaction. *Cancer Res* **59**: 3899–3903.

Maekawa M, Sugano K, Kashiwabara H, Ushiama M, Fujita S, Yoshimori M, Kakizoe T (1999) DNA methylation analysis using bisulfite treatment and PCR-single-strand conformation polymorphism in colorectal cancer showing microsatellite instability. *Biochem Biophys Res Commun* **262**: 671–676.

Mikovits JA, Young HA, Vertino P, Issa JP, Pitha PM, Turcoski-Corrales S, Taub DD, Petrow CL, Baylin SB, Ruscetti FW (1998) Infection with human immunodeficiency virus type 1 upregulates DNA methyltransferase, resulting in *de novo* methylation of the gamma interferon (IFN-gamma) promoter and subsequent downregulation of IFN-gamma production. *Mol Cell Biol* **18**: 5166–5177.

Muegge K, Young H, Ruscetti F, Mikovits J (2003) Epigenetic control during lymphoid development and immune responses: aberrant regulation, viruses, and cancer. *Ann N Y Acad Sci* **983**: 55–70.

Nakayama M, Wada M, Harada T, Nagayama J, Kusaba H, Ohshima K, Kozuru M, Komatsu H, Ueda R, Kuwano M (1998) Hypomethylation status of CpG sites at the promoter region and overexpression of the human MDR1 gene in acute myeloid leukemias. *Blood* **92**: 4296–4307.

Rakyan VK, Hildmann T, Novik KL, Lewin J, Tost J, Cox AV, Andrews TD, Howe KL, Otto T, Olek A, Fischer J, Gut IG, Berlin K, Beck S (2004) DNA methylation profiling of the human major histocompatibility complex: a pilot study for the human epigenome project. *PLoS Biol* **2**: e405.

Robertson KD, Wolffe AP (2000) DNA methylation in health and disease. *Nat Rev Genet* **1**: 11–19.

Rund D, Azar I, Shperling O (1999) A mutation in the promoter of the multidrug resistance gene (MDR1) in human hematological malignancies may contribute to the pathogenesis of resistant disease. *Adv Exp Med Biol* **457**: 71–75.

Schaeffeler E, Eichelbaum M, Brinkmann U, Penger A, Asante-Poku S, Zanger, UM, Schwab M (2001) Frequency of C3435T polymorphism of MDR1 gene in African people. *Lancet* **358**: 383–384.

Shao W, Tang J, Dorak MT, Song W, Lobashevsky E, Cobbs CS, Wrensch MR, Kaslow RA (2004) Molecular typing of human leukocyte antigen and related polymorphisms following whole genome amplification. *Tissue Antigens* **64**: 286–292.

Tang J, Wilson CM, Meleth S, Myracle A, Lobashevsky E, Mulligan MJ, Douglas SD, Korber B, Vermund SH, Kaslow RA (2002) Host genetic profiles predict virological and immunological control of HIV-1 infection in adolescents. *AIDS* **16**: 2275–2284.

Tost J, Dunker J, Gut IG (2003a) Analysis and quantification of multiple methylation variable positions in CpG islands by Pyrosequencing. *BioTechniques* **35**: 152–156.

Tost J, Schatz P, Schuster M, Berlin K, Gut IG (2003b) Analysis and accurate quantification of CpG methylation by MALDI mass spectrometry. *Nucleic Acids Res* **31**: e50.

Xiong Z, Laird PW (1997) COBRA: a sensitive and quantitative DNA methylation assay. *Nucleic Acids Res* **25**: 2532–2534.

Zeschnigk M, Bohringer S, Price EA, Onadim Z, Masshofer L, Lohmann DR (2004) A novel real-time PCR assay for quantitative analysis of methylated alleles (QAMA): analysis of the retinoblastoma locus. *Nucleic Acids Res* **32**: e125.

See also Tooke N and Pettersson M (2004): http://www.devicelink.com/ivdt/archive/04/11/002.html

Protocol 10.1

Genomic DNA samples

1a. Cell line DNA – as a source of external reference material, several genomic DNA samples (e.g. KT17) established for the 10th International Histocompatibility Workshops were purchased from Fred Hutchinson Cancer Research Center (Shao *et al.*, 2004). The pre-determined concentration of 200 ng/μL corresponds to ~28,571 genome copies/μL

1b. Patient DNA samples – these samples were extracted from whole blood using the QIAamp DNA Blood Mini kit (Qiagen) (Tang *et al.*, 2002); DNA concentration was initially estimated by optical density at 260 nm; more reliable quantification was done using the PicoGreen dsDNA Assay (Molecular Probes) that stains double-stranded DNA

Digestion of gDNA with *EcoR* I (New England Biolabs)

2a. *EcoR* I reaction mix (60 μL each)

10x NE buffer H	6.0 μL
EcoR I (10 U/μL)	0.5 μL
gDNA (~300 ng/μL)	20.0 μL
Sterile water	33.5 μL

2b. Incubate sample at 37°C for 4 hr before inactivating enzyme at 65°C for 20 min (best done in a thermocycler with heated lid)

2c. Separate *EcoR* I-digested gDNA from salt and enzyme by size exclusion, using 96-well filter plate (Millipore) attached to a manifold vacuum

2d. Recover *EcoR* I-digested gDNA in 120 μl sterile water. The digested DNA is enough for many downstream applications; with an estimated 95% recovery rate, DNA concentration becomes ~45 ng/μL

Digestion of gDNA with *BstU* I (New England Biolabs)

3a. *BstU* I reaction mix (40 μL each)

10x NE buffer 2	4.0 μL
BstU I (10 U/μL)	0.5 μL
EcoR I-treated gDNA (~45 ng/μL)	20.0 μL (one-sixth of the total)
Sterile water	15.5 μL

3b. Incubate sample at 60°C for 2 hr

3c. Separate *BstU* I-digested gDNA from salt and enzyme by size exclusion, using 96-well filter plate (Millipore) attached to a manifold vacuum

3d. Recover *BstU* I-digested gDNA in 120 µl sterile water for qPCR (DNA concentration becomes ~7 ng/µL)

qPCR in 96-well plates

4a. Prepare qPCR reaction mixture, with each having a total volume of 25 µL:

2x QuantiTect master mix (Qiagen)*	12.5 µL
Forward primer (20 µM)	0.5 µL
Reverse primer (20 µM)	0.5 µL
100x fluorescein	0.25 µL
BstU I-digested gDNA	1.0 µL (~7 ng)
PCR water	10.25 µL

*Reagents from several other suppliers do not work consistently without further optimization

4b. Thermocycling in the iCycler® (with heated lid) (BioRad)
Step 1. 95°C, 15 min (to active the HotStarTaq)
Step 2. 35 cycles of 95°C for 20 sec (denaturing), 60°C (annealing, temperature decreasing at 0.2°C per cycle until it reached 58°C) for 30 sec, 72°C for 36 sec (extension), with optics activated after each extension,
Step 3. Plot melting curve from 65°C to 95°C

4c. Characterization of qPCR amplicon (optional) by conventional techniques including
• Run PCR products on 2.0% agarose gel to visualize DNA bands (with ethidium bromide staining and UV transillumination)
• Determine the size of qPCR amplicons by comparing with bands in the 100-bp DNA ladder (New England Biolabs)
• Confirm the DNA sequence identity by nested PCR, Southern blot, or direct cycle sequencing

Design and documentation of qPCR assays

5a. For each experiment, the first 10 wells are reserved for duplicates of five serially diluted (e.g. 1:8, 1:64, 1:512, and 1:4096), positive control DNA (high molecular weight genomic DNA or plasmid DNA containing the target sequence) with a known copy number. Other wells are used for testing samples (also in duplicates)

5b. Verify that each qPCR test yields specific amplicons (e.g. 4c) and that all melting curves overlap to form a single peak; qPCR assays containing an internal probe (e.g. TaqMan®) should be used if these conditions are not met

5c. Use the threshold cycle numbers (C_t) of control samples to generate a standard curve in each test, with known input DNA copy numbers transformed to $\log_{10}$ (e.g. *Table 2*); experiments are deemed reliable if the r^2 for the standard curve is greater than 0.95

5d. With DNA samples treated with *EcoR* I and *BstU* I, qPCR for *MDR1* promoter fragment 2 (no *EcoR* I site; no *BstU* I site) was used to quantify the total input DNA copy numbers in each sample (see *Figure 10.1*)

5e. qPCR for *MDR1* promoter fragment 1 (no *EcoR* I site; 1 *BstU* I site) and *MDR1* intron 1 (no *EcoR* I site; 6 *BstU* I sites) defined the DNA copies resistant to *BstU* I (due to CpG methylation)

5f. The ratio of *BstU* I-resistant *MDR1* DNA to total input DNA serves as a crude measure (see discussion in the text and *Figure 10.1*) of methylation level in test samples

6. Verification of CpG methylation by pyrosequencing

6a. Design PCR and sequencing primers using the Pyrosequencing™ Assay Design Software

6b. Treat gDNA samples with bisulfite to convert unmethylated cytosine to uracil (U), following procedures outlined for the EZ DNA methylation kit (Zymo Research)

6c. Perform pyrosequencing using the Pyro Gold reagents, as recommended by the manufacturer (Biotage); proportion of methylcytosine (^{m}C) is calculated automatically in the allele frequency mode

Mitochondrial DNA analysis

11

Steve E. Durham and Patrick F. Chinnery

11.1 Introduction

The field of mitochondrial genetic research is an emerging and rapidly expanding area of interest. A recent OVID (Medline) search revealed 300 references between 1975–1979 citing mitochondria as a keyword in some format (mitochondria, mitochondrial or mtDNA), but by 2000–2004 this number has increased to 14,000. Mitochondria are essential intracellular organelles that are the primary source of adenosine triphosphate (ATP), the high energy molecule required for all active intracellular processes. ATP is generated by oxidative phosphorylation which is carried out by the respiratory chain on the inner mitochondrial membrane. The respiratory chain is composed from over 100 different polypeptides, making up five enzyme complexes, some of which are synthesized within the mitochondria from their own DNA (the mitochondrial genome, mtDNA). In addition to their fundamental role in energy metabolism, mitochondria are involved in intracellular calcium signaling and apoptosis. It therefore comes as no surprise that diverse disciplines have developed an interest in mitochondria, from inherited neurological disorders to cancer and the ageing process. The mtDNA molecule forms the focus of this chapter.

11.2 Mitochondrial genetics

mtDNA is the only source of DNA outside the nucleus in mammals and has a number of distinct properties. The human mitochondrial genome is a double-stranded circular DNA molecule, 16,569 bp in length, accounting for approximately 1% of total cellular DNA. The complete sequence was determined in 1981 (Anderson *et al.*, 1981) (Cambridge Reference Sequence – CRS) and was re-sequenced in 1999 (Andrews *et al.*, 1999) enabling the correction of some sequence errors and common polymorphic variants. The revised Cambridge Reference Sequence (rCRS) is available through the MITOMAP web site (MITOMAP, 2005). Expression of almost the complete genome is required for respiratory chain function, in comparison to only 7% of the nuclear genome being expressed at any one time. mtDNA has a different codon usage, is strictly maternally inherited and undergoes little if any intermolecular recombination. mtDNA encodes 37 genes; 13 essential polypeptides of the mitochondrial respiratory chain as well as 22 transfer RNA (tRNA) and 2 ribosomal RNA (rRNA) genes (*Figure 11.1*) that are

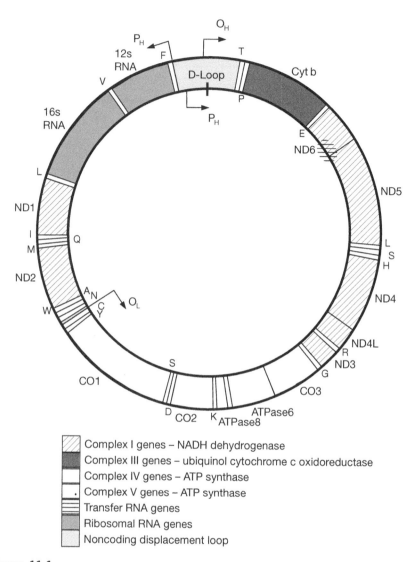

Figure 11.1

Sequence arrangement of the human mitochondrial genome. Genes are coded as the differently shaded inset. O_H and O_L are the origins of replication for the heavy and light strands respectively. P_H and P_L are the promoters for the heavy and light strands respectively.

essential for intra-mitochondrial translation and transcription. The mitochondrial genome contains no introns and is so compact that there are regions of gene overlap. The displacement loop (D-loop), approximately 1.1 kb in length, is not directly involved in protein synthesis but plays a crucial role in regulating translation and transcription of the mtDNA molecule.

11.2.1 mtDNA mutations

Mutations of mtDNA can be present as large-scale rearrangements (deletions or duplications) and point mutations. Small deletions and inversions have been described but these are rare and will not be considered further. Duplications are not considered pathogenic, although they may predispose to deletion formation. Many point mutations do not alter the corresponding protein sequence or have no demonstrable phenotypic effect but some do cause a defect of the respiratory chain. This is either through a direct effect on a structural subunit, or indirectly through an effect on intra-mitochondrial protein synthesis. Similar effects are seen from most large-scale deletions as they often affect both structural subunits and protein synthesis.

11.2.2 mtDNA copy number and heteroplasmy

Multiple copies of the mitochondrial genome can be found in individual mitochondria, with many mitochondria present in a single cell. The precise number of genomes per mitochondrion is unknown and is likely to vary between cell types; the generally accepted range is between 2 to 10 molecules per organelle. The number of mitochondria per cell is dependent on the metabolic demands of that cell type and can vary from 100s to 10,000s. In most individuals at birth, all of the mtDNA molecules are identical (called homoplasmy). However, the presence of mutated mtDNA usually generates a mixture of wild-type and mutated molecules within the same cell. This phenomenon is referred to as heteroplasmy. The percentage level of mutated mtDNA often varies from cell to cell, and from organ to organ within an individual and may change over time.

11.2.3 The threshold effect

Single-cell and cybrid (cytoplasmic hybrid) studies have shown that the proportion of mutated mtDNA is directly related to the expression of a biochemical defect within the cell. There is a critical ratio between wild-type and mutant mtDNA that must be exceeded before phenotypic expression of the disease occurs. For the majority of mtDNA mutations this percentage mutation load threshold correlates with the histochemically demonstratable activity of cytochrome *c* oxidase (COX, complex IV of the respiratory chain). Above the mutation threshold cells appear COX negative, while at lower mutation loads COX activity can still be demonstrated. The threshold level is variable and dependent on the type of mutation and its location, with some tRNA mutations requiring 85% mutant mtDNA while some protein coding mutations and deletions can express a mutant phenotype at much lower levels. Although the percentage level of mutated mtDNA appears to be important, for most mutations it is not clear whether the respiratory chain defect is actually due to the ratio of mutated to wild-type mtDNA, or related to the absolute amount of wild-type mtDNA (analogous to haplo-insuffiency), or mutated mtDNA (through a dominant negative effect).

11.2.4 Mutation rate of mtDNA

The mutation rate of the mitochondrial genome is estimated as being some ten times faster than that of the nuclear genome. The high mutation rate seen with mtDNA is likely to be caused by a cumulative effect of the high levels of reactive oxygen species (ROS), the absence of histones and the limited DNA repair mechanisms available. This has three major consequences; 1) mtDNA has accumulated mutations throughout evolution, leading to many polymorphic differences between unrelated individuals; 2) mtDNA mutations are a relatively common cause of sporadic and inherited human disease; and 3) mtDNA accumulates somatic mutations throughout human life.

11.2.5 Mitochondrial DNA, ageing and disease

Ageing

Histochemical studies have shown that a small proportion of non-dividing cells from healthy humans acquire a respiratory chain defect throughout life. Single-cell studies have identified pathogenic mutations of mtDNA within the majority of these cells, explaining the accumulation of potentially pathogenic mtDNA mutations with age in many human tissues (Michikawa *et al.*, 1999b; Attardi, 2002; Fayet *et al.*, 2002). Both point mutations and deletions have been detected, but deletions appear to be the most common cause of age-related COX deficiency. The majority of mtDNA deletions appear to occur between sites of direct DNA sequence repeats of 4 to 15 bp. This suggests that the mechanism by which deletions occur is by slippage during replication or by homologous recombination. The majority of deletions are found on the major arc of the mitochondrial genome, a region defined as being between the origin of light strand replication (O_L) and the origin of heavy strand replication (O_H). Approximately 95% of reported deletions occur in this region and remove genes encoding subunits of COX, ATP synthase and NADH dehydrogenase. The first mtDNA deletion reported to accumulate with age was a deletion of 4,977 bp. The so-called 'common deletion' occurs between two 13 bp repeat sites, at nucleotide positions 8,470 to 8,483 and 13,447 to 13,459, removing approximately 5 kb of mtDNA between the ATPase8 and ND5 genes. In accordance with the MITOMAP website, all base positions are numbered from the light strand (L strand). *Table 11.1* lists some of the mtDNA mutations associated with ageing.

Primary mitochondrial DNA diseases

Primary mitochondrial diseases are those caused by a mutation, either a point mutation or deletion, in the mitochondrial genome. Mitochondrial disease is a term which covers a broad spectrum of multi-systemic disorders. The pathogenic mutations compromise respiratory chain function leading to cellular dysfunction and ultimately cell death. *Table 11.2* summarizes some of the more common mtDNA disorders.

Table 11.1 Examples of mtDNA deletions and base substitutions which have been reported at higher prevalences in healthy aged individuals with the tissues each has been seen in. Deletion start and end points are shown. CSB2 – Conserved Sequence Block 2; O_H and O_L - origins of replication for the heavy and light strand respectively; P_H and P_L – promoters for heavy and light strand respectively.

Mutation	Location	Affected tissue
mtDNA deletions		
4977 bp	8468 to 13446	Heart, liver, brain, skeletal muscle and other tissues
6063 bp	7842 to 13905	Liver
7400 bp	8648 to 16085	Myocardium, brain cortex and putamen
3895 bp	548 to 4442	Skeletal muscle
Multiple deletions	Various	Various
mtDNA point mutations		
8344A>G	tRNA^{Lys}	Extraocular muscle
3243A>G	tRNA$^{Leu(UUR)}$	Extraocular muscle
4460T>C,	tRNA^{Met}	Skeletal muscle
4421G>A	tRNA^{Met}	Skeletal muscle
Cins 310	D-Loop (CSB2)	Skin fibroblasts
414T>G	D-Loop (P_L)	Skin fibroblasts
189A>G	D-Loop (O_H)	Skeletal muscle
408T>A	D-Loop (P_L)	Skeletal muscle
150C>T	D-Loop (O_H)	Leukocytes

Table 11.2 Common mitochondrial disorders and associated mtDNA mutations.

Disorder	Clinical features	mtDNA mutation
Chronic Progressive External Ophthalmoplegia (CPEO)	External ophthalmoplegia and bilateral ptosis	Large Scale deletions or
Kearns Sayre Syndrome (KSS)	Early PEO onset (<20 yrs) with pigmentary retinopathy with one of; elevated CSF, cerebellar ataxia and heart block	duplications
Pearson Syndrome (PS)	Sideroblastic anemia in childhood, pan-cytopenia and exocrine pancreatic failure	
Lebers Hereditary Optic Neuropathy (LHON)	Subacute painless bilateral loss of vision, onset ~20 yrs. Male affected bias	3460G>A 11778G>A 14484T>C
Neurogenic weakness with ataxia and retinitis pigmentosa (NARP)	Early adult/late childhood onset peripheral neuropathy with ataxia and pigmentary retinopathy	8993T>G/C
Mitochondrial encephalomyopathy with lactic acidosis and stroke-like episodes (MELAS)	Stroke like episodes, onset <40 yrs. Seizures, dementia, ragged red fibers and lactic acidosis	3243A>G 3251A>G 3271T>C
Myoclonic epilepsy with ragged red fibers (MERRF)	Myoclonus, seizures, cerebellar ataxia and myopathy	8344A>G 8356T>C
Leigh Syndrome (LS)	Subacute relapsing encephalopathy with cerebellar and brain stem signs, onset in infancy	8993T>G/C
Non-syndromic sensorineural deafness		7445A>G
Aminoglycoside induced non-syndromic deafness		1555A>G

mtDNA point mutations

Designing as assay for the detection of mtDNA point mutations is, in essence, no different to detecting nDNA mutations by allelic discrimination. Examples of allelic discrimination assays are presented in Chapters 8, 9, and 17 and will not be repeated here. The major difference is that when measuring nuclear alleles the percentage values are either 50% heterozygous or 100% homozygous. When measuring the mutant load in the mtDNA, values of heteroplasmy from 1 to 99% can be present. These need to be accurately measured when investigating a pathogenic point mutation, e.g. 8344A>G (Szuhai et al., 2001), as the level of heteroplasmy has important implications on the clinical phenotype and disease progression. Validation across the full range of heteroplasmy values is required to enable meaningful data to be obtained from the assay and for valid conclusions to be drawn.

Nuclear mitochondrial disorders

Nuclear mitochondrial disorders are those conditions where the primary genetic defect is encoded on the nuclear genome but the effect of this defect is realized in mitochondria (Spinazzola and Zeviani, 2005). This effect is either directly on mtDNA maintenance or on mitochondrial processes, e.g. OXPHOS and protein import. Some examples of these disorders where the effect is on the mtDNA are shown in *Table 11.3*. When considering real-time PCR for these nuclear mutations the primary genetic defect is either homozygous or heterozygous, so quantification of the mutant load is unnecessary. Detection of the presence of the nuclear mutation can be easily achieved by another means, e.g. sequencing, RFLP, primer extension assays. It is the secondary defect of the mtDNA which is of relevance here, be it the level of heteroplasmy (point mutations or deletions, single or multiple) or the absolute quantity of mtDNA (depletion).

mtDNA mutations and cancer

mtDNA substitutions have been detected in a number of different human cancers (Penta et al., 2001). It is currently unclear whether these mutations are a secondary phenomenon, and are of no direct relevance to carcinogenesis, or whether they are primarily involved in the disease (Coller et al., 2001). In

Table 11.3 Nuclear-mitochondrial disorders where the primary genetic defect is on the nuclear genome and results in a secondary mtDNA genetic defect.

Disorder	Nuclear defect	mtDNA mutation
Autosomal dominant PEO	Adenine Nucleotide Translocator (*ANT1*) mtDNA polymerase γ (*POLG*) Twinkle helicase (*C10orf2*)	Multiple mtDNA deletions
Mitochondrial neuro-gastrointestinal encephalomyopathy	Thymidine Phosphorylase (*TP*)	
Myopathy with mtDNA depletion	Thymidine Kinase (*TK2*)	mtDNA depletion
Myopathy with hepatic failure	Deoxyguanosine Kinase (*DGOUK*)	

most cases the mutations are homoplasmic (i.e. all of the mtDNA is mutated within individual cells). Each tumor contains a specific point mutation (or mutations), and allele detection is usually achieved by sequencing.

Toxin-induced mtDNA depletion

Shortly after the introduction of nucleoside analogue treatment for acquired immunodeficiency due to HIV infection, it became clear that some patients receiving AZT (azydothymidine) developed a mitochondrial myopathy due to mtDNA depletion. mtDNA is replicated by pol γ, which is inhibited by a variety of nucleoside analogues (part of the HAART, or highly active anti-retroviral therapy regime) (Lewis *et al.*, 2003). Acute and subacute nucleoside toxicity can be life-threatening, due to the lactic acidosis and liver failure. A number of studies have described mtDNA depletion in peripheral blood immediately preceding the clinical deterioration, raising the possibility that measuring mtDNA levels (relative to a single copy nuclear gene), might be used to monitor therapy and prevent fatal complications (de Mendoza *et al.*, 2004) or the more chronic complications such as lipodystrophy (McComsey *et al.*, 2005b). However, not all studies are in agreement, and it may be necessary to study a specific group of blood monocytes to gain clinically useful data (Hoy *et al.*, 2004). Moreover, with the advent of new generation therapies, nucleoside toxicity is less of a problem (Lewis *et al.*, 2003; McComsey *et al.*, 2005a). This chapter will not discuss mtDNA assays in HAART in greater detail, but the basic principle of measuring mtDNA copy number will be considered in greater depth and has broad applications across this and other disciplines.

11.3 Mitochondrial DNA analysis by real-time PCR

The focus of this chapter is the detection and quantification of mtDNA deletions and the measurement of total mtDNA copy number in individual cells. We will cover some of the considerations required when measuring changes in mtDNA copy number from tissue homogenate samples, or blood extractions where mtDNA:nDNA assays are required.

11.3.1 Detection method

Each real-time system differs in its configuration, however they are all similar in so much as they detect emitted fluorescence for each reaction from the specific fluorophore(s) being used. For the purpose of this discussion we will simply be referring to two methods, DNA binding dyes and probes. We are aware that there are subgroups within these categories but they are discussed elsewhere in detail in this book. For detailed guidance on the selection of an appropriate detection method (binding dye or probe) the reader is referred to the relevant chapters of this book.

11.3.2 Oligonucleotide fluorescent probes

We appreciate that there are a variety of different types of probe technologies available to the investigator. Detailed information of the different probes can be found in Chapters 1 and 5. For the purpose of this chapter we will simply refer to all types of them under the umbrella term 'probes.'

Probes have two major advantages over DNA binding dyes. First they add an extra layer of specificity to a reaction. The correct binding of both primers and the probe are required to generate a signal. This combination of primer/probe binding makes it less likely to generate a signal from a non-specific product or from misannealing.

Second, probes facilitate multiplex reactions. Oligonucleotide probes can be labeled with fluorophores which emit at different wavelengths, this means different probes can be used to detect multiple targets within a single reaction tube. Using the machines currently available it is possible to perform a four color multiplex. It should be noted however, that the optimization required to perform this is extensive and expensive for both time and reagents. A large proportion of the genetic analysis performed in the mitochondrial field is carried out at the single-cell level, limiting DNA quantity, pressing the investigator into detecting several targets in a single reaction.

11.3.3 DNA binding dyes

SYBR Green® is the DNA binding dye most commonly used in real-time PCR applications. Chapters 1, 7, 8, and 9 all contain reviews and comparisons of this technology to others available and again, the reader is directed to these chapters for detailed information.

DNA binding dyes are molecules which bind non-specifically to double-stranded DNA molecules. This includes specific products, misprimed non-specific products and primer-dimer molecules. Due to this non-specific binding, amplicon selection and primer design are of utmost importance with this technology. Self–self primer binding, dimer formation and loop structures can all introduce non-specific signal. DNA binding dye technology allows the investigator to view the number of products contributing to the observed fluorescent signal. This is achieved by the gradual increase of the temperature, post-amplification, causing the amplicon to dissociate to become single-stranded. As this occurs, the DNA binding dye is released and the fluorescence signal decreases. Based on the DNA sequence of the amplicon there is a specific temperature at which it will be completely dissociated. However, as different regions of the amplified product will dissociate at different temperatures before the whole product is single-stranded, this means there will be a graded decrease in fluorescence. Software packages translate this data into a 'melting curve', represented as a single peak per product in the negative first-derivative curve of the actual melting curve. As informative as this is, when first setting up a method the melting curve data should not be relied on alone for determining whether you have a single product or not. The nature of the melting curve means that the product peaks are broad. It is very possible that this could mask the presence of secondary products not that dissimilar in length or thermal profile. The degree to which this problem presents itself is dependent on the melting curve parameters. All primer pairs should be checked for specificity by separating the amplification products through a high percentage agarose gel.

11.3.4 Considerations when designing a mtDNA real-time assay

Primer and probe design

The presence of mitochondrial DNA pseudogenes in the nuclear genome has implications for primer and probe design. Approximately 0.016% of the nuclear genome is nonfunctional mitochondrial DNA (Woischnik and Moraes, 2002). This issue is more pronounced with real-time PCR amplicons as they are short in length and this increases the possibility of pseudogenic amplification. To ensure that the primers you have designed are amplifying a product of genuine mitochondrial amplification it is useful not only to check that the primers do not anneal to the nuclear genome by *in silico* methods (using BLAST) but to also amplify extracted DNA from rho0 cells. Rho0 cells are devoid of mtDNA therefore no amplification products should be seen.

There are specific features of mtDNA which can interfere with assay design. Strand asymmetry, the uneven distribution of purine and pyrimidine nucleotides across the heavy and light strands can give rise to large regions with undesirable primer characteristics. The presence of C-tract length variants and the highly polymorphic nature of the D-loop, make this region particularly difficult for robust assay design. It is suggested that investigators avoid this region unless they have good D-loop characterization for the cohort of samples they are investigating and are aware of these variations. Web based systems, such as MFold (Zuker, 2003), determine the three dimensional structure of a target DNA sequence. By understanding the structure of the region of interest, primers can be selectively placed in accessible locations. As with any PCR based assay, good primer design initially will save time and money in the long term.

Primer and probe design is covered by Wang and Seed in Chapter 5 of this book with additional considerations given in Chapter 7. The considerations when designing a mtDNA assay are the same as those required when designing any real-time assay. Most real-time PCR machines are supplied with primer design software in some format, however there are good programs freely available on the web to aid in amplicon selection.

When designing primers for a DNA binding dye real-time reaction the investigator can apply the same design parameters as for any normal PCR reaction. Amplicon size is less restrictive with binding dye reactions as the polymerase does not have to displace a probe. Amplicons in the size range 100 bp to 250 bp can be readily amplified using short, two step amplification protocols (denaturing then combined annealing/extension) with no visible effect on efficiency. With longer amplicons it may be necessary to resort to three step PCR amplification (denaturing, annealing and then extension). Wherever possible primers should be selected which have no self binding, loop or dimer formation. Where this is not possible we suggest that primers with dG binding values greater than –2 kCal/mol are avoided. From experience, primers adhering to this restriction give more reproducible results.

In probe-based assays the polymerase uses its 5' endonuclease activity to displace a bound probe, this slows the extension of the amplicon. It is with that in mind that probe based assays are not recommended to exceed

150 bp, amplicons of less than 100 bp are recommended. Restrictions on primer design follow similar parameters to those applied to primers for SYBR® Green assays. These criteria should also be applied to the probe, ensuring that no binding occurs between either primer or the probe.

When designing multiplex reactions the consideration of primer/probe interaction becomes of greater importance. For example, in a singleplex TaqMan® probe reaction there are three oligonucleotides that can interact with each other. With a four color multiplex there are now 12 oligonucleotides to optimize in the same reaction, posing a particular challenge during assay development.

Amplicon selection

When looking at a mtDNA assay, the choice of target and/or control region will vary depending on the questions being asked by the investigator and by the origin of the DNA samples. One of the unique problems encountered when analyzing mtDNA is the presence of so many different reported DNA deletions. When designing the assay, it is important to bear in mind that mtDNA deletions accumulate with age, and that some deletion species are found more prevalently in some tissue types than others. What has been used as a reference target in one study may not necessarily work as a reference in a different tissue or patient population. The distribution of mtDNA deletions around the genome can be seen in *Figure 11.2*. The data have been taken from the MITOMAP web site (MITOMAP, 2005) and only deletions with reported breakpoints have been included. It can be clearly seen that the only region free from deletions is the D-loop. mtDNA species with deleted D-loop regions do not appear to persist in cells, most likely due to the removal of the regulatory regions preventing accumulation to detectable levels. It should be noted the D-loop is highly polymorphic and, although free from deletions, it is therefore not be the best choice as a reference region for an assay.

Quantification of mtDNA deletions

When investigating mtDNA deletions there are two major approaches. The first will allow the measurement of the total mutation load in a sample, whereas the second is mtDNA deletion specific. The difference arises from the selection of amplicons. Both approaches require the selection of a control region which is not deleted in the cohort of samples under investigation. With the first approach the second amplicon is situated within the region deleted. This allows the measurement of wild-type mtDNA, enabling calculation of the total mutant load present by subtraction. This approach does not take into account whether this is from a single deletion or from multiple deletions which remove the wild-type amplicon region. The second approach is deletion specific. The second amplicon is located across the deletion breakpoint and so a product will only be amplified when a specific deletion is present. Both approaches can be used with DNA binding dyes or probe technologies. If selecting to use probe technology then having a primer on either side of the deletion with the probe sequence actually crossing the breakpoint will increase the reaction specificity to that single deletion.

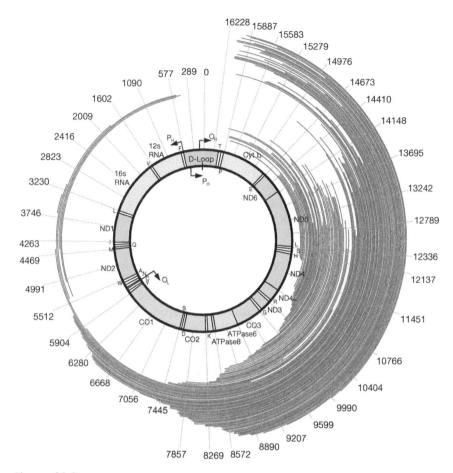

Figure 11.2

Distribution of reported mtDNA deletions around the genome. Each single line represents a single DNA deletion species, the regions covered by the line being those removed by the deletion. Data taken from the MITOMAP web site. Only deletions with reported breakpoints are illustrated. Numbers show the base positions relative to the revised Cambridge Reference Sequence (rCRS). The bias towards the major arc can clearly be seen.

Relative or absolute quantification?

Relative quantification is discussed by Pfaffl in Chapter 3. In its simplest form, this is the comparison of the amplicon of interest to a control amplicon. For a mtDNA assay, either the mtDNA region (as above) or nDNA can act as a control. When selecting a nDNA region it is useful to select a single copy gene to allow for your experimental data to be easily adjusted to a value per nuclear gene copy.

Relative quantification of an mtDNA reference gene/region to a sample mtDNA gene/region allows for the calculation of the percentage of one species to another. This in essence represents the percentage heteroplasmy of a mtDNA sample. For the design of such experiments it is important that both amplification reactions are linear over the expected experimental range of DNA concentrations. Ideally, when analyzed over a DNA dilution series, both reactions should generate equations for the line that are the same. From the slope (gradient) of this line it is possible to show that the efficiency of amplification for both reactions is the same. The intercept value of the equation shows whether for the same number of copies the reactions generate the same C_t. Perhaps the more important of these two pieces of data is the slope, or PCR amplification efficiency. Assuming that a PCR reaction is 100% efficient, each molecule should be copied once each cycle, leading to a doubling of the copy number each cycle. Over a log dilution series this equates to a 3.3 cycle difference between log dilutions. The slope of the line through the dilution series should therefore be as close to –3.3 as possible. Slopes less than –3.4 indicate a reaction that is less than 100% efficient. Conversely gradients greater than –3.2 indicate an efficiency of more than 100%, most likely due to primer-dimers formation or non-specific product formation. Consistency in the slope indicates that the reaction is optimized, slope values between –3.2 and –3.4, show a well optimized reaction. If the two reactions have the same slope this shows that, over the linear range, the two reactions are consistent. If the intercept is the same, then they are consistently the same. If the intercept is different but the slope the same, and this is a reproducible difference, the two reactions can still be compared and this difference can be accounted for mathematically in the post-run data analysis.

Comparing the quantity of mtDNA to a nDNA reference allows correction for the difference in the number of mononuclear cells present in each sample. As we are not performing a gene expression study and simply a DNA investigation, it does not matter if there is more than one copy of the gene, or if it is differentially expressed in different cells. Any region or gene can therefore be used. The same numbers of copies of each piece of nDNA are present in every cell, so the fluorescent signal will be consistent between samples containing the same number of cells. This is true in most circumstances, however not in cancer-derived cell lines, for example. Selection of any region will allow the normalization of your mtDNA data. However, it is more beneficial and we would recommend the selection of a single copy gene as the nDNA reference target. This allows for the quantity of mtDNA in the sample to be expressed as copies per diploid genome without the need for an mtDNA copy number standard curve. A mtDNA : nDNA ratio in this manner is of most use when investigating DNA extracted from bloods or homogenate tissue samples, where the number of cells present is unknown. If the study is on individual micro-dissected cells, then comparison to a nuclear target may not be necessary.

Unlike relative quantification methods, the absolute method requires a known standard curve to generate meaningful data. This means, therefore, that a series of standards needs to be generated. There are several means for doing this and considerations that need to be taken into account for each.

Generating copy number standards

PCR generated DNA templates: by using a pair of nested primers it is quick and easy to generate and quantitate a template to act as a standard for your reactions. The first primer pair generates a large PCR product, containing the region of mtDNA for investigation. This is then purified before being quantified. The second nested primer pair is used for the real-time reaction. We have found from experience that there are some regions that simple will not perform well as real-time templates. This is probably due to sequence characteristics leading to unwanted secondary structure within the PCR product.

There are significant differences between the available methods for PCR product purification and the subsequent performance of these products as real-time PCR templates. PCR purification spin columns are one quick and easy method for removing salt and unused PCR components from a product. However, we have found that despite using no buffers, these columns have a severely detrimental effect on real-time performance (*Figure 11.3B*). It may be

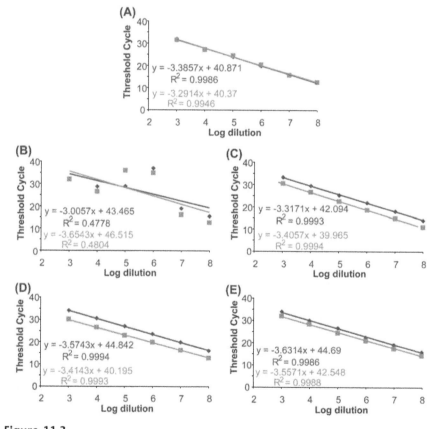

Figure 11.3

Different template preparations behave differently when used as absolute quantification standard curves for two mtDNA assays. Reaction One detects wild-type using a region of ND4 (black), the second detecting total mtDNA using a region of ND1 (grey). Each graph shows the standard curve for both reactions performed on 1:10 serial dilutions from different templates. **A.** whole genomic DNA extract from blood; **B.** PCR amplified product, purified using PCR purification spin columns; **C.** PCR amplified product, purified using gel extraction; **D.** PCR product cloned into vector, circular; **(E)** PCR product cloned into vector, linearized.

through an interaction with the spin column membrane, mechanical disruption or some chemical leeching from the membrane into the purified DNA sample. We have routinely used spin columns to purify PCR products for DNA sequencing with no problems, and continue to do so. However, we have been unable to use this technology to purify templates for real-time PCR reactions.

For the purification of PCR templates to be used as real-time standards we would recommend the use of a gel extraction method. This process removes all unincorporated PCR reaction components and leaves a single clean template for use as a standard. However, when compared to genomic DNA dilution series PCR template reactions do not always perform the same (*Figure 11.3C*) and so caution is advised.

Cloned PCR templates: to ensure a continuing and reproducible supply of your template it is convenient to use the nested primer pair approach, with the addition of cloning the primary, large product into a vector before transforming it into competent cells, extracting and purifying the DNA. This has the added advantage of ensuring you have a homoplasmic supply of your target region. For rare or low level heteroplasmic mutations this may be of particular interest. Two lines of thought exist regarding the use of cloned PCR products as real-time templates. The first suggests that for maximum PCR efficiency the vector needs to be linearized by restriction digest prior to real-time amplification. The second is that the vector be used in its closed circular form. We have investigated both variations and compared to the data obtained from genomic mtDNA. We have found that for mitochondrial studies, reactions behave differently on cloned templates, both closed (*Figure 11.3D*) and linear (*Figure 11.3E*), than on genomic mtDNA.

Genomic mtDNA: there is something reassuring about being able to compare like with like. Copy number standard curves based on PCR products are not the same as 16.5 kb closed circular mtDNA genome molecules (*Figure 11.3A*). However, there is the difficulty of quantifying mtDNA in a DNA extract due to the presence of nDNA. We have investigated several approaches to this. Ultra-centrifugation through a caesium chloride gradient enables the separation of mtDNA from nDNA by the differential movement through the gradient. We have found that through using nDNA specific primers a product can be amplified from the 'clean' mtDNA fraction.

Southern blot analysis is another approach that could be used to quantify mtDNA in a DNA extract. By running a known set of standards, quantified pure PCR products, along with several dilutions of the homogenate DNA sample and probing with an mtDNA specific probe the quantity of mtDNA can be calculated. Quantification through this approach should be viewed as inaccurate, in our opinion. There appears to be preferential binding of the probe to the purified PCR products. This may be due to the PCR products being short linear molecules that the probe can access more readily.

The third approach requires a reliable real-time assay to already be in place using one of the other two methods of generating a standard curve. A genomic DNA sample can be quantified by real-time using either a PCR template or cloned PCR product standard curve. Experience has shown that for difficult amplicons the option of PCR template or cloned product as a standard is not viable. However, other regions prove to be remarkably reproducible by these methods. Awareness of the potential to introduce errors in the calculation is inherent with this approach. Back calculation of the quantification by independent methods is advised, either spectrophotometry or fluorimetry. Quantification of a genomic mtDNA sample by this template real-time route has great benefits when investigating a difficult region. A second benefit to this is the provision of a quantified standard for any mitochondrial assays you design in the future. An example of absolute quantification in determination of viral load is given in Chapter 14, Protocol 3.

DNA extraction from single cells

As with every DNA based assay the quality of the DNA used in the assay has a massive effect on the quality of the resulting data. When dealing with DNA extracted from blood or tissue homogenates this does not pose a real problem. There is a large selection of commercially available kits and protocols which result in a high concentration of good quality DNA. DNA obtained through this route performs very well in real-time reactions.

However, DNA quality can be a major problem when investigating individual cells. Many commercially available kits do not cater for the low quantity of DNA and small volumes involved when working with individual cells. When we applied a standard single cell lysis protocol for real-time PCR amplification, considerable variability was observed in the absolute copy number measurement. Partial lysis of the cell and poor DNA extraction are the most likely explanation. If the study requires the relative quantification, then this does not pose such a huge problem, because inefficient extraction is unlikely to be template-specific. When measuring the absolute amount of mtDNA in a sample, partial lysis poses a real problem. If not all mtDNA molecules are available for amplification then it becomes difficult to draw meaningful conclusions.

To examine the true effect of lysis conditions on the measurement of absolute copy number, different lysis conditions should be considered. Cell lysis buffers can contain EDTA, as this is a magnesium chelating agent its effect on subsequent real-time reactions should be understood. Historically individual cell lysis reactions have been kept at as small a volume as possible to maximize the final DNA concentration. DNA purification methods are not 100% efficient. When measuring the quantity of mtDNA in a sample, a percentage loss can mean any subsequent conclusions are flawed. As such, post-lysis purification is not regularly performed. *Table 11.4* shows

Table 11.4 Components and concentrations of different lysis buffers used to determine the effect of lysis efficiency on the detectable mitochondrial copy number.

Lysis buffer	Lysis buffer without EDTA	ABI GeneAmp® PCR buffer plus
50 mM Tris-HCl, pH8.5	50 mM Tris-HCl, pH8.5	ABI GeneAmp® PCR buffer (1X)
1mM EDTA, pH8	0.5% Tween 20	0.5% Tween 20
0.5% Tween 20	100 μg Proteinase K	100 μg Proteinase K
100 μg Proteinase K		

Table 11.5 mtDNA copy number measurements from skeletal muscle fibers dissected from 20 μm cryostat sections and lyzed using different lysis methods. TOT shows the total mtDNA copy number measured using a region of ND1 and WT shows the wild-type mtDNA copy number measured using a region of ND4. The % Deln value shows the percentage deletion calculated using both measurements. Percentage deletion values should be zero as the fibers are from a control individual without an mtDNA deletion.

(A) Lysis buffer			(B) Lysis buffer without EDTA		
Copy number:			Copy number:		
TOT	WT	% Deln	TOT	WT	% Deln
1530	32200	−2008.4	18000	19800	−9.9
538	27800	−5070.9	51200	13400	73.9
757	1950	−157.1	37800	11500	69.4
19400	2050	89.4	19000	50400	−164.7
6350	42800	−573.9	15800	12900	18.3
6210	12200	−1866.6	18600	10100	45.7
13900	1300	90.6	34600	13200	61.7
23200	5010	78.4	30000	12500	58.5
20000	957	95.2	46800	13300	71.5
8830	42100	−376.5	48600	14800	69.5
25500	8350	67.2	14500	45200	−212.1
1790	33800	−1784.6	12100	11100	8.2

(C) ABI GeneAmp® PCR buffer plus			(D) Lysis buffer then purification		
Copy number:			Copy number:		
TOT	WT	% Deln	TOT	WT	% Deln
766	761	0.36	8190	8300	−1.4
1600	9430	−488.2	33800	6030	82.1
1120	2420	−116.0	59	2740	−4548.4
−	−	−	9890	7910	20.0
2670	2990	−12.1	7270	7640	−5.1
2550	3630	−42.3	29700	4230	85.7
9780	3200	67.3	10100	17300	−71.6
1320	15800	−1095.5	375	2740	−630.1
−	−	−	91100	16700	−83.3
3270	4210	−28.9	11000	40500	−268.9
2440	11500	−373.6	8460	7520	11.1
1290	1530	−18.3	55900	64600	−15.6

different lysis buffers and *Table 11.5* show the effect the use of each buffer has on the mtDNA copy number measured from skeletal muscle fibers taken from 20 μm cryostat sections. What can be clearly observed is that the inclusion of EDTA in the buffer increases the variability observed, whereas a post-lysis purification markedly decreases the copy number measurement.

Lysis temperature and incubation times are yet another variable that can have a large effect on the quantity and quality of DNA extract from a single cell. Using the lysis buffer without EDTA (*see Table 11.4*), lysis conditions over two temperatures and two incubation times have been investigated.

Table 11.6 Effect of lysis temperature and incubation time on mtDNA copy number measured from control skeletal muscle fibers taken from 20 μm cryostat sections.

2 hour incubation at:		16 hour incubation at:	
37°C	55°C	37°C	55°C
15200	18500	17800	23000
7640	19500	20700	20400
22100	16300	12400	18000
15400	15400	17100	37200
20700	16800	27200	29800
15400	24400	15900	19500
21400	19800	51200	22400
15300	15400	19800	24300
15600	14900	23200	25100
15600	22200	22500	27900
17800	16900	14900	26000
16600	18500	21600	31300

Incubation at 37°C and 55°C and for both 2 hour and 16 hour-time periods showed some variation in the copy number measured from single skeletal muscle fibers taken from 20 μm cryostat sections. Incubating at 55°C for 16 hours appears to generate a higher and more reproducible copy number value (*Table 11.6*).

Replicate analysis

When considering single-cell analysis and designing a real-time PCR assay, it does not follow that the more concentrated the DNA sample the more accurate the measurement. Many investigators may favor making a single measurement on the complete lysis product, rather than replicate measurements on multiple aliquots, either for data quality or cost reduction. It is essential when analyzing mtDNA, or any DNA, by real-time PCR to make replicate measurements on each sample/standard to have confidence in the data obtained. In the majority of cases the replicates will be similar, questioning the need and expense of replicate measurements. However, there are times when an outlier occurs and the inclusion of this in the analyzed data seriously affects the conclusions. When analyzing replicates these outliers can be identified and removed from the analysis. However, when making replicate measurements it is important to maintain the pipetted volume at a sensible value. If trying to replicate small volumes, a small percentage error in volume has a much greater effect on the final measured quantity. To this end we would recommend that the minimum final volume should be 5 ul, so the cell lysis volume should be adjusted to allow for this. If using a SYBR® Green reaction this may limit the investigator to two targets per cell before the starting quantity of DNA is too dilute to amplify. This raises the circumstances mentioned earlier when the benefits of a multiplex probe based real-time assay would enable the investigation of more targets per cell.

Table 11.7 Replicate vs. single measurements. **A.** Total (TOT) and wild-type (WT) mtDNA copy numbers measured in five single skeletal muscle fibers from a control individual. Four replicate measurements were made per fiber. Highlighted in bold are potential outlier values, approximately a factor of ten greater than the other three measurements; **B.** Average copy number values for both TOT and WT including all four values from each fiber, shown also is the calculated percentage deletion; **C.** Average copy number values for both TOT and WT excluding the potential outliers, shown also is the calculated percentage deletion. Exclusion of the potential outliers puts the average TOT and WT copy numbers more in agreement with each other, as expected in control fibers.

(A)

	Copy number		(B)	Average values		
	TOT	*WT*		*TOT*	*WT*	*% Deln*
Fiber 1	30600	*162000*	Fiber 1	32400	64600	49.9
	32500	32200	Fiber 2	65400	30500	−114.4
	31800	31400	Fiber 3	18700	19800	5.69
	34600	32600	Fiber 4	28800	29000	0.6
			Fiber 5	63300	28100	−125.2
Fiber 2	*172000*	28600				
	28800	27800	(C)			
	29300	33700		Average values excluding outliers		
	31700	31900		TOT	WT	% Deln
			Fiber 1	32400	32000	−1.0
Fiber 3	14600	21100	Fiber 2	29900	30500	1.8
	18400	19700	Fiber 3	18700	19800	5.7
	20300	19300	Fiber 4	28800	29000	0.7
	21300	19000	Fiber 5	26300	28300	7.3
Fiber 4	29200	30300				
	29200	28100				
	27700	28600				
	28900	28800				
Fiber 5	24800	25000				
	27300	28600				
	26700	31500				
	174000	27400				

In skeletal muscle fibers from control individuals we have investigated replicate versus single measurements. We performed two reactions per fiber using a single measurement for each reaction (i.e. measuring the percentage deletion). In control fibers, the copy number value obtained from each reaction should be identical and the calculated percentage deletion should be zero. The calculation of the ratio magnifies the observed error however, high percentage deletion values and large negative values can be seen (*see % Deln values in Table 11.5*). When the investigation was repeated making quadruplicate measurements for two reactions there was often a single replicate which stands out as an obvious outlier (*Table 11.7*). With all replicate values included in that analysis, the percentage deletion values can be seen to be more variable. Statistically, the outlier values can be legitimately removed if they fall outside of two standard deviations of the remaining three replicates. When this screening is applied we can see that the percentage deletion values lie much closer to the zero value expected.

Real-time PCR reaction optimization

Optimization of a real-time assay is covered in more detail in Chapters 1 and 3, however it cannot be over-stressed how important the careful optimization of all reaction components and conditions is. Primer and probe concentrations should be optimized to give the earliest signal for a specific amplicon, this ultimately equates to the sensitivity of the reaction. The annealing temperature for any given primer set, or primer/probe combination can be calculated by any number of different means. These calculated temperatures are not always the temperatures at which the best performance is achieved. Standard PCR on a gradient block PCR machine can be used to check the specificity of real-time primers. To check the specificity of a primer/probe combination then a gradient on a real-time PCR machine will allow visualization of the fluorescence signal from the probe. The melting curve from a SYBR® Green real-time reaction gives a reasonable indication as to the number of products amplified in a reaction. However, as the amplicon 'melts' at a range of temperatures along its length until completely single stranded, these peaks are quite broad. Products very similar in size may not be distinguishable by this approach.

Magnesium chloride concentration affects the activity of the polymerase in the reaction. Polymerases from different manufacturers will all require different magnesium chloride concentrations to enable them to perform maximally. The best indicator for optimization is the PCR efficiency and reaction reproducibility. The PCR efficiency is taken from the number of PCR cycles between log dilutions of a standard template and was discussed earlier. Most manufacturers supply reagents with an optimized buffer. This should perform well for the majority of amplicons. However, magnesium chloride concentration should be optimized for each reaction by amplifying a dilution series at a range of magnesium chloride concentrations. Comparison of the PCR efficiencies will enable the investigator to select the optimum concentration for each assay they design. *Figure 11.4* shows a magnesium chloride optimization experiment for an mtDNA real-time assay in the ND1 region. Magnesium chloride concentrations of 1.5 mM, 2.5 mM, 3.0 mM, 3.5 mM, 4.0 mM and 4.5 mM are shown in *Figure 11.4 parts A to F* respectively. As mentioned previously investigators should aim to have reactions with a slope as close to –3.3 as possible. An important issue is the consistency of efficiency. An efficiency of 96% on one run and then 65% on a subsequent run highlights an assay that is not well optimized.

11.4 Discussion

What we hope to have presented here is a coherent flow through the considerations when designing an mtDNA real-time PCR assay. Many of these considerations are no different to when designing any real-time assay, nuclear or mitochondrial, and where this is the case we have directed the investigator to the relevant chapter. As a large amount of mitochondrial research is carried out at the level of the individual cell, we feel that it is these considerations that should strongly influence the design and

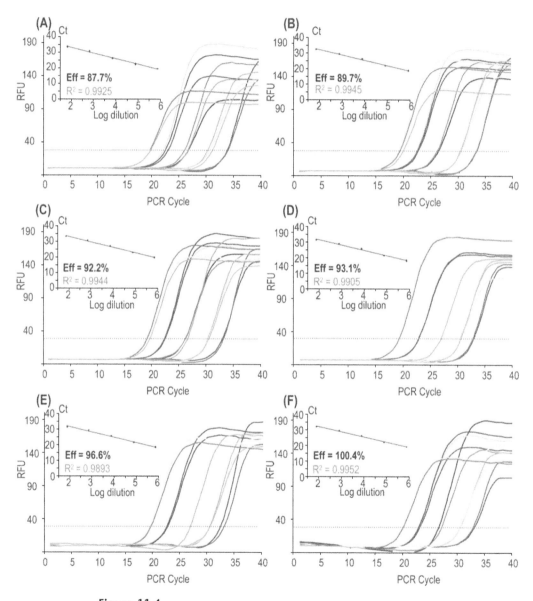

Figure 11.4

Effect of final Magnesium Chloride concentration of real-time PCR efficiency. Each graph shows a 1:10 serial dilution series, amplified for a region of ND1. Inset on each graph is the standard curve at that magnesium chloride concentration showing the equation for the line, the R^2 value and the PCR efficiency. **A.** 1.5 mM; **B.** 2.5 mM; **C.** 3.0 mM; **D.** 3.5 mM; **E.** 4.0 mM; **F.** 4.5 mM. Improved reaction efficiency can be seen as the magnesium chloride concentration increases.

optimization of a real-time assay and have focused the attention of this chapter in that direction.

The variability in the real-time PCR machines and in the different manufacturers' reagents, together with the numerous different targets make it impossible to generate a generic protocol for an mtDNA real-time assay. We have covered what we believe to be the main considerations when working in the mtDNA field and feel it is sufficient to say that any real-time assay has the potential to generate strong data if correctly optimized. We would advise strongly against trying to cut corners and costs by minimizing the optimization and replicate analysis. Optimization and validation is essential to be able to draw conclusions from real-time data and should be carried out in accordance with guidelines laid out in the other chapters of this book. The conditions found to work for one assay will not necessarily work well for all.

References

Anderson S, Bankier AT, Barrell BG, de Bruijn MH, Coulson AR, Drouin J, Eperon IC, Nierlich DP, Roe BA, Sanger F, Schreier PH, Smith AJ, Staden R, Young IG (1981) Sequence and organization of the human mitochondrial genome. *Nature* **290**(5806): 457–465.

Andrews RM, Kubacka I, Chinnery PF, Lightowlers RN, Turnbull DM, Howell N (1999) Reanalysis and revision of the Cambridge Reference Sequence for human mitochondrial DNA. *Nature Genetics* **23**(2): 147.

Attardi G (2002) Role of mitochondrial DNA in human aging. *Mitochondrion* 2: 27–37.

Coller HA, Khrapko K, Bodyak ND, Nekhaeva E, Herrero-Jimenez P, Thilly WG (2001) High frequency of homoplasmic mitochondrial DNA mutations in human tumors can be explained without selection. *Nat Genet* **28**(2): 147–150.

de Mendoza C, Sanchez-Conde M, Ribera E, Domingo P, Soriano V (2004) Could mitochondrial DNA quantitation be a surrogate marker for drug mitochondrial toxicity? *AIDS Rev* **6**(3): 169–180.

Fayet G, Jansson M, Sternberg D, Moslemi AR, Blondy P, Lombes A, Fardeau M, Oldfors A (2002) Ageing muscle: clonal expansions of mitochondrial DNA point mutations and deletions cause focal impairment of mitochondrial function. *Neuromuscul Disord* **12**(5): 484–493.

Hoy JF, Gahan ME, Carr A, Smith D, Lewin SR, Wesselingh S, Cooper DA (2004) Changes in mitochondrial DNA in peripheral blood mononuclear cells from HIV-infected patients with lipoatrophy randomized to receive abacavir. *J Infect Dis* **190**(4): 688–692.

Lewis W, Day BJ, Copeland WC (2003) Mitochondrial toxicity of NRTI antiviral drugs: an integrated cellular perspective. *Nat Rev Drug Discov* **2**(10): 812–822.

McComsey G, Bai RK, Maa JF, Seekins D, Wong LJ (2005a) Extensive investigations of mitochondrial DNA genome in treated HIV-infected subjects: beyond mitochondrial DNA depletion. *J Acquir Immune Defic Syndr* **39**(2): 181–188.

McComsey GA, Paulsen DM, Lonergan JT, Hessenthaler SM, Hoppel CL, Williams VC, Fisher RL, Cherry CL, White-Owen C, Thompson KA, Ross ST, Hernandez JE, Ross LL (2005b) Improvements in lipoatrophy, mitochondrial DNA levels and fat apoptosis after replacing stavudine with abacavir or zidovudine. *Aids* **19**(1): 15–23.

Michikawa Y, Laderman K, Richter K, Attardi G (1999a) Role of nuclear background and in vivo environment in variable segregation behavior of the aging-dependent T414G mutation at critical control site for human fibroblast mtDNA replication. *Somat Cell Mol Genet* **25**(5–6): 333–342.

Michikawa Y, Mazzucchelli F, Bresolin N, Scarlato G, Attardi G (1999b) Aging-dependent large accumulation of point mutations in the human mtDNA control region for replication. *Science* **286**(5440): 774–779.

MITOMAP (2005) A Human Mitochondrial Genome Database. http://www.mitomap.org

Penta JS, Johnson FM, Wachsman JT, Copeland WC (2001) Mitochondrial DNA in human malignancy. *Mutat Res* **488**(2): 119–133.

Spinazzola A, Zeviani M (2005) Disorders of nuclear-mitochondrial intergenomic signaling. *Gene* **18**: 162–168.

Szuhai K, Ouweland J, Dirks R, Lemaitre M, Truffert J, Janssen G, Tanke H, Holme E, Maassen J, Raap A (2001) Simultaneous A8344G heteroplasmy and mitochondrial DNA copy number quantification in myoclonus epilepsy and ragged-red fibers (MERRF) syndrome by a multiplex molecular beacon based real-time fluorescence PCR. *Nucleic Acids Res* **29**(3): E13.

Woischnik M, Moraes CT (2002) Pattern of organization of human mitochondrial pseudogenes in the nuclear genome. *Genome Res* **12**(6): 885–893.

Zuker M (2003) Mfold web server for nucleic acid folding and hybridization prediction. *Nucleic Acids Res* **31**(13): 3406–3415.

Real-time immuno-PCR

Kristina Lind and Mikael Kubista

12

12.1 Introduction

12.1.1 Immunoassays

Immunoassays make use of the excellent binding specificity of antibodies and have been used for protein detection since the 1960s (Yalow and Berson, 1960). Antibodies can be directed with high specificity to a large variety of antigens. The still most popular and used technique for protein quantification, ELISA, was developed in the 1960s (Engvall and Perlman, 1971). Sandwich ELISA uses one antibody to capture the analyte and an enzyme immobilized by a second antibody for detection. The enzyme catalyzes a colorimetric reaction which product is readily detected for quantification. Through the reaction the signal from the analyte increases linearly in time and low concentrations of the antigen can be detected.

12.1.2 Immuno-PCR

In 1992 Sano *et al.* developed immuno-PCR, which is an immunoassay similar to ELISA but uses PCR for the detection instead of a colorimetric enzyme reaction. They constructed a streptavidin-protein A chimera that was bound to a biotinylated linear plasmid. The protein A-streptavidin-DNA-conjugate bound the detection antibody which was connected to the analyte in the microtiter plate. The DNA was then amplified by PCR. The PCR product was finally detected by gel electrophoresis (Sano *et al.*, 1992). In an alternative setup the amplified PCR product was detected by PCR ELISA (Niemeyer *et al.*, 1997). Here biotin labeled PCR primers and a digoxigenin labeled nucleotide was used. Following PCR the products were immobilized in streptavidin coated microtiter plates and analyzed by ELISA using digoxigenin IgG-alkaline phosphatase conjugate. These post-PCR analysis methods are time consuming and cross-contamination is a major problem. Further drawback is that the PCR is run to completion resulting in loss of linearity. A much easier, faster and more convenient approach is to use real-time PCR for direct quantification of the DNA (*Figure 12.1*). This was first demonstrated by Sims *et al.*, 2000. Compared to ELISA, real-time immuno-PCR is more sensitive and has much larger dynamic range (*Figure 12.2*). These advantages are due to the fundamentally different signal generation and detection processes. In ELISA the detection antibody is coupled to an enzymatic system that generates linear signal increase with time, while in real-time immuno-PCR the detection antibody is coupled to a nucleic acid template that generates an exponential signal growth in time

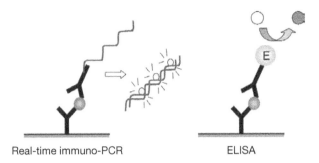

Figure 12.1

Schematic set-up of real-time immuno-PCR and ELISA.

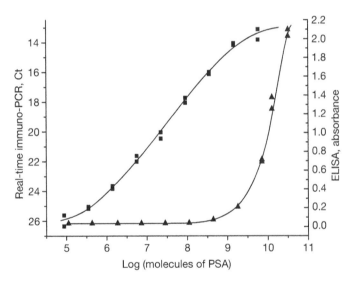

Figure 12.2

Comparison of real-time immuno-PCR (squares) and ELISA (triangles) on standard samples. Reprinted from *Journal of Immunological Methods*, **304**, 107–116, Figure 5 with permission from Elsevier B.V.

when amplified by PCR. Further, ELISA signal is read out as intensity value at the end of the reaction, while in real-time PCR the number of amplification cycles to reach a particular signal level is registered.

12.2 Assemblages for real-time immuno-PCR

The immunoassay can be assembled in several ways that differ mainly in two aspects: how the capture antibody is bound to the surface, and how the DNA-label is attached to the detection antibody (Lind and Kubista, 2005).

12.2.1 Attaching capture antibody

A straightforward way to attach the capture antibody to the vessel surface is by physical adsorption. No special tubes or linking groups are needed. A solution containing the antibody is added to the wells and incubated overnight allowing the antibodies to adsorb to the surface. Some will have their antigen binding site facing upwards being accessible and some will have the antigen binding site hidden. The capture antibody can also be attached through a streptavidin-biotin link. This requires that the capture antibody is biotinylated and that the surface is coated with streptavidin. With this setup the capture antibody can either be attached to the surface before being exposed to the antigen or it can be incubated in the test solution to bind antigen and then immobilized to surface. The assay time with this approach is much shorter than with physical adsorption. Drawback is that streptavidin coated plates are many times more expensive than regular plates.

12.2.2 Labeling detection antibody with DNA

There are currently two established ways to attach a DNA-label to the detection antibody. One is through streptavidin-biotin linkage and the other by covalent coupling using a heterobifunctional cross-linking agent. The streptavidin-biotin link is usually assembled stepwise by incubating one component at a time during the assay. This is rather straight forward but requires additional incubation and washing steps, which increase technical variation. The covalent antibody/DNA conjugate is prepared in advance and purified. This reduces the number of reaction components, making the assay easier to optimize.

12.3   Real-time immuno-PCR details

12.3.1 Reaction containers and instruments

Immunoassays, like ELISA, are performed in plates optimized for protein binding. These are typically made of polystyrene or polycarbonate. For real-time immuno-PCR containers that fit, real-time PCR instruments are needed. Regular ELISA plates do not fit PCR instruments because they have flat bottoms and their size does not match regular PCR blocks. Immunoassays are most conveniently performed in 96-well plates or in strips of eight or twelve tubes. A multichannel pipettor can then be used to add assay components and for washing. Nunc® TopYield polycarbonate strips are flat bottomed and fit normal PCR blocks. These have been used in real-time immuno-PCR attaching the capture antibody by physical adsorption to the wells (Adler et al., 2003). Regular polypropylene plates for real-time PCR, such as the AB-0600 from ABgene, that fit, for example, the iCycler iQ®, from Biorad, can also be used. When we compared the polypropylene plates with the polycarbonate strips we obtained higher reproducibility with the polypropylene container. The polypropylene plates are also available coated with streptavidin for attachment of capture antibody via biotin.

A number of real-time PCR instruments on the market are suitable for real-time immuno-PCR. We find the 96-well platforms most convenient. Our lab has tested the iCycler iQ® from Biorad, the 5700 and 7700 from ABI, the Mx3005p from Stratagene, and the Quantica from Techne, with very good results when using the 96-well microtiter plates recommended by the respective vendors. Also the capillary based LightCycler® has been used for real-time immuno-PCR (McKie *et al.*, 2002). Here the assay was first built up in an ELISA compatible 96-well plate. The DNA-label was then cut by a restriction enzyme and the solution was transferred to LightCycler® capillaries for real-time PCR analysis.

12.3.2 DNA-label

The DNA-label serves as template for the PCR and should be carefully designed. Most important is that the DNA-label and the primer pair used produce negligible amounts of non-specific primer-dimer products. Its length does not seem to be crucial. We have tested lengths between 66–1098 bp and found no important dependence. However, we have not performed any systematic study. Usually we design the PCR system to give short amplicons (60–150 basepairs), to have high PCR efficiency and to keep cost down. If streptavidin-biotin linkage is going to be used to connect the DNA to the detection antibody, biotinylated primers can be used to produce the DNA-label. For chemical conjugation suitably modified polynucleotides are available from most oligohouses, and the shorter they are, the less expensive. The modification can be either an amino- or a thiol-group depending on the chemistry that will be used. If succinimidyl-4-(N-maleimidomethyl) cyclohexane-1-carboxylate (SMCC) is used, the polynucleotide should be amino modified. We always modify the 5′ end of the polynucleotide because it is cheaper, but 3′ end modifications should also be fine. To minimize risk of contamination, we recommend using a different DNA sequence for real-time immuno-PCR from those used by others in the lab. Ideally a non-natural DNA sequence can be designed. Such DNA sequences including primers and protocol are available from our center (www.tataa.com).

12.3.3 Blocking agents

In immunoassays, non-specifically bound reagents give rise to background signal that limits assay sensitivity. To reduce the background, blocking agents are used. These are often inert protein blends, such as bovine serum albumin (BSA) and milk powder. Other common additives are detergents, such as Tween 20. The blocking agents cover any empty surface preventing direct binding of the reaction components to the vessel.

12.3.4 Controls

The performance of the real-time immuno-PCR assay can be tested by running a number of control reactions in parallel with the test samples. The background signal from non-specifically bound assay components is assessed by a background control (BC) that contains the same amount of antibodies and DNA as positive samples, but without antigen. The BC signal is caused by non-specific adsorption of reaction components and should have as high C_t as possible. The second control is the no template control (NTC). The NTC contains the real-time PCR mastermix only. It reveals any contamination of the mastermix and its C_t reflects primer-dimer formation. If the NTC and BC have similar C_t, primer-dimer formation limits sensitivity, and it is probably a good idea to redesign the real-time PCR assay.

12.3.5 Optimizing concentrations

When setting up an assay for the first time, it is important to optimize the capture and detection antibodies' of concentrations. In the following, we describe a basic procedure of how to optimize a standard set-up based on adsorbed capture antibody and chemically conjugated detection antibody/DNA. The optimal concentrations depend on the type of antibodies used, but are typically in the range 1–20 µg ml^{-1} for the capture antibody and 1–100 ng ml^{-1} of detection antibody. Using antibodies with higher affinity and specificity for the antigen usually results in more sensitive assay. To determine optimal concentrations of capture and detection antibodies, an evaluation set-up with serial dilutions of both antibodies and antigen, as shown in *Figure 12.3*, can be performed. In a first attempt the real-time immuno-PCR protocol below can be used.

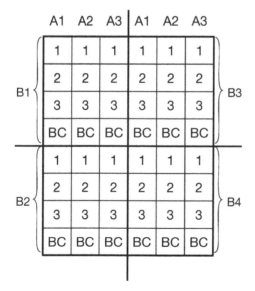

Figure 12.3

Microtiter plate set-up for capture and detection antibody concentration optimization. 1–3 represent different concentrations of antigen covering the range of interest. BC is the background control. A1–A3 is a ten-fold dilution series of capture antibody, and B1–B4 is a five-fold dilution series of detection antibody/DNA conjugate.

The combination of capture and detection antibodies' concentrations that give largest C_t difference between the lowest antigen concentration and the BC sample is the best choice, of course, provided they are within the linear range of C_t versus the logarithm of antigen concentration. If this is not the case one must choose different concentrations of the antibodies or replace them.

References

ACS Committee on Environmental Improvement (1980) Guidelines for data acquisition and data quality evaluation in environmental chemistry. *Anal Chem* **52**: 2242–2249.

Adler M, Wacker R, Niemeyer CM (2003) A real-time immuno-PCR assay for routine ultrasensitive quantification of proteins. *Biochem Biophys Res Commun* **308**: 240–250.

Engvall E, Perlman P (1971) Enzyme-linked immunosorbent assay (ELISA). Quantitative assay of immunoglobulin G. *Immunochemistry* **8**: 871–874.

Lind K, Kubista M (2005) Development and evaluation of three real-time immuno-PCR assemblages for quantification of PSA. *J Immunol Methods* **304**: 107–116.

McKie A, Samuel D, Cohen B, Saunders NA (2002) Development of a quantitative immuno-PCR assay and its use to detect mumps-specific IgG in serum. *Journal of Immunological Methods* **261**: 167–175.

Niemeyer CM, Adler M, Blohm D (1997) Fluorometric polymerase chain reaction (PCR) enzyme-linked immunosorbent assay for quantification of immuno-PCR products in microplates. *Analytical Biochemistry* **246**: 140–145.

Sano T, Smith CL, Cantor CR (1992) Immuno-PCR: Very sensitive antigen detection by means of specific antibody-DNA conjugates. *Science Washington DC* **258**: 120–122.

Sims PW, Vasser M, Wong WL, Williams PM, Meng YG (2000) Immunopolymerase chain reaction using real-time polymerase chain reaction for detection. *Anal Biochem* **281**: 230–232.

Yalow RS, Berson SA (1960) Immunoassay of endogenous plasma insulin in man. *J Clin Invest* **39**: 1157–1175.

Protocol 12.1: Basic real-time immuno-PCR protocol

Chemicals:

Adsorption buffer: 0.2 M NaH_2PO_4
Wash buffer: 0.154 M NaCl, 5 mM Tris pH 7.75, 0.005% Tween 20, and 0.1% Germall
Blocking/incubation buffer: 0.5% BSA, 0.05% Tween 20 in phosphate buffered saline pH 7.3 (137 mM NaCl, 2.7 mM KCl, 4.3 mM Na_2HPO_4, and 1.4 mM KH_2PO_4)
Capture antibody
Detection antibody/DNA conjugate
Antigen standard
PCR chemicals

Equipment:

For washing: use either a multichannel pipette or Nunc® immunowash. Other automatic washing devices that handle V-shaped PCR-tubes can also be used.
Real-time PCR instrument with 96-well plate format.
Precision pipettes
PCR microtiter plate
Adhesive seal
Vortex
Microcentrifuge

1. **Adsorption of capture antibody**. Dilute capture antibody to desired concentration in 0.2 M NaH_2PO_4. A volume of 2,500 µl is needed per 96-well plate. Add 25 µl of the capture antibody solution to each well and cover the plate with adhesive seal. Incubate at 4 °C overnight (see *Figure 12.4*)

2. **Blocking**. Wash the antibody coated wells three times with wash buffer. Add 200 µl blocking buffer to each well. Cover with adhesive seal and incubate at 37°C for 1 h or at 4°C overnight

3. **Prepare analyte and detection antibody/DNA conjugate solution**. Prepare an analyte dilution series by diluting stock

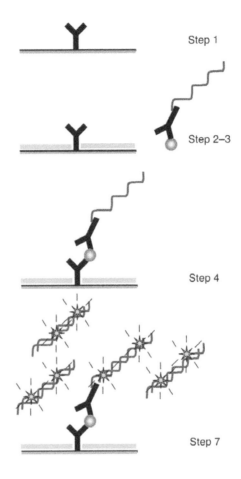

Step 1

Step 2–3

Step 4

Step 7

Figure 12.4

Steps in basic real-time immuno-PCR.

antigen solution in incubation buffer. The standard concentrations should cover the concentrations of the samples that will be analyzed. Never attempt to quantify antigen at concentrations outside the range of the standards. Prepare also an antibody/DNA conjugate solution of suitable concentration in incubation buffer. For duplicate samples mix 11 µl of either the analyte standard or the test sample with 44 µl of diluted conjugate solution in a clean microtiter plate. Prepare BC samples by mixing 11 µl of incubation buffer with 44 µl of the conjugate solution in a separate tube. Incubate all samples for 1 h at room temperature

4. **Add samples to the wells.** Wash the blocked wells ones with wash buffer and add 25 µl of the incubated samples to each well. Seal with adhesive seal and incubate at room temperature for 1 h

5. Wash the wells six times with wash buffer and ten times with milli-Q water

6. Tap the microtiter plate upside down on paper towels to remove as much of the water as possible

7. **Run real-time PCR.** Add 25 µl PCR mastermix to each well. Prepare also an NTC by adding PCR mastermix to a clean well. Seal the microtiter plate with an adhesive seal that is compatible with the real-time PCR instrument. Start the PCR. Use the same cycling protocol as for the regular real-time PCR

Data analysis

Construct a standard curve from the standard samples by plotting C_t versus the logarithm of analyte concentration. Fit a line by linear regression to the data points that follow linear response. Frequently the lowest and highest concentrations deviate from linearity. To include those one may use more advanced models, such as 3rd degree polynomial to fit the data points. To determine the sensitivity of the assay, run replicates of the BC together with the standard curve samples (Plate scheme in *Figure 12.5*). Estimate the apparent concentrations the BC signals correspond to using the standard curve, and calculate the standard deviation (SD) and mean. The limit of detection is defined as the mean of the apparent concentrations of the BC samples plus three times the SD (ACS Committee on Environmental Improvement, 1980). The reproducibility of the assay is determined from at least eight replicates of standard samples with high, medium, and low analyte concentration that are run together with a complete standard curve (*Figure 12.5*). The sample concentrations should be within the linear range of the standard curve. Estimate the concentrations from the repeat samples from the standard curve and calculate their mean and the SD. Finally calculate the relative standard deviation at the three concentrations as the SD/mean. Typically the relative standard deviation decreases with increasing analyte concentration.

BC	BC	BC	L	M	H
1	1	BC	L	M	H
2	2	BC	L	M	H
3	3	BC	L	M	H
4	4	BC	L	M	H
5	5	BC	L	M	H
6	6	BC	L	M	H
7	7	BC	L	M	H

Figure 12.5

Microtiter plate set-up to estimate the sensitivity and reproducibility of the real-time immuno-PCR. Numbers 1 to 7 represent standard antigen samples of different concentrations. BC is the background control. Repeats of standard samples with low (L), medium (M), and high (H) concentrations of antigen.

Preparing and purifying antibody/DNA conjugates

There are many heterobifunctional cross-linking agents suitable for conjugation of a DNA-label to a detection antibody. They have two reactive groups, one that reacts with the antibody, and a second that reacts with the DNA-label joining them together. We usually use SMCC (*Figure 12.6*), which reacts with an amino-group and a thiol-group. Thiol-groups can be added to antibodies by treating them with 2-iminothiolane (Traut's reagent), which reacts with primary amines introducing the sulfhydryl groups. The polynucleotide can be synthesized with an artificial amino-group in either end for coupling to SMCC. Conjugation is performed by first reacting the polynucleotide with SMCC and then reacting the SMCC-polynucleotide conjugate with the sulfhydryl activated antibody (*Figure 12.6*).

Conjugation protocol

Chemicals:

Detection antibody
5' aminomodified polynucleotide
TSE pH 8.5 (0.1 M triethanolamine, 0.1 mM sodium chloride, 1 mM EDTA, pH 8.5)
2-iminothiolane
Glycine pH 7.3
TSE pH 12.9 (0.1 M triethanolamine, 0.1 mM sodium chloride, 1 mM EDTA, pH 12.9)
SMCC
DMSO
TSE pH 7.3 (0.1 M triethanolamine, 0.1 mM sodium chloride, 1 mM EDTA, pH 7.3)

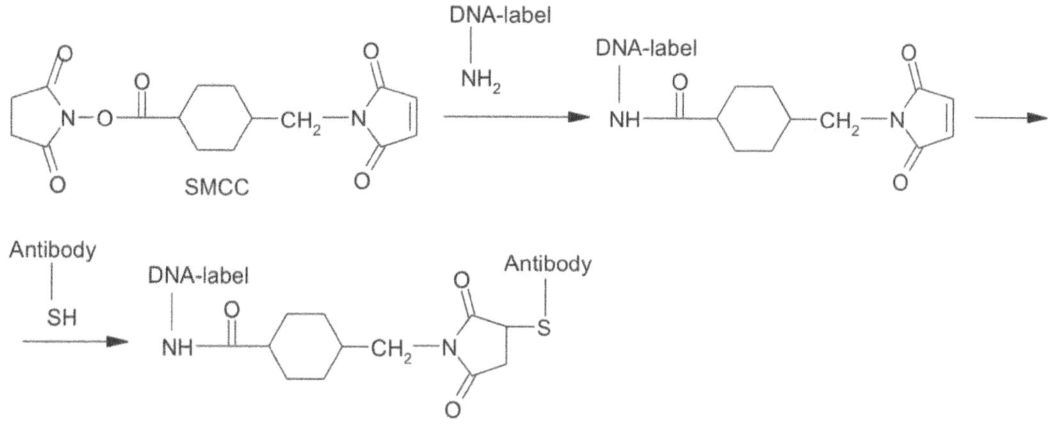

Figure 12.6

Conjugation of DNA and antibody with SMCC.

Equipment:	Precision pipettes
	NAP-5, NAP-10, or PD-10 columns (Amersham Biosciences)
	Centricon YM-30 (Millipore)
	Vortex
	Centrifuge
	Glass vial

1. Change buffer of the detection antibody solution to TSE (pH 8.5) and concentrate to 3–5 mg antibody per ml using, for example, Centricon YM-30 (Millipore)

2. Prepare 50 mM solution of 2-iminothiolane in TSE (pH 12.9) and mix immediately with the antibody solution in proportion 1:33. For example, if the antibody volume is 1 ml, add 30 μl 2-iminothiolane. Incubate for 30 min at room temperature

3. Quench the reaction with 1 M glycine (pH 7.3) using the same volume as of 2-imino-thiolane

4. Purify the activated antibody with a small gel filtration column such as NAP-5, NAP-10, or PD-10 using TSE (pH 7.3). The activated antibody is not very stable and should be used within 2 h

5. Dilute the polynucleotide to about 70 nmol ml^{-1} in 0.1 M TS (pH 7.7). The molar amount of the polynucleotide should be approximately two times the molar amount of antibody used

6. Dissolve 3–5 mg of SMCC in DMSO to 20 mM in a small glass vial

7. Add SMCC to the polynucleotide solution to a concentration of 2 mM. Incubate for 20 min at room temperature

8. Quench the reaction by adding 1/50 volume of 1 M glycin (pH 7.3) to the tube

9. Purify the activated polynucleotide with a small gel filtration column, such as NAP-5, NAP-10, or PD-10. Elute with TSE (pH 7.3). The activated polynucleotide should be used within 2 h

10. Determine the concentrations of antibody and polynucleotide spectroscopically

11. Mix the polynucleotide and antibody in a 2:1 molar ratio

12. Incubate 1 h at room temperature, and store the reaction tube in the refrigerator for at least 2 h and at most 72 h before purification

Purification protocol

Rigorous purification of the antibody/DNA conjugate is crucial for the sensitivity of the assay. Different methods to remove non-conjugated antibody and free DNA can be used. We recommend first removing free antibody with anion-exchange chromatography followed by gel filtration to separate the free DNA from the conjugate.

Ion-exchange column: Resource® Q 1 ml (Amersham Biosciences)
Buffers: 10 mM Tris pH 8.0 in 0 M NaCl and in 1.5 M NaCl for gradient elution.
Collect eluate with fraction collector.
Identify the fractions containing free DNA and antibody/DNA conjugate by absorption spectroscopy. Pool and concentrate these fractions using a Centricon YM-100 (Millipore), or similar, to a volume (typically about 0.5 ml) suitable for the gel filtration column to be used.
Gel filtration column: Superdex 200 HR10/30 (Amersham Biosciences)
Buffer: 50 mM Tris, 1 mM EDTA, 0.15 M NaCl, pH 8.0.
Collect eluate with fraction collector
Fractions to use in real-time immuno-PCR are best identified in test runs. The amount to use in the test runs can be estimated by regular real-time PCR. Fractions containing active conjugate are

pooled and stored either in refrigerator or freezer. About 5 mg ml^{-1} of BSA can be added to stabilize the solution. Although it may depend on the antibodies, we have found stock conjugate solutions to be stable upon storage at +4°C or –20°C for at least 1 year.

Clinical microbiology

Burcu Cakilci and Mehmet Gunduz

13

13.1 Introduction

13.1.1 Importance of detection and quantification in microbiology

The contribution of clinical microbiology laboratories to the effective treatment of patients with bacterial infections depends on early and accurate identification, and rapid susceptibility testing of bacteria. The rapid, specific, and sensitive detection of the pathogens is essential for providing immediate therapeutic intervention, subsequent control of potential epidemics, and has a great impact on the outcome of individual patients especially those with systemic infections or immune disorders. Although detection and identification of pathogenic microorganisms is the essence of culture-based assays in clinical microbiology; importance of quantitative measurements of pathogens has increased and extended to monitoring effectiveness of the antibiotics. Recently, quantification of pathogenic microorganisms and their activities has become almost as important as their rapid detection and identification.

13.1.2 From traditional methods to real-time PCR in microbiology

Prior to development of the recent molecular diagnostic technologies and their applications in microbiology laboratories, traditional diagnostic methods were used for qualification and quantification of the microorganisms. Direct counting of individual cells by microscopy from samples or culture plates was probably the oldest technique used for microbial quantification. DNA binding stains were also used for the same purpose (Jansson and Prosser, 1997). As enumeration has been of interest, many different cell detection techniques have been developed including fluorescent tagged antibodies, rRNA targeted fluorescent oligonucleotides, confocal laser scanning microscopy and flow cytometric methods (Jansson and Prosser, 1997). Among the traditional methods culture and culture-based methods were considered as reliable approaches and mostly named as the 'gold standard' for the identification and quantification of bacteria. Besides being time-consuming, traditional methods have some limitations such as poor sensitivity, slow growing or poorly viable organisms, narrow detection ranges, complex interpretation, high levels of background, and non-specific cross-reactions (Mackay, 2004). Traditional microbial culture methods supply valuable data especially for new, uncharacterized or atypical organisms for further studies and epidemiological assessments (Mackay, 2004).

There has been a sharp increase in molecular biological techniques over the last few decades, which has resulted in developments in many areas of the life sciences, including bacteriology (Millar and Moore, 2004). The field of molecular biology has been revolutionized by amplifying as few as one copy of a gene into billions of copies, after the introduction of PCR. PCR-based methods offer high sensitivity, specificity and simplification of testing by process automation and incorporation of non-isotopic detection (Sintchenko *et al.*, 1999). In the early stages, the use of PCR methodology was limited by the qualitative nature of PCR-based applications. The logarithmic amplification of the target sequence in PCR reveals with no correlation between the initial and final amounts of DNA product. Due to this limitation, conventional PCR is considered a semiquantitative technique.

Recently, real-time PCR technology appeared as a major development among the PCR-based systems. In these assays, fluorescent signals are generated as the PCR takes place and thus real-time monitoring during the amplification is possible. This useful combination of nucleic acid amplification and signal detection has reduced the time required for nucleic acid detection. In addition, as a quantitative method for detection of bacterial species, sensitivity of real-time PCR enables detection of less than 10 cells with sensitive background regulations (Maeda *et al.*, 2003; Ott *et al.*, 2004). Today, real-time PCR is used to detect nucleic acids from food, vectors used in gene therapy protocols, genetically modified organisms, and areas of human and veterinary microbiology and oncology (Mackay, 2004).

13.2 Real-time PCR studies in microbiology

13.2.1 Basics for microbial quantitation

In real-time PCR, it is possible to measure the amount of PCR product at any time point. During the exponential phase in real-time PCR experiments, a cycle threshold (C_t) is determined at which point the fluorescence signal generated from a sample is significantly greater than background fluorescence. These C_t values are directly proportionate to the amount of starting template and are the basis for calculating mRNA expression levels or DNA copy number measurements. The quantification of the PCR product at specific cycle numbers allows complete understanding of the PCR process with truly quantitative estimations.

External and internal controls

All real-time PCR assays should include positive and negative controls and reliability of an assay is intimately associated with the quality of the assay controls. Some substances found in clinical samples can inhibit PCR, possibly via effecting binding and/or polymerization activity of DNA polymerases, such as bile salts and complex polysaccharides of feces and heme of blood (Nolte, 2004; Monteiro *et al.*, 1997; Al-Soud and Radstrome, 2001). Degradation of target nucleic acid, sample processing errors, thermal cycler malfunction, and carry-over contamination are some of the other reasons that may cause false-negative results (Nolte, 2004).

External controls are simple approaches to detect inhibitors but addition of an external control to a separate reaction increases the costs and does not always work for large batch sizes (Nolte, 2004). Usually, an external control is created by using a cloned amplicon, a portion of the target organisms' genome, or simply the purified amplicon itself (Mackay, 2004). The data generated from individual amplification of the known dilution series of external control create the basis of the standard curve which is used for analyzing the unknown sample content. Internal controls, which are spiked into clinical patient sample, provide an accurate way to check the integrity of all amplification steps. They should imitate the target sequence and if they are not amplified, that should indicate the inhibition of the PCR analysis of clinical patient sample (Nolte, 2004; Cockerill, 2003). Melting curve analyses can be another important quality control feature that confirms PCR amplicons as the correct amplification product and can discriminate base pair differences in target DNA, such as mutations (Cockerill, 2003; Wittwer and Kusukawa, 2004).

Absolute and relative quantification

Absolute quantification determines the exact number of the target present in the sample, by relating the PCR signal to a standard curve, therefore requiring that the absolute quantities of the standard be known. Absolute quantification makes it easier to compare data from different assays and necessary when exact transcript copy number need to be determined, e.g. viral or bacterial load in microbiology clinics (Mackay, 2004; Livak and Schmittgen, 2001).

The relative quantification method quantifies the target sequence amount normalized to an endogenous reference control (Livak and Schmittgen, 2001). This reference can be co-amplified separately or in the same tube in a multiplex assay; therefore with this method, rather than using a known amount of standard, comparison of the relative amount of target sequence to the reference values is possible (Livak and Schmittgen, 2001). The result is given as relative to that reference value and reference can be any transcript, as far as its sequence is known. Housekeeping genes are often used for this purpose, as they are widely expressed in various cells and tissues. The comparative C_t method is similar to the relative quantification method, except the usage of arithmetic formulas to achieve results for relative quantification. The relative standard or comparative C_t methods are simpler and easier to establish than absolute quantification as well as providing sufficient information in many cases. Details of quantification methods are given in Chapters 3 and 6.

13.2.2 Bacteria

The real-time PCR systems brought significant advantages to clinical microbiology. Real-time PCR applications are more sensitive, time and cost effective for both detection and quantification of the pathogens and for identification of the specific genes or mutations in microorganisms (Pahl *et al.*, 1999; Maeda *et al.*, 2003; Boutaga *et al.*, 2003; Saukkoriipi *et al.*, 2004; Sloan *et al.*, 2004; Huletsky *et al.*, 2004). Real-time PCR applications allow fast confirmations of conventional detection and serology results, correction of non-specific serological false positive results, identification of the genes that encode virulence factors or antimicrobial resistance, recognition of the mutations of genes that are associated with antimicrobial resistance. Thus, it may be possible to commence appropriate antibiotic therapy sooner, or modify the therapy earlier in the presence of a molecular identification from a culture-negative or serology-negative specimen, or quantitative monitoring of the pathogen during therapy (Millar and Moore, 2004). Although not yet perfect, they have the potential to revolutionize the way in which diagnostic tests are performed.

Bacterial detection and quantification

Rapid detection by real-time PCR has been an important advantage where early diagnosis and appropriate antibiotic therapy are vital for survival, and traditional methods are often time-consuming. For instance, the early detection of bacterial DNA in the blood of critically ill patients with traditional culture diagnostic is still technically difficult (Cursons *et al.*, 1999). Real-time PCR assays have been developed for quantification of different bacteria, including *Chlamydia pneumoniae* in human atherosclerotic plaques (Ciervo *et al.*, 2003), intestinal bacterial populations (Ott *et al.*, 2004) and *Streptococcus pneumoniae* in nasopharyngeal secretions (Saukkoriipi *et al.*, 2004). *Helicobacter pylori* is considered to be the major causative agent of gastritis in acute or chronic forms and an important factor for etiology of peptic ulcer and gastric cancer. Real-time PCR technique based on the amplification of a fragment of the 23S rRNA gene has been developed (Lascols *et al.*, 2003). This system aimed at detecting *H. pylori* directly from gastric biopsy specimens, quantifying the bacterial density and testing the susceptibility of the strain to clarithromycin. Real-time PCR also proved its high sensitivity for assessment of CagA protein, which is one of the main virulence factors of *H. pylori*, as a diagnostic tool for the pathogenity of *H. pylori* infection (Yamazaki *et al.*, 2005). The detection of *H. pylori* colonization has been performed using a sensitive and specific quantitative test based on the amplification of the fragment 26-kDa *Helicobacter*-specific gene by real-time PCR (SYBR® Green I) and sensitivity of the developed system is reported as five bacterial cells per PCR sample (Mikula *et al.*, 2003)

The protection of populations against a bioterrorism act has become a major concern for the governments. For such threats, a rapid, specific and sensitive detection of the microbial agent is of paramount importance. *Yersinia pestis*, is enzootic in many countries and the causative agent for bubonic plague, can be viewed as a potential biological weapon, according to its ability to cause high rates of morbidity and mortality in humans, with fatality rates of 50–100% if untreated (Brubaker, 1991). The intentional dissemination of plague would most likely occur by aerosolization, causing fulminant pneumonia in exposed individuals (Loiez *et al.*, 2003). Out of 11 species of *Yersinia*, only three are pathogenic for humans. This may cause false identification in commercially available biochemical identification systems (Tomaso *et al.*, 2003). With recently developed real-time PCR systems for detection, identification and quantification of *Yersinia pestis*

(Tomaso *et al.*, 2003; Loiez *et al.*, 2003), it is possible to achieve sensitive and reliable results in rapid diagnosis of the pneumonic plague which is the form of plague most likely to result from a biological warfare attack.

Real-time PCR has also improved the detection of *Bacillus anthracis*, another potential biological terrorism tool. *Bacillus anthracis* is etiologic agent for zoonotic disease anthrax, and recognized as a potential agent of bioterrorism. Because of non-specific early symptoms of disease *Bacillus anthracis* infection and diagnosing anthrax in humans is difficult (Makino and Cheun, 2003). Identification of the organism in cultures, clinical specimens and environmental samples are essential for ensuring that patients infected with *B. anthracis* are treated as soon as possible (Ryu *et al.*, 2003). Therefore, detecting and monitoring anthrax spores in the environment, especially in the air, can improve prognosis due to earlier initiation of therapies and prevent spread of the infection. Routine culture and biochemical testing methods are useful for the identification of *Bacillus anthracis* but it takes 24 to 48 hours or longer (Bell *et al.*, 2002). Conventional PCR-based techniques were considered as a useful tool for detection of *Bacillus anthracis* but they had limitations. Rapid-cycle real-time PCR instruments, like LightCycler®, can be used for reliable and fast detection without carry-over contamination (Makino and Cheun, 2003; Bell *et al.*, 2002).

Examining antibiotic susceptibilities

Successful applications of real-time PCR have been developed for detection of drug-resistant bacterial strains or antimicrobial testing of bacteria (Rolain *et al.*, 2004). Examples include *Helicobacter pylori* (Lascols *et al.*, 2003), *Staphylococcus aureus* (Huletsky *et al.*, 2004), Enterecocci (Sloan *et al.*, 2004), *Enterococcus faecalis and Enterococcus faecium* (Woodford *et al.*, 2002). One of the most harmful human pathogens, *Mycobacterium tuberculosis*, may be one of the best examples for underlining the importance of rapid antimicrobial susceptibility testing with real-time PCR. *M. tuberculosis* strains resistant to anti-tuberculosis agents are being detected more frequently and emergence of these strains threatens the capability of controlling the disease worldwide (Wada *et al.*, 2004). Approximately 3.2% of the new tuberculosis cases are reported to be caused by multidrug-resistant strains of *M. tuberculosis* (Viedma *et al.*, 2003). Although the culture method still remains as the gold standard for susceptibility testing, it takes at least two weeks to get results (Wada *et al.*, 2004). New real-time PCR systems for detection and identification of multidrug-resistant *M. tuberculosis* strains bring new opportunities to clinics for effective treatment of tuberculosis (van Doorn *et al.*, 2003, Rindi *et al.*, 2003).

Detection of specific bacterial gene regions

The high sensitivity gained by real-time PCR allows detection of specific gene or gene regions. Target-specific bacterial genes can be detected even when they are present in very low concentrations, thus permitting detection and/or quantification of minor bacterial subpopulations, bacterial mutations, and drug-resistance strains (Sloan *et al.*, 2004; Costa *et al.*, 2005).

Resistance in *Mycobacterium tuberculosis* is mostly conferred by point mutations in genes coding for drug targets or drug-converting enzymes (van Doorn *et al.*, 2003; Rindi *et al.*, 2003). The catalase peroxidase gene (*katG*) is the most commonly targeted gene region for the detection of isoniazid-resistance, with the majority of mutations occurring at codon 315 in 30–90% of isoniazid-resistant strains. Genotypic real-time PCR assays that detect mutations within such regions are predictive of clinical drug resistance and have potential to provide rapid detection of resistance in bacterial isolates (Rindi *et al.*, 2003). Recently developed real-time PCR genotypic assays for detection of isoniazid resistance in *Mycobacterium tuberculosis* simply detect the mutations C (-15) T and G (-24) T in the regulatory region of the *inhA* gene (Rindi *et al.*, 2003).

13.2.3 Fungi and parasites

Definitive diagnosis of fungal infections depends on a positive blood culture or histopathological evidence of deep-tissue invasion but the sensitivity of the blood culture, especially in invasive aspergillosis, is very low. Culture techniques are also time-consuming, taking up to several weeks – an unacceptable time for the initiation of specific treatment of fungal infection. Alternative techniques to the traditional methods, such as enzyme linked immunosorbent assay detection of the galactomannan antigen in serum and bronchoalveolar lavage (BAL) fluid, and a plasma-β-D-glucan assay, have recently been introduced. Although these techniques demonstrate good sensitivity and varying degrees of specificity in high-risk patients, they may only give positive results at advanced stages of the infection (Ascioglu *et al.*, 2002). Real-time PCR technology, however, is a simple, quick, highly specific and sensitive method for the detection of infectious agents.

Besides fungal detection, quantification of the fungal burden is also important in monitoring the effectiveness of treatment. Several investigators recently applied and showed the effectiveness of real-time PCR technology. White *et al.* (2003) devised a method for rapid identification of 39 systemic infections with Candida species (within one working day) using the LightCycler® real-time PCR system. Similarly, *Aspergillus fumigatus* DNA in both BAL and blood samples from high-risk patients can be rapidly and specifically detected and quantified by a LightCycler® based real-time PCR assay (Spiess *et al.*, 2003). Using an iCyler iQ® real-time PCR system, pulmonary aspergillosis has been identified in over 90% of BAL samples of high-risk patients (Sanguinetti *et al.*, 2003). Several studies reported the application of real-time PCR in susceptibility of *Aspergillus* and Candida species to antifungal agents (Trama *et al.*, 2005; Chau *et al.*, 2004; Balashov *et al.*, 2005).

As one of the major infectious diseases in developing world, malaria has also been targeted by rapid diagnostic testing by real-time PCR methods. Lee *et al.*, (2002) developed a TaqMan® real-time PCR technique for malaria diagnosis that can be adapted for high-throughput rapid screening of hundreds of samples during an outbreak to prevent further transmission of malaria with a better sensitivity and specificity than those of the microscopic method. The method is also shown to be useful in monitoring the effectiveness of antimalarial therapy, especially in situations where drug-resistant strains of the parasites are prevalent (Lee *et al.*, 2002). Recently a single reaction real-time PCR assay has been described to detect and identify all plasmodium species in 3 hours, including standard DNA sample preparation, amplification and detection, with sensitivities equivalent to microscopy (Mangold *et al.*, 2005).

The intestinal protozoan parasite *Entamoeba histolytica* is endemic in a large part of the world and is considered responsible for millions of cases of dysentery and liver abscess each year. Traditionally, the laboratory detection of *E.histolytica* in human feces has depended on the microscopic examination of fresh stool samples. However, recently identified *Entamoeba dispar* is a nonpathogenic type with no requirement of treatment and cannot be differentiated morphologically from *E. histolytica*. In this regard, Blessmann *et al.* (2002) developed a sensitive, specific and quick real-time PCR method which can distinguish the two species. In another study using multiplex real-time PCR, simultaneous detection of *E. histolytica*, *Giardia lamblia* and *Cryptosporidium parvum* in fecal samples is successfully performed (Verweij *et al.*, 2004).

In conclusion, real-time PCR technology offers a sensitive, specific and fast detection of fungal infections as well as parasites. Routine clinical applications of the technique in near future will result in the better control and treatment of infectious diseases.

References

Al-Soud WA, Radstrom P (2001) Purification and characterisation of PCR-inhibitory components in blood cells. *J Clin Microbiol* **39**: 485–493.

Ascioglu S, Rex JH, de Pauw B, Bennett JE, Bille J, Crokaert F, Denning DW, Donnelly JP, Edwards JE, Erjavec Z, Fiere D, Lortholary O, Maertens J, Meis JF, Patterson TF, Ritter J, Selleslag D, Shah PM, Stevens DA, Walsh TJ (2002) Defining opportunistic invasive fungal infections in immunocompromised patients with cancer and hematopoietic stem cell transplants: an international consensus. *Clin Infect Dis* **1**:34(1): 7–14

Balashov SV, Gardiner R, Park S, Perlin DS (2005) Rapid, high-throughput, multiplex, real-time PCR for identification of mutations in the cyp51A gene of Aspergillus fumigatus that confer resistance to itraconazole. *J Clin Microbiol* **43**(1): 214–222.

Bell CA, Uhl JR, Hadfield TL, David JC, Meyer RF, Smith TF, Cockerill III FR (2002) Detection of *Bacillus anthracis* DNA by LightCycler PCR. *J Clin Microbiol* **40**(8): 2897–2902.

Blessmann J, Buss H, Nu PA, Dinh BT, Ngo QT, Van AL, Alla MD, Jackson TF, Ravdin JI, Tannich E (2002) Real-time PCR for detection and differentiation of Entamoeba histolytica and Entamoeba dispar in fecal samples. *J Clin Microbiol* **40**(12): 4413–4417.

Boutaga K, van Winkellhoff AJ, Vandenbroucke-Grauls CMJE, Savelkoul PHM (2003) Comparison of real-time PCR and culture for detection of *Porphyromonas gingivalis* in subgingival plaque samples. *J Clin Microbiol* **41**(11): 4950–4954.

Brubaker RR (1991) Factors promoting acute and chronic diseases caused by *Yersinia*. *Clinical Microbilogy Reviews* **4**: 309–324.

Chau AS, Mendrick CA, Sabatelli FJ, Loebenberg D, McNicholas PM (2004) Application of real-time quantitative PCR to molecular analysis of Candida albicans strains exhibiting reduced susceptibility to azoles. *Antimicrob Agents Chemother* **48**(6): 2124–2131.

Ciervo A, Petrucca A, Cassone A (2003) Identification and quantification of *Chlamydia pneumoniae* in human atherosclerotic plaques by LightCycler real-time PCR. *Molecular and Cellular Probes* **17**: 107–111.

Cockerill FR (2003) Application of RapidCycle real-time polymerase chain reaction for diagnostic testing in the clinical microbiology laboratory. *Arch Pathol Lab Med* **127**: 1112–1120.

Costa AM, Kay I, Palladino S (2005) Rapid detection of *mecA* and *nuc* genes in staphylococci by real-time multiplex polymerase chain reaction. *Diagnostic Microbiology and Infectious Disease* **51**: 13–17.

Cursons RT, Jeyerajah E, Sleigh JW (1999) The use of polymerase chain reaction to detect septicaemia in critically ill patients. *Crit Care Med* **27**: 937–940.

Huletsky A, Giroux R, Rossbach V, Gagnon M, Vaillancourt M, Bernier M, Gagnon F, Truchon K, Bastien M, Picard FJ, van Belkum A, Ouellette M, Roy PH, Bergeron MG (2004) New real-time PCR assay for rapid detection of methicillin-resistant staphylococcus aureus directly from specimens containing a mixture of staphylococci. *J Clin Microbiol* **42**(5): 1875–1884.

Jansson JK, Prosser JI (1997) Quantification of the presence and activity of specific microorganisms in nature. *Molecular Biotechnology* **7**(2): 103–120.

Lascols C, Lamarque D, Costa JM, Copie-Bergman C, Le Glaunec JM, Deforges L, Soussy CJ, Petit JC, Delchier JC, Tankovic J (2003) Fast and accurate quantitative detection of *Helicobacter pylori* and identification of Clarithromycin resistance mutations in H. pylori Isolates from gastric biopsy specimens by real-time PCR. *J Clin Microbiol* **41**(10): 4573–4577.

Lee MA, Tan CH, Aw LT, Tang CS, Singh M, Lee SH, Chia HP, Yap EP (2002) Real-time fluorescence-based PCR for detection of malaria parasites. *J Clin Microbiol* **40**(11): 4343–4345.

Livak KJ, Schmittgen TD (2001) Analysis of relative gene expression data using real-time quantitative PCR and the $2^{-\Delta\Delta Ct}$ method. *Methods* **25**: 402–408.

Loiez C, Herwegh S, Wallet F, Armand S, Guinet F, Courcol RJ (2003) Detection of Yersinia pestis in sputum by real-time PCR. *J Clin Microbiol* **41**(10): 4873–4875.

Mackay IM (2004) Real-time PCR in the microbiology laboratory. *Clin Microbiol Infect* **10**: 190–212.

Maeda H, Fujimoto C, Haruki Y, Maeda T, Kokeguchi S, Petelin M, Arai H, Tanimoto I, Nishimura F, Takashiba S (2003) Quantitative real-time PCR using TaqMan and SYBR Green for Actinobacillus actinomycetemcomitans, Porphyromonas gingivalis, Prevotella intermedia, tetQ gene and total bacteria. *FEMS Immunol Med Microbiol* **39**(1): 81–86.

Makino S, Cheun H (2003) Application of the real-time PCR for the detection of airborne microbial pathogens in reference to the anthrax spores. *J Microbiol Methods* **53**: 141–147.

Mangold KA, Manson RU, Koay ES, Stephens L, Regner M, Thomson RB Jr, Peterson LR, Kaul KL (2005) Real-time PCR for detection and identification of Plasmodium spp. *J Clin Microbiol* **43**(5): 2435–2440.

Mikula M, Dzwonek A, Jagusztyn-Krynicka K, Ostrowski J (2003) Quantitative detection for low levels of *Helicobacter pylori* infection in experimentally infected mice by real-time PCR. *J Microbiol Methods* **55**(2): 351–359.

Millar BC, Moore JE (2004) Molecular diagnostics. current options. FROM: methods in molecular biology. In: 266: Genomics, proteomics, and clinical bacteriology: methods and reviews. Woodford N, Johnson A (eds) Humana Press Inc., Totowa, NJ.

Monteiro L, Bonnemaison D, Vekris A, Petry KG, Bonnet J, Vidal R, Cabrita J, Megraud F (1997) Complex Polysaccharides as PCR Inhibitors in Feces: Helycobacter pylori model. *J Clin Microbiol* **35**(4): 995–998.

Nolte FS (2004) Novel internal controls for real-time PCR assays. *Clinical Chemistry* **50**(5): 801–802.

Ott SJ, Musfeldt M, Ullmann U, Hampe J, Schreiber S (2004) Quantification of intestinal bacterial populations by real-time PCR with a universal primer set to minor groove binder probes: a global approach to the enteric flora. *J Clin Microbiol* **42**(6): 2566–2572.

Pahl A, Kuhlbrandt U, Brune K, Rollinghoff M, Gessner A (1999) Quantitative detection of *Borrelia burgdorferi* by real-time PCR. *J Clin Microbiol* **37**(6): 1958–1963.

Rindi L, Bianchi L, Tortoli E, Lari N, Bonanni D, Garzelli G (2003) A real-time PCR assay for detection of isoniazid resistance in *Mycobacterium tuberculosis* clinical isolates. *Journal of Microbiological Methods* **55**: 797–800.

Rolain JM, Mallet MN, Fournier PE, Raoult D (2004) Real-time PCR for universal antibiotic susceptibility testing. *Journal of Antimicrobial Chemotherapy* **54**: 538–541.

Ryu C, Lee K, Yoo C, Seong WK, Oh HB (2003) Sensitive and rapid quantitative detection of anthrax spores isolated from soil samples by real-time PCR. *Microbiol Immunol* **47**(10): 693–699.

Sanguinetti M, Posteraro B, Pagano L, Pagliari G, Fianchi L, Mele L, La Sorda M, Franco A, Fadda G. Comparison of real-time PCR, conventional PCR, and galactomannan antigen detection by enzyme-linked immunosorbent assay using bronchoalveolar lavage fluid samples from hematology patients for diagnosis of invasive pulmonary aspergillosis. *J Clin Microbiol* **41**(8): 3922–3925.

Saukkoriipi A, Leskela K, Herva E, Leinonen M (2004) *Streptococcus pneumoniae* in nasopharyngeal secretions of healthy children:comparison of real-time PCR and culture from STGG-transport medium. *Molecular and Cellular Probes* **18**: 147–153.

Sintchenko V, Iredell JR, Gilbert GL (1999) Point of view: is it time to replace the petri dish with PCR? Application of culture-independent nucleic acid amplification in diagnostic bacteriology: Expectations and reality. *Pathology* **31**: 436–439.

Sloan LM, Uhl JR, Vetter EA, Schleck CD, Harmsen WS, Manahan J, Thompson RL, Rosenblatt JE, Cockerill III FR (2004) Comparison of the LightCycler van A/van B detection assay and culture for detection of vancomycin-resistant entereococci from perianal swabs. *J Clin Microbiol* **42**(6): 2636–2643.

Spiess B, Buchheidt D, Baust C, Skladny H, Seifarth W, Zeilfelder u, Leib-Mosch C, Morz H, Hehlmann R (2003) Development of a LightCycler PCR assays for detection and quantification of *Aspergillus fumigatus* DNA in clinical samples from neutropenic patients *J Clin Microbiol* **41**(5): 1811–1818.

Tomaso H, Reisinger EC, Al Dahouk S, Frangoulidis D, Rakin A, Landt O, Neubauer H (2003) Rapid detection of Yersinia pestis with multiplex real-time PCR assays using fluorescent hybridisation probes. *FEMS Immunol Med Microbiol* **38**(2): 117–126.

Trama JP, Mordechai E, Adelson ME (2005) Detection of Aspergillus fumigatus and a mutation that confers reduced susceptibility to itraconazole and posaconazole by real-time PCR and pyrosequencing. *J Clin Microbiol* **43**(2): 906–908.

van Doorn HR, Claas ECJ, Templeton KE, van der Zanden AGM, Vije AK, Jong MD, Dankert J, Kuijper EJ (2003) Detection of a point mutation associated with high-level isoniazid resistance in *Mycobacterium tuberculosis* by using real-time PCR technology with 3'-minor groove Binder-DNA probes. *J Clin Microbiol* **41**(10): 4630–4635.

Verweij JJ, Blange RA, Templeton K, Schinkel J, Brienen EA, van Rooyen MA, van Lieshout L, Polderman AM (2004) Simultaneous detection of *Entamoeba histolytica*, *Giardia lamblia*, and *Cryptosporidium parvum* in fecal samples by using multiplex real-time PCR. *J Clin Microbiol* **42**(3): 1220–1223.

Viedma G (2003) Rapid detection of resistance in *Mycobacterium tuberculosis*: a review discussing molecular approaches. *Clinical Microbiol Infect* **9**: 349–359.

Wada T, Maeda S, Tamaru A, Imai S, Hase A, Kobayashi K (2004) Dual-Probe assay for rapid detection of drug-resistant *Mycobacterium tuberculosis* by real-time PCR. *J Clin Microbiol* **42**(11): 5277–5285.

White PL, Shetty A, Barnes RA (2003) Detection of seven Candida species using the Light-Cycler system. *J Med Microbiol* **52**(Pt 3): 229–238.

Wittwer CT, Kusukawa N (2004) Real-time PCR. In: Persing DH, Tenover FC, Versalovic J, Tang YW, Unger ER, Relman DA, White TJ (eds). Diagnostic molecular microbiology; principles and applications. ASM Press, Washington, DC, 71–84.

Woodford N, Tysall L, Auckland C, Stockdale MW, Lawson AJ, Walker RA, Livermore DM (2002) Detection of oxazolidinone-resistant *Enterococcus faecalis* and *Enterococcus faecium* strains by real-time PCR and PCR-restriction fragment length polymorphism analysis. *J Clin Microbiol* **40**(11): 4298–4300.

Yamazaki S, Kato S, Matsukura N, Ohtani M, Ito Y, Suto H, Yamazaki Y, Yamakawa A, Tokudome S, Higashi H, Hatakeyama M, Azuma T (2005) Identification of *Helicobacter pylori* and the *cagA* genotype in gastric biopsies using highly sensitive real-time PCR as a new diagnostic tool. *FEMS Immunology and Medical Microbiology* **44**(3): 261–268.

Protocol 13.1: Detection of *Prevotella intermedia* by real-time PCR assay

Sample preparation

After removal of supragingival plaque above the gingival line, isolate the sampling sites with sterile cotton roll, and then air-dry.

Remove the subgingival plaque samples from the deepest pockets from the same patients with individual sterile Gracey curettes.

Immerse the plaque sample in 1.0 ml PBS (-) (GibcoBRL, MD, USA), and vortex for 30 sec, followed by centrifugation for 20 min at 15,000 g. Remove the supernatant.

Extract the bacterial DNA from cultivated strains and clinical plaque samples by using InstaGene® Matrix (Bio-Rad Lab., CA, USA), according to manufacturer's instructions.

Oligonucleotides

For the identification of unique bacterium, primers designed from the species-specific region on the 16S rDNA are used, based on their known 16S rDNA sequences (GenBank). The sequences of primers for real-time PCR can be designed by using Primer Express® (Applied Biosystems) and OLIGO Primer Analysis Software (Version 4.0, Molecular Biology Insights, Inc.).

One set of primer sequences for amplification of *Prevotella intermedia* are as follows:

Forward	5'-AATACCCGATGTTGTCCACA -3'
Reverse	5'-TTAGCCGGTCCTTATTCGAA -3'

For specificity testing of primers, extracted DNAs from 10^7 cells of *Prevotella intermedia* can be examined as templates with the primers. PCR amplifications should be observed when the target strains are used as templates. There should be no increase in the fluorescence with the mismatch strains. Melting curves of all PCR products should show sharp peaks at the expected Tm of the products. These preliminary results indicate that each real-time PCR specifically amplified the target DNA.

Reaction mixture

The total volume used in this protocol is 50 μl, however if necessary, each reaction can be made in a total volume of 25 μl maintaining the correct concentrations of reagents:

2 × SYBR® Green Master Mix	26.0 µL
Forward primer (10 µM)	2.0 µL
Reverse primer (10 µM)	2.0 µL
Molecular biology grade H$_2$O	15.0 µL
Specimen or control nucleic acid extract	5.0 µL
Total	50.0 µL

Carefully pipette the reaction mixture into wells of MicroAmp® Optical 96-well Reaction Plate (Applied Biosystems). Use 10^7–10^2 cell dilutions of *P. intermedia* to make a standard curve in each individual reaction plate and use 5 µL of molecular biology grade H$_2$O as negative control. Add 5 µL DNA in each sample well.

Place the reaction plate into GeneAmp® 5700 Sequence Detection System (PE Applied Biosystems).

Cycling conditions

Thermocycling program for this pair of primers consists of 40 cycles of 95°C for 15s and 60°C for 1 min with an initial cycle of 95°C for 1 min. Accumulations of the PCR products are detected in each cycle by monitoring the increase in fluorescence of the reporter dye double-strand DNA-binding SYBR® Green.

Analysis

Dissociation curve (melting curve, Tm) analysis should be performed within the temperature range of 60°C to 95°C to detect specific amplification of the target sequence. GeneAmp® 5700 SDS software (PE Applied Biosystems) can be used for melting curve analysis. Details of melting analysis are provided in Chapter 9.

Clinical virology

David M. Whiley and Theo P. Sloots

14

14.1 Introduction

There has been a marked improvement in the diagnosis of viral infections over the last decade with the application of the polymerase chain reaction (PCR), and more recently real-time PCR. The employment of real-time PCR for virus detection offers the advantage of high sensitivity and reproducibility combined with an extremely broad dynamic range. However, the greatest impact of real-time PCR has undoubtedly been its ability to quantify the viral load in clinical specimens. The results can be expressed in absolute terms with reference to quantified standards, or in relative terms compared to another target sequence present in the sample. In addition, quantitative PCR (qPCR) tests offer the possibility to determine the dynamics of viral proliferation, monitor the response to treatment, and distinguish between latent and active infections.

Many qPCR protocols for the detection of viral targets have now been published and commercial assays are available for a number of clinically important viruses, including human immunodeficiency virus (Kumar *et al.*, 2002; O'Doherty *et al.*, 2002), hepatitis B and C viruses (Berger *et al.*, 1998), and cytomegalovirus (Boeckh and Boivin, 1998). However, the number of commercial assays available is still limited and this has led to the development and introduction of in-house real-time qPCR protocols.

Current approaches to the development of viral real-time PCR assays commonly employ fluorescent chemistry to effect the kinetic measurement of product accumulation. These may include non-specific compounds such as the DNA intercalating dye SYBR® Green or sequence specific oligoprobes which employ fluorescent resonance energy transfer (FRET). In practice, our experience has been that the specific approach has yielded better results for the detection of viral pathogens in clinical specimens, and particularly so in the determination of viral loads.

Recently, technical improvements in sequence detection systems have enabled extensive characterization of the viral genome, including determination of viral subtype, genotype, variants mutants and genotypic resistance patterns. However, such characterization has demonstrated that many regions of the viral genome are heterogeneous, displaying sequence variation not only between genera, but also variants of the same virus isolated at a single geographic location. Such variation has profound implications for the design of sensitive and specific real-time PCR protocols, particularly for RNA viruses where genomic polymorphism is widespread.

Also, in order to obtain meaningful results, it is important that the efficiency of the PCR does not vary greatly due to minor differences between samples, and careful optimization of the PCR conditions are required to obtain consistent results. This is particularly important in the design and performance of qPCR. Even so, other factors, inherent to the target virus, also can affect the results of qPCR. Recent studies in our laboratory have demonstrated that sequence variation in the primer or probe target sites has a significant and major impact of the validity of qPCR results in determining viral load (Whiley and Sloots, 2005b). This chapter will examine the effect of sequence variation of primer and probe target sites on assay performance, and its effect on quantification of viral loads.

14.2 Qualitative real-time PCR for viral disease

14.2.1 Sequence variation and assay performance

It must be recognized that the success of any real-time PCR assay is highly dependent on the nature of the primer and probe target sequences utilized in the assay. These sequences need to be specific for the target virus and must be well conserved across different strains or types of the virus, otherwise the sensitivity of the assay may be compromised. However, for many viruses it can be difficult to identify such sequences. There are several reasons for this, including sequence polymorphism of the viral genome, a lack of sequence information as well as perceived changes in the clinical relevance of certain viral subtypes. Parainfluenza virus type 3, respiratory syncytial virus, human metapneumovirus, and influenza virus A provide good examples of this phenomenon.

Sequence polymorphism and false-negative results

Sequence polymorphism occurs throughout the genomes of many viruses, and any one viral species can comprise multiple genotypes exhibiting considerable sequence variation. Furthermore, within any single patient population or between different populations different strains of a viral species may circulate. If polymorphism occurs in the primer or probe target sites on the viral genome, then the PCR assay may have a lower sensitivity in its ability to detect these strains or at worse fail to detect the virus strain altogether.

Recently, we compared a 5′ nuclease assay targeting the parainfluenza 3 nucleocapsid protein gene (WhSl-para3–5N; unpublished data) with a previously described 5′ nuclease assay targeting the parainfluenza 3 hemagglutinin-neuraminidase gene (HN-para3–5N; Watzinger et al., 2004) for the detection of parainfluenza 3 in nasopharyngeal aspirate specimens collected from a local patient population. During the course of the study, 34 positive samples were identified with both assays detecting 33 of these. The cycle threshold (C_t) values for these 33 positive samples were comparable in both assays, with an average difference of only 1.1 cycles. However, one sample was negative by the WhSl-para3–5N assay but was positive by the HN-para3–5N assay, providing a C_t value of 25. Such a low C_t value indicates that the viral load was relatively high in this specimen and therefore the false-negative result could not be explained by the differences in the detection limit of each assay. The presence of parainfluenza 3 in this specimen was confirmed by a direct fluorescence assay (DFA), yet the specimen consistently provided negative results in the WhSl-para3–5N assay. Sequencing of the viral genome that was targeted by the TaqMan® probe showed four mismatches. These were almost certainly responsible for the false-negative result (see *Troubleshooting guide 14.1*).

The WhSl-para3–5N assay was designed using the parainfluenza 3 sequences for the nucleocapsid protein gene currently available on the Genbank sequence database. Based on this information, the primer and probe targets appeared to be well conserved and certainly the mismatches identified above were not represented in Genbank sequences. Quite simply, this demonstrates that the use of sequence information from public databases has limitations for viral assay design. Furthermore, the above results highlight that sequence polymorphism between viral strains circulating in a population can lead to false-negative results in PCR assays, and that these results may occur irrespective of viral load. Overall, our experience indicates that sequence variation is a more widespread problem for the design and performance of viral PCR assays. However, the occurrence of false-negative results due to sequence variation can usually be identified by the use of a second assay and confirmed by gene sequencing.

Sequence polymorphism and fluorescent response

The impact of sequence variation upon real-time PCR assays is not simply restricted to producing false-negative results, but can have more subtle effects, including the reduction of the overall fluorescent signal produced in an assay. Recently, we developed a real-time PCR for the detection of respiratory syncytial virus (RSV) using a 3' minor groove binder (MGB) TaqMan® probe. These MGB probes are particularly suited for the detection of single nucleotide polymorphisms (Kutyavin et al., 2000). However, they are also quite useful where only limited target sequence is available, given their shorter length compared to standard DNA TaqMan® probes. It is for the latter reason that we designed a MGB probe or the detection of respiratory syncytial virus sequences. This RSV MGB assay used RSV L-gene sequences available on the Genbank public database. Based on these sequences, two primers (forward AGTAGACCATGTGAATTCCCTGC; reverse GTCGATATCTTCATCACCATACTTTTCTGTTA) and one MGB probe (FAM-TCAATACCAGCTTATAGAAC-MGB-BHQ1) were designed.

During the initial evaluation of the assay, significant differences were observed between the linear amplification curves produced by positive specimens. In particular, the fluorescent signal for some positive specimens deviated only slightly from the baseline fluorescence (see *Figure 14.2*). Sequence analysis of the amplification products from several of these positives samples showing varying fluorescent signal revealed that mismatches with the MGB probe were responsible for the differences in fluorescence (see *Troubleshooting guide 14.2*). Briefly, no mismatches between the target and MGB probe gave an optimal fluorescent signal, whereas a single mismatch significantly lowered the fluorescent signal, the extent of which was dependent on where the mismatch occurred. Notably, specimens providing the lowest fluorecent signal had a mismatch nearer the 3' end of the MGB probe (Whiley and Sloots, 2006.)

These results highlight the usefulness of MGB TaqMan® probes for the detection of mismatches. However, they suggest that MGB TaqMan® probes may have limited utility for viral diagnostics, particularly where the probes are used to screen for uncharacterized viral strains. It should be noted that we have been using standard DNA TaqMan® probes for many of our respiratory virus assays and have not observed similar differences in fluorescent signal. As a result, our laboratory now preferentially adopts standard DNA TaqMan® probes, rather than MGB TaqMan® probes, for routine viral screening.

Limited sequence information and assay development

The issues concerning sequence variation can be further complicated by the fact that sequence information on public databases may comprise only a limited number of viral strains usually from discrete geographic locations. Therefore, the development of PCR assays based on such limited information may affect the sensitivity of assay performance. This can be most pronounced for newly detected viruses. Following the initial characterisation of human metapneumovirus (van den Hoogen et al., 2001), our laboratory developed a PCR assay to detect this virus in clinical specimens. This assay used a 5' nuclease probe targeting the nucleocapsid protein gene (hMPV-N-5N; Mackay et al., 2003), and was developed using the limited sequence data present on Genbank at the time. However, subsequent studies revealed that human metapneumovirus comprised at least four genetic lineages (A1, A2, B1 and B2) with considerable sequence variation between these lineages (van den Hoogen et al., 2004). As a result we found that the hMPV-N-5N assay failed to detect some human metapneumovirus strains of the B lineage. Sequencing revealed that mismatches were present between the primers and probe and the nucleocapsid gene sequences of some human metapneumovirus B viral strains, and that these were responsible for the limitations of the hMPV-N-5N assay. Subsequently another hMPV 5' nuclease assay targeting the human metapneumovirus nucleocapsid protein gene was described by Maertzdorf et al., (2004), and reportedly was capable of detecting human metapneumovirus from all known genetic lineages. We have since used this assay for the detection of human metapneumovirus lineage B strains in our local patient population. Overall, the initial limitations of our hMPV-N-5N assay was entirely due to the fact that the assay was designed using limited sequence data available on Genbank at the time, and predominantly comprised human metapneumovirus type A sequences.

The changing clinical significance of influenza A subtypes

In recent years, influenza type A has provided a good example of the challenges confronting real-time PCR detection of viral agents comprising genetically heterogeneous populations. Influenza A exists as multiple subtypes showing considerable sequence variation between and within subtypes. However, the problems concerning the PCR detection of influenza A are not due to a lack of available sequence information. Genbank, for example, contains an extensive amount of sequence data for numerous genes from a wide range of viral subtypes representing various regions around the world.

Rather, the recent problems concerning PCR detection of influenza A have involved a change in the genetic nature of the viral subtypes circulating in a population. Following the Hong Kong pandemic of 1968, the influenza A types infecting humans have remained fairly stable with the common subtypes causing disease including H1N1, H1N2, H2N2 and H3N2. As a consequence, numerous PCR assays have been developed to ensure the detection of these subtypes. However, in the late 1990s avian strains of influenza were shown to infect humans (Hammel and Chiang, 2005), and several incidents of infection were reported, including H5N1, H7N2, H7N3, H7N7, and H9N2, in countries including Canada, the United States, the Netherlands as well as numerous Asian countries. Established PCR protocols were unable to detect these avian strains due to the genetic heterogeneity of their genomes, and, consequently, there was a growing need to review diagnostic capabilities for influenza A in many countries.

These limitations of the established PCR protocols were addressed by the development of a conventional PCR assay capable of detecting influenza A viruses from multiple animal species (Fouchier *et al.*, 2000). More recently, our laboratory developed a 5′ nuclease real-time RT-PCR (WhSl-FluA-5N; *protocol 1*) for the detection of both the common human influenza A subtypes as well as the avian subtypes now known to infect humans. To validate our WhSl-FluA-5N assay, it was compared to a previously described influenza A 5′ nuclease RT-PCR assay (FluA-TM; van Elden *et al.*, 2001), which also targeted the influenza A matrix protein gene. Overall, the results showed good agreement between the two assays when applied to the detection of influenza type A in nasopharyngeal aspirate samples obtained from patients in our local Australian population. However, the WhSl-FluA-5N assay could detect a greater number of influenza A subtypes than the FluA-TM.

Briefly, nucleic acids were extracted from 23 influenza type A cultures comprising subtypes H1N1 (6 strains), H3N2 (12), H5N1 (2), H5N3 (1), H7N7 (1), and H9N2 (1), and were tested by both assays. All H1N1 and H3N2 strains were detected by both assays, whereas, the H5N1, H5N3, H7N7, and H9N2 strains were positive by the WhSl-FluA-5N assay only (Whiley and Sloots, submitted for publication). It should be noted that the FluA-TM assay has been used successfully in our laboratory for the detection of influenza A for a number of years, and only recently have we adopted the WhSl-FluA-5N assay given the potential introduction of avian strains into our population.

On the whole, the above results highlight that the range of influenza A subtypes affecting humans can change and that, given the sequence variation existing between such subtypes, laboratory diagnostic assays need to be reviewed as new subtypes infecting humans are identified. Potentially, the use of the WhSl-FluA-5N assay may need to be reassessed if other influenza A subtypes are identified or if genomic sequence variation of local strains leads to mismatches in the primer or probe target sites.

Additional comments

It should be noted that the successful validation of a PCR assay against specimens collected from any one patient population at a particular time may give only limited indication to the performance of that assay in another patient population, or even on the same patient population at a different time. Essentially this is due to differences and mutation of circulating viral strains in those populations, and highlights the need for assay validation on virus populations that represent local virus types and subtypes. Following validation there is then a clear need for ongoing quality control of assay performance.

Many of the limitations due to sequence variation can often be complicated by a lack of sequence information for local viral subtypes during assay development. This may be overcome by laboratories conducting their own sequencing studies, but this increases assay development costs considerably, as a broad range of viral strains need to be sequenced to provide adequate information. In general, we use sequence information on public databases as a tentative guide only and develop two PCR assays, targeting different genes, which are then evaluated in parallel.

Although these issues highlight some notable pitfalls concerning the application of molecular assays in viral diagnostics, they detract from the enormous potential of PCR over traditional techniques, such as DFA and virus isolation by cell culture. The major benefit of PCR is its potential for greater sensitivity and specificity compared to the traditional methods. We believe there is a need for acceptance that some viral strains may be missed by PCR in the same manner that the traditional methods may fail to detect low viral loads or non-viable organisms. Nonetheless, the problems concerning strain variation may be overcome by using two PCR assays with different gene targets in parallel, although this would inevitably increase the cost of testing.

14.3 Virus typing using sequence-specific probes

An additional advantage of real-time PCR technology in viral diagnostics is its ability to detect and characterize viral types simultaneously within a single reaction, thereby further reducing cost and hands-on time. In our laboratory, two formats have been adopted for virus typing. These are consensus hybridization probes combined with melting curve analysis, and type-specific TaqMan® probes labeled with different fluorophores.

14.3.1 Hybridization probes

The use of hybridization probes with melting curve analysis provides significant benefits for diagnostic virology. The approach adds an extra element of specificity to any viral PCR assay. This is because mismatches with the probes will decrease the melting temperature of the hybridization complex formed between the probes and the PCR product during melting curve analysis. The extent of this decrease is dependent on the number and position of the mismatches with the probes. This may serve as a quality control mechanism, where shifts in observed melting temperature from that of expected melting temperature are indicative of sequence variation within the target and may warrant further investigation of the assay performance. On the other hand, such shifts in melting temperature can be advantageous, enabling characterization and typing of two closely related viruses. Briefly, this involves designing a set of hybridization probes capable of hybridizing to both viral types, but having a limited and defined number of mismatches with the DNA of one type and no mismatches with the other type. It is these mismatches and subsequent difference in melting temperature that then enable the two viruses to be distinguished during melting curve analysis.

To date, we have used hybridization probe protocols coupled with melting curve analysis for the routine detection and differentiation of herpes simplex virus (HSV) types 1 and 2 and human polyomaviruses JC and BK (JCV and BKV). Overall, the assays have high clinical sensitivity and specificity. However, two limitations have become evident. First, sequence variation within the viral types can interfere with virus characterization, and second, the assays can fail to detect one viral type when both are present in a single sample (Whiley and Sloots, 2005a).

The effect of sequence variation in hybridization probe target sites

The herpes simplex virus PCR assay initially used in our laboratory was the LightCycler® hybridization probe method originally described by Espy *et al.*, (2000), using an asymmetric primer concentration for the improved resolution of melting temperatures (*Protocol 2*). Briefly, the assay targets the herpes simplex virus DNA polymerase gene and works on the basis that both hybridization probes are complementary to target sequences on the herpes simplex virus type 2 DNA genome. However, one of the probes has two mismatches with herpes simplex virus type 1 DNA target sequences, thereby theoretically lowering the melting temperature for herpes simplex virus type 1 by approximately 7°C. In applying this method we found that the majority of our local herpes simplex virus strains can be differentiated as herpes simplex virus 1 or 2, but that approximately 5% of herpes simplex virus positive specimens produce melting temperatures that are not characteristic of either the type 1 or 2 melting curves. Sequence analysis of the PCR products from several of these problematic samples revealed unexpected sequence variation in the probe binding sites, which was responsible for the uncharacteristic melting temperatures. In particular, a range of melting temperatures was observed for herpes simplex virus type 2 strains circulating in our local population, all attributable to polymorphism of the probe binding sites. (see *Figure 14.3* and *Troubleshooting guide 14.3*; Whiley *et al.*, 2004)

To circumvent the problems associated with sequence variation of our local viral strains, our laboratory developed a herpes simplex virus hybridization probe (LightCycler®) assay targeting the glycoprotein D gene (Whiley *et al.*, 2004). The assay proved to be more successful for characterizing our local herpes simplex virus strains, yet recently variations in melting temperature have again been observed using this assay in a small number of herpes simplex virus positive samples (unpublished data). Overall, this suggests that the use of hybridization probes for herpes simplex virus genotyping could be inherently problematic due to genomic sequence variation between viral strains. This is supported by other studies (Anderson *et al.*, 2003; Smith *et al.*, 2004). Smith *et al.*, (2004) has emphasized that polymorphism occurs throughout the herpes simplex virus genome and has recommended that at least two real-time PCR assays be used for accurate typing of herpes simplex virus strains.

Interestingly, our laboratory has used a polyomavirus hybridization probe assay for a number of years, yet no shifts in melting temperature have been observed for our local JCV or BKV strains. However, the number of herpes simplex virus -positive samples detected greatly exceeds those for polyomavirus, and similar sequence variation may yet be detected in the future, as sequencing studies have revealed that polymorphism does occur near the primer and probe target sites of the assay in our locally circulating polyomavirus strains. Therefore, the fact that melting shifts have not yet been observed in the polyomavirus assay may merely be a statistical anomaly (Whiley and Sloots, 2005a).

The presence of two viral types

Another serious limitation of melting curve analysis is the potential failure to detect two viral types when both are present in a single specimen. This was observed in the polyomavirus hybridization probe assay when the results were compared to those from a conventional PCR using JCV and BKV specific probes (Whiley et al., 2001). In the validation process, both assays provided 100% agreement in detecting the presence of polyomaviruses in urine samples. However, the hybridization probe assay failed to detect the presence of both JCV and BKV in three of five specimens containing both viral types. Our subsequent results show that the relative viral loads of JCV and BKV in a single sample need only differ by one log for the assay to fail in the detection of both viruses (Whiley and Sloots, 2005a).

Similar observations were also made for specimens collected from patients dually infected with HSV 1 and HSV 2. Recently, we compared the HSV hybridization probe-based assay, targeting the HSV glycoprotein D gene (Whiley et al., 2004), with a previously described nuclease probe (TaqMan®) duplex real-time PCR assay, utilizing HSV type-specific primers and probes also targeting the glycoprotein D gene (Weidmann et al., 2003). Briefly, this investigation tested 10-fold dilutions of HSV type 2 spiked with HSV type 1 DNA (see Table 14.1). Overall, the duplex TaqMan® assay provided significantly

Table 14.1 Ten-fold dilutions of HSV type 2 DNA, spiked with HSV type 1 DNA, tested in the D glycoprotein gene hybridization probe and duplex TaqMan® assays.

Dilutions	1	2	3	4	5	6
Copies of HSV type 2	10^7	10^6	10^5	10^4	10^3	10^2
Copies of HSV type 1	10^4	10^4	10^4	10^4	10^4	10^4
Results:						
Hybridization probe assay						
HSV type 2	POS	POS	POS	POS	POS	neg
HSV type 1	neg	neg	POS	POS	POS	POS
Duplex TaqMan® assay						
HSV type 2	POS	POS	POS	POS	POS	POS
HSV type 1	POS	POS	POS	POS	POS	POS

neg = negative, POS = positive

better discriminatory power, detecting the presence of both viruses in all dilutions. The results of the HSV hybridisation probe assay largely reflected those previously observed for the polyomavirus assay, and in three of six specimens, only the HSV type at the higher concentration was detected whereas both types were detected where the relative loads were within one log.

14.3.2 Additional comments

Failure to detect the presence of certain viral types can have a significant impact on patient management. For example, in transplant patients early detection of BK virus (BKV) reactivation is important as this is often the precursor to BKV associated disease, including hemorrhagic cystitis. Notably JCV and BKV are often shed simultaneously in human urine. Therefore, the presence of JCV strains or otherwise sequence polymorphism of the target binding sites could confound the detection of BKV in the urine of transplant patients, reducing the potential to predict the onset of disease.

Co-infections of HSV types 1 and 2 are less common and the failure to detect one of these viruses is arguably of little importance to clinical management and treatment. Nonetheless, incorrect typing of HSV can have significant psychological and social implications, even though the clinical management of the patient is unaffected.

These issues highlight the limitations of hybridization probes used in melting curve analysis for viral characterization in a clinical setting. If the accurate identification of a specific virus type carries clinical importance then careful consideration must be given to the detection methods used in the assay. Preferably, such formats should not rely on consensus oligonucleotide sequences but use type-specific primers and probes.

14.4   Quantification of viral load

Quantitative PCR analysis of viral nucleic acid is now used by diagnostic laboratories worldwide and is particularly useful for monitoring viral loads in patients to assess the effect of anti-viral therapy and potentially to detect drug-resistant viral strains. In clinical virology, qPCR is commonly used for the detection and quantification of blood-borne viruses, including hepatitis B and C and human immunodeficiency viruses. However, it is also increasingly used to monitor viral pathogens of transplant patients, including Epstein–Barr virus, cytomegalovirus and BK virus. In recent years, our laboratory has used quantitative real-time PCR to investigate BK viral loads in a transplant patient population (see *Protocol 14.3*). BKV can cause several clinical manifestations in immunocompromised patients including hemorrhagic cystitis and allograft nephropathy. Since adopting qPCR, several limitations of this technology have been identified, including the impact of PCR inhibitors, poor nucleic acid extraction and the effect of sequence variation in PCR primer target sites. Each one of these will impact upon the accuracy of the qPCR results.

14.4.1 The use of an internal control in clinical molecular virology

As discussed in previous chapters, the generation of accurate qPCR results is dependent on the strict control and standardization of many assay processes. In qPCR applied to clinical virology, the main sources of error include the nucleic acid isolation process and the presence of PCR inhibitors in clinical specimens. In our opinion, the best way to control for these factors is by using an internal control strategy previously described by Niesters (2004). Briefly, this involves spiking specimens and standards prior to extraction with a known concentration of an unrelated virus, in this case seal herpesvirus (PhHV-1), which is then extracted with the test specimen. The specimen extracts are subsequently each tested by a PhHV-1 TaqMan® assay and the C_t values compared to a predetermined range of C_t values. Specimens providing C_t values for PhHV-1 that are outside this expected range are then repeated (*Protocol 14.4*). The benefit of this approach is that it simultaneously controls for both the presence of PCR inhibitors as well as inefficiencies in the extraction protocol. Further, the extraction of the seal herpesvirus is more likely to mimic the extraction of the target virus of interest, than, for instance, the use of a plasmid as an internal control. This procedure is suitable for both quantitative and qualitative viral PCR applications and can also be used for quality control of new reagents or test procedures. Almost any unrelated virus may be used in a similar manner, so long as the virus is unlikely to be found in clinical samples, does not cross-react with any of the primers and probes used in the clinical assays, and preferably is not harmful to humans. The disadvantages of using such a strategy are the increased costs and extra processing time of running the internal control assay in addition to the test assay. Nonetheless, with increasing regulation and accreditation of *in-vitro* diagnostic devices such protocols should be considered as best practice.

14.4.2 Impact of target sequence variation on qPCR

Again, sequence variation may also have a significant impact upon the accuracy of viral qPCR assays. Recently, we examined the impact of sequence mismatches in the primer target sites on qPCR by constructing a number of primers incorporating various mismatches with a target sequence on the BKV T-antigen gene. Essentially these primers were modifications of the primers used in the BKV-Tag-qPCR assay described previously (see *Protocol 14.3*). The results showed that as few as two mismatches in the 3′ end of a single primer could introduce considerable error, underestimating viral load by up to 3 logs (Whiley and Sloots, 2005b). The overall impact this has on qPCR results is dependent on the conditions used in the assay and may hinge on the nucleotide composition of the primers, the PCR annealing temperature, the reaction mix composition and the specific nucleotides that are the subject of mismatch.

The concern of sequence variation and its impact on qPCR led us to examine a second BKV qPCR assay (BK-qLC) previously developed in our laboratory. This assay was based on the original polyomavirus hybridization probe assay, described above, using consensus primers and probes to detect both human polyomaviruses, JCV and BKV, followed by melting curve differentiation (Whiley *et al.*, 2001). To make the assay specific for BKV, the consensus forward primer (PoL1s; Whiley *et al.*, 2001) was replaced with a BKV specific forward primer. This new assay methodology (BK-qLC assay) was then validated against the BKV-Tag-qPCR assay (see *Protocol 14.3*) for the detection of BKV in urine samples obtained from transplant patients. The results showed good agreement between the two assays, with the detection of 19 BKV positive samples by both methods. In addition, there was good agreement between the C_t values and estimated viral loads for all but one of these 19 positive specimens. For 18 of these the average difference in C_t values between the two assays was only 1.0 cycles and differences in viral load were within one log. However, one positive specimen gave a C_t value of 30.5 in the BK-qLC and 20.7 in the BKV-Tag-qPCR assay. This difference of 9.8 cycles represented a difference in viral load of three logs and therefore was a significant discrepancy between the two assays. Sequencing the PCR product obtained in the BK-qLC from this specimen revealed a single base mismatch of the nucleotide targeted by the extreme 3' end of the forward primer (see *Troubleshooting guide 14.4*).

Overall, these results show that sequence variation in the primer binding sites can be the source of significant error in the estimation of viral load using qPCR even when other sources of error are standardized. This has considerable implications for viral quantification studies, particularly those targeting viruses that exhibit extensive sequence variation between strains and subtypes, such as the RNA viruses. Human metapneumovirus, respiratory syncytial virus and adenoviruses are just a few examples where sequence heterogeneity has been demonstrated. In fact, by performing real-time PCR assays targeting different genes in parallel for each of these viruses, we have found significant discrepancies between C_t values for positive specimens (data not published), suggesting that the impact of sequence variation on qPCR may be significant.

Overall, these results show that the impact of sequence variation is quantitative as well as qualitative. Interestingly, the BK-qLC performed well as a qualitative assay in the initial evaluation, demonstrating that a good qualitative PCR assay may not necessarily be suitable for use as a quantitative assay. Clearly, a different procedure must be followed to validate a quantitative assay to that employed for a qualitative assay. Such validation may involve sequencing the primer and probe target sites and determining if mismatches occur. Nevertheless, before accepting a viral qPCR protocol, we recommend performing two separate assays in parallel, validating one against the other, and comparing the C_t values to identify any discrepancies in performance. Not withstanding the above issues, we have found the BKV-Tag-qPCR assay to be reliable for quantifying BKV DNA and to date, have not found sequence variation in the primer or probe target sites of any local BKV strains.

14.4.3 Additional comments

There are many other factors concerning viral qPCR standards that should be considered. First, the use of plasmids as external standards has been criticized, because, although these may be well characterized, little may be known about the composition of the clinical samples being analyzed. Thus, the use of plasmids may erroneously assume comparable amplification characteristics to the viral target in test samples. Niesters and Puchhammer-Stockl (2004) emphasize that standards should use whole virus diluted in the same appropriate matrix as the clinical sample, and nucleic acids must be isolated using the same extraction procedure. Multiple replicates of the standards should be analyzed to determine confidence limits based on inter-assay and intra-assay variation. The problems concerning standards are further complicated by a lack of universally accepted international standards for most clinically significant viruses (Niesters and Puchhammer-Stockl, 2004), and thus, qPCR results may vary between different laboratories. It is essential therefore, that clinicians are aware of these limitations and their potential for variation in the qPCR results. Recognition of these does not devalue the overall utility of qPCR, but rather will allow for a proper assessment of the results within a relevant clinical context.

14.5 Conclusions

Real-time PCR methodologies for the detection of viruses are now firmly entranced in many clinical laboratories, offering the major advantage of improved sensitivity over more traditional methods. The application of this technology has been most successful in those areas where conventional virological techniques did not exist or were inadequate. Its introduction into the virology laboratory has led to the recognition and improved diagnostics of several viruses that may not be isolated, allowed for quantification of viral loads during the infection process and the subsequent monitoring of antiviral drug therapy, as well as characterization of mutations in viral genes that are responsible for drug resistance.

The number of commercially available assays is increasing, but many laboratories still design, develop and validate assays in house. As these in-house, real-time PCR methods have evolved there has been an increasing awareness of the need for standardized reagents and quality control programs in order to produce clinically significant results. In addition, the targeted nature of PCR brings with it some inherent limitations that are difficult to control, particularly in virology, where the heterogeneous nature of the viral genome may lead to difficulties in assay design and performance. An awareness of these issues will ultimately result in a better understanding of this new technology, and enable us to fully explore its potential as a modern diagnostic tool.

References

Anderson TP, Werno AM, Beynon KA, Murdoch DR (2003) Failure to genotype herpes simplex virus by real-time PCR assay and melting curve analysis due to sequence variation within probe binding sites. *J Clin Microbiol* **41**: 2135–2137.

Berger A, Braner J, Doerr HW, Weber B (1998) Quantification of viral load: clinical relevance for human immunodeficiency virus, hepatitis B virus and hepatitis C virus infection. *Intervirology* **41**: 24–34.

Boeckh M, Boivin G. (1998) Quantitation of cytomegalovirus: methodologic aspects and clinical applications. *Clin Microbiol Rev* **11**: 533–554.

Espy MJ, Uhl JR, Mitchell PS, Thorvilson JN, Svien KA, Wold AD, Smith TF (2000) Diagnosis of herpes simplex virus infections in the clinical laboratory by LightCycler™ PCR. *J Clin Microbiol* **38**: 795–799.

Fouchier RA, Bestebroer TM, Herfst S, Van Der Kemp L, Rimmelzwaan GF, Osterhaus AD (2000) Detection of influenza A viruses from different species by PCR amplification of conserved sequences in the matrix gene. *J Clin Microbiol* **38**: 4096–4101.

Hammel JM, Chiang WK (2005) Update on emerging infections: news from the Centers for Disease Control and Prevention. Outbreaks of avian influenza A

(H5N1) in Asia and interim recommendations for evaluation and reporting of suspected cases – United States, 2004. *Ann Emerg Med* **45**: 88–92.

Hirsch HH, Mohaupt M, Klimkait T (2001) Prospective monitoring of BK virus load after discontinuing sirolimus treatment in a renal transplant patient with BK virus nephropathy. *J Infect Dis* **184**: 1494–1495.

Kumar R, Vandegraaff N, Mundy L, Burrell CJ, Li P (2002) Evaluation of PCR-based methods for the quantitation of integrated HIV-1 DNA. *J Virol Methods* **105**: 233–246.

Kutyavin IV, Afonina IA, Mills A, Gorn VV, Lukhtanov EA, Belousov ES, Singer MJ, Walburger DK, Lokhov SG, Gall AA, Dempcy R, Reed MW, Meyer RB, Hedgpeth J (2000) 3'-minor groove binder-DNA probes increase sequence specificity at PCR extension temperatures. *Nucleic Acids Res* **28**: 655–661.

Mackay IM, Jacob KC, Woolhouse D, Waller K, Syrmis MW, Whiley DM, Siebert DJ, Nissen M, Sloots TP (2003) Molecular assays for detection of human metapneumovirus. *J Clin Microbiol* **41**: 100–105.

Maertzdorf J, Wang CK, Brown JB, Quinto JD, Chu M, de Graaf M, van den Hoogen BG, Spaete R, Osterhaus AD, Fouchier RA (2004) Real-time reverse transcriptase PCR assay for detection of human metapneumoviruses from all known genetic lineages. *J Clin Microbiol* **42**: 981–986.

Niesters HG (2004) Molecular and diagnostic clinical virology in real time. *Clin Microbiol Infect* **10**: 5–11.

Niesters HG, Puchhammer-Stockl E (2004) Standardisation and controls, why can't we overcome the hurdles? *J Clin Virol* **31**: 81–83.

O'Doherty U, Swiggard WJ, Jeyakumar D, McGain D, Malim MH (2002) A sensitive, quantitative assay for human immunodeficiency virus type 1 integration. *J Virol* **76**: 10942–10950.

Smith TF, Uhl JR, Espy MJ, Sloan LM, Vetter EA, Jones MF, Rosenblatt, JE, Cockerill FR 3rd (2004) Development, implementation, and trend analysis of real-time PCR tests for the clinical microbiology laboratory. *Clin Micro News* **26**: 145–153.

van den Hoogen, BG, de Jong J, de Groot R, Fouchier RA, Osterhaus AD (2001) A newly discovered human pneumovirus isolated from young children with respiratory tract disease. *Nat Med* **7**: 719–724.

van den Hoogen BG, Herfst S, Sprong L, Cane PA, Forleo-Neto E, de Swart RL, Osterhaus AD, Fouchier RA (2004) Antigenic and genetic variability of human metapneumoviruses. *Emerg Infect Dis* **10**: 658–666.

van Elden LJ, Nijhuis M, Schipper P, Schuurman R, van Loon AM (2001) Simultaneous detection of influenza viruses A and B using real-time quantitative PCR. *J Clin Microbiol* **39**: 196–200.

Watzinger F, Suda M, Preuner S, Baumgartinger R, Ebner K, Baskova L, Niesters HG, Lawitschka A, Lion T (2004) Real-time quantitative PCR assays for detection and monitoring of pathogenic human viruses in immunosuppressed pediatric patients. *J Clin Microbiol* **42**: 5189–5198.

Weidmann M, Meyer-Konig U, Hufert FT (2003) Rapid detection of herpes simplex virus and varicella-zoster virus infections by real-time PCR. *J Clin Microbiol* **41**: 1565–1568.

Whiley DM, Mackay IM, Sloots TP (2001) Detection and differentiation of human polyomaviruses JC and BK by LightCycler™ PCR. *J Clin Microbiol* **39**: 4357–4361.

Whiley DM, Mackay I, Syrmis MW, Witt MJ, Sloots TP (2004) Detection and differentiation of herpes simplex virus types 1 and 2 by a duplex LightCycler™ PCR that incorporates an internal control PCR reaction. *J Clin Virol* **30**: 32–38.

Whiley DM, Sloots TP (2005a) Melting curve analysis using hybridization probes: limitations in microbial molecular diagnostics. *Pathology* **37**: 254–256.

Whiley DM, Sloots TP (2005b) Sequence variation in primer targets affects the accuracy of viral quantitative PCR. *J Clin Virol* **34**: 104–107.

Whiley DM, Sloots TP (2006) Sequence variation can affect the performance of minor groove binder TaqMan probes in viral diagnostic assays. *J Clin Virol* **35**: 81–83.

Protocol 14.1: Detection of influenza A by a 5' nuclease real-time RT-PCR assay

Sample preparation

Nasopharyngeal aspirate specimens are placed in 1mL of sterile saline and vigorously mixed on a Vortex. Nucleic acids are extracted from 0.2 mL of this suspension using the Roche MagNAPure™ instrument (Roche Diagnostics, Australia) as per manufacturer's instructions.

Influenza A oligonucleotides (targeting the influenza A matrix gene)

Forward CTTCTAACCGAGGTCGAAACGTA[a]
Reverse GGTGACAGGATTGGTCTTGTCTTTA[a]
Probe Fam-TCAGGCCCCCTCAAAGCCGAG-BHQ1[a]

Reaction mix (Qiagen One-Step RT-PCR Kit; Qiagen, Australia)

Qiagen One-Step RT-PCR Enzyme mix:	1.0 µL
Qiagen One-Step RT-PCR Buffer (5x):	5.0 µL
Qiagen One-Step RT-PCR dNTP mix (10 mM):	1.0 µL
Forward primer (10 µM):	2.0 µL
Reverse primer (10 µM):	2.0 µL
Probe (20 µM):	0.2 µL
RNase-free water:	8.8 µL
Specimen or control nucleic acid extract:	5.0 µL
Total:	25.0 µL

Cycling conditions (ABI Prism® 7500; Applied Biosystems, USA)

RT step	1 cycle: 50°C 20 min
Activation/denaturation	1 cycle: 95°C 15 min
Cycling	45cycles: 95°C 15 sec 60°C 60 sec acquisition

Result interpretation: Specimens providing exponential amplification
                             curves are considered positive for influenza A
                             RNA.

                             [a] These sequences matched the matrix protein
                             genes of a broad range on influenza A subtypes,
                             including H1N1 (Genbank accession number
                             AY619976), H1N2 (AY233392), H2N2 (M63531),
                             H3N2 (AY210264), H5N1 (AY646180), H5N3
                             (AY300973), H7N2 (AY241624), H7N3
                             (AY611525), H7N7 (AJ619676), H9N2
                             (AF523497).

Protocol 14.2: Detection and differentiation of herpes simplex virus types 1 and 2 using a hybridization probe real-time PCR assay

Sample preparation

Swab specimens are placed in 1 mL of sterile saline and vortexed. Nucleic acids are extracted from 0.2 mL of each suspension using the Roche MagNAPure™ instrument (Roche Diagnostics, Australia).

HSV oligonucleotides (targeting the HSV DNA polymerase gene; Espy *et al.*, 2000).

Forward GCTCGAGTGCGAAAAAACGTTC
Reverse CGGGGCGCTCGGCTAAC
Probe 1 GTACATCGGCGTCATCTGCGGGGGC
AAG-fluorescein
Probe 2 LC-Red640–TGCTCATCAAGGGCGTG
GATCTGGTGC-Phosphate

Reaction mix (LightCycler® FastStart DNA Master Hybridization Probes kit; Roche Diagnostics, Australia)

Roche kit Master reagent (10 x)	2.0 µL
$MgCl_2$ (25 mM)	2.4 µL
Forward primer (10 µM)	0.4 µL
Reverse primer (10 µM)	0.8 µL
Probe 1 (20 µM)	0.2 µL
Probe 1 (20 µM)	0.2 µL
RNase-free water	9.0 µL
Specimen or control nucleic acid extract	5.0 µL
Total	20.0 µL

Cycling conditions (LightCycler®; Roche Diagnostics, Australia)
Activation/denaturation   1 cycle:
   95°C 10 min
Cycling   55 cycles:
   95°C 10 sec
   55°C 10 sec single acquisition
   72°C 20 sec
Melting curve analysis   1 cycle:
   95°C 5 sec
   40°C 5 sec
   95°C 5 sec continuous acquisition
   (ramp at 0.2°C/sec)

Result interpretation:

Specimens providing a melting temperature of 67°C are considered positive for HSV type 1 DNA whereas specimens providing a melting temperature of 74°C are considered positive for HSV type 2 DNA.

Protocol 14.3: Quantitative analysis of BKV load by a 5′ nuclease real-time PCR (BKV-Tag-qPCR) assay

Sample and standard preparation

Specimens and standards are prepared in the same manner, and should also include internal control nucleic acid (see *Protocol 14.4*). Standards must be diluted in the appropriate matrix, for example when urine samples are tested, 10-fold dilutions of the standard are prepared in negative urine. Briefly, 100 μL of BKV standard at 1×10^8 copies per mL is added to 900 μL of negative urine. Subsequent dilutions are performed to provide seven standards ranging from 1×10^7 to 10 copies per mL of urine. Nucleic acids are extracted from 0.2 mL of each sample and standards using the Roche MagNAPure™ instrument (Roche Diagnostics, Australia).

BKV oligonucleotides (targeting the BKV T antigen gene; Hirsch *et al.*, 2001)

Forward primer	AGCAGGCAAGGGTTCTATTAC TAAAT
Reverse primer	GAAGCAACAGCAGATTCTCAACA
Probe	FAM-AAGACCCTAAAGACTTTCC CTCTGATCTACACCAGTTT-TAMRA

Reaction mix (TaqMan® Universal PCR Master Mix; Applied Biosystems, Australia)

TaqMan® Universal PCR Master Mix	12.5 μL
Forward primer (10 μM)	1.0 μL
Reverse primer (10 μM)	1.0 μL
Probe (20 μM)	0.2 μL
RNase-free water	5.3 μL
Specimen or control nucleic acid extract	5.0 μL
Total	25.0 μL

Cycling conditions (ABI Prism® 7500; Applied
Biosystems, United States)
Activation/denaturation   1 cycle:
    95°C 10 min
Cycling   5 cycles:
    95°C 15 sec
    60°C 60 sec acquisition

Result interpretation: The BKV concentration in each specimen is determined using a standard curve. Standard curves
are generated from the range of controls included
in the assay protocol, by software provided with
the instrument (ABI 7500 system V.1.2.3f2).
Values are expressed in copies per mL of specimen. Results are interpreted subject to the internal control results (see *Protocol 14.4*).

Protocol 14.4: Real-time PCR internal control (Niesters, 2004)

Sample preparation:

The concentration of seal herpesvirus (PhHV-1) added to each patient specimen should be sufficient to provide an average C_t value of approximately cycle 30 in a correctly extracted sample containing no PCR inhibitors. This equates to a final concentration of approximately 5,000 copies per mL of specimen. Briefly, 10µL of PhHV-1 at 500 copies per µL should be added to 1mL of each specimen.

PhHV-1 oligonucleotides
Forward primer   GGGCGAATCACAGATTGAATC
Reverse primer   GCGGTTCCAAACGTACCAA
Probe   TTTTTATGTGTCCGCCACCAT CTGGATC

The reaction mix, fluorophores and cycling conditions used for the PhHV-1 assay can readily be adapted to those utilized by the testing laboratory. Conditions must be standardized so that the assay achieves a C_t value of approximately cycle 30, with minimal inter-assay and intra-assay variability, for a correctly extracted sample containing no PCR inhibitors.

Result interpretation:

Specimens providing C_t values exceeding two standard deviations from the expected C_t value are considered to be unsuitable. They may contain PCR inhibitors, have been inefficiently extracted, or have been subject to other conditions that have compromised the PCR. These specimens should be repeated.

Specimens providing C_t values within two standard deviations from the expected C_t value can be used in the calculation of viral load of the target test virus using a standard curve.

Troubleshooting guide 14.1: False-negative result in the WhSl-para3–5N assay

The specimen that was not detected in the WhSl-para3–5N assay was investigated further to identify the cause of the assay failure. Following thermocycling, the WhSl-para3–5N PCR reaction mix for this specimen was examined by gel electrophoresis and a PCR amplification product of the expected size (105 base pairs) was observed (see *Figure 14.1*). This demonstrates that amplification had occurred and suggests that the

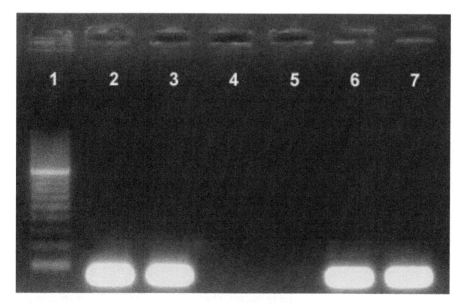

Figure 14.1

Gel electrophoresis image of the reaction mix from the specimen providing a false-negative result in the WhSl-para3–5N assay. A 105 base pair product was observed for the specimen (run in duplicate in lanes 6 and 7) and positive controls (lanes 2 and 3) but not in the negative controls (lanes 4 and 5). A 200 base pair ladder was included in lane 1.

source of the failure was not in amplification but in the detection of the amplification product. The PCR product from this specimen was then sequenced to identify the oligonucleotide target sequences in this parainfluenza type 3 strain.

The fact that an amplification product was produced indicated that both primers hybridized to the target, therefore, mismatches between the probe and its target appeared to be the likely source of the problem. Thus, the same primers that were used in the WhSl-para3–5N assay could be used for the DNA sequencing reaction. On the other hand, had a PCR product not been obtained, then mismatches between the primers and their targets would be the more probable explanation. Thus, a second primer pair, external to those utilized by the assay would need to be designed and employed to sequence the primer targets. Nevertheless, for this investigation a second external primer was used to illustrate the process.

Briefly, the two external primers (forward ACCAGGAAACTATGCTGCAGAACGGC and reverse CCATACCTGATTGTATTGAAGAATGAAGC) were used in a standard RT-PCR reaction to amplify a 409 base pair sequence from the above specimen. The amplification products were separated by electrophoresis through a 1% agarose gel. The specific band (409 bp) was excised and purified using a QIAquick gel extraction kit (Qiagen, Germany). DNA sequencing was performed using the above forward primer in an ABI Prism Dye Terminator Cycle Sequencing kit (Applied Biosystems, USA) and was analyzed on an ABI 377 DNA sequencer (Applied Biosystems, USA). The sequence obtained from this specimen was then compared to the primer and probe sequences used in the WhSl-para3–5N assay.

Interestingly, the sequencing results showed that mismatches (bold underlined) were present with all three oligonucleotides. For both primers, there were mismatches towards the 5′ end. It should be noted that the length and Tm of primers typically used in standard TaqMan® cycling conditions will usually allow them to accommodate one or two 5′ mismatches with little impact on the PCR. This is consistent with the results observed here, where amplification product was present for this specimen. However, in this parainfluenza type 3 strain there were four mismatches within the binding site of the WhSl-para3–5N TaqMan™ probe. This many mismatches most likely prevented the probe from adequately hybridizing to the target and therefore caused the detection failure.

F Primer   CGGTGACACAGTGGATCAGATT
Specimen   TGGTGACACAGTGGATCAGATT

R Primer   AGGTCATTTCTGCTAGTATTCATTGT
           TATT
Specimen   AGATCATTCCTGCTAGTATTCATTGT
           TATT

Probe   TCAATCATGCGGTCTCAACAGAGC
           TTG
Specimen   TCAATTATGCGATCCCAACAGAGC
           TTA

Troubleshooting guide 14.2: Variation in fluorescent signal from positive specimens in the RSV MGB assay

Three RSV-positive specimens providing representative fluorescent signals (see *Figure 14.2*) were investigated further to identify the cause of the variation in fluorescent response. Briefly, the PCR products from these specimens were

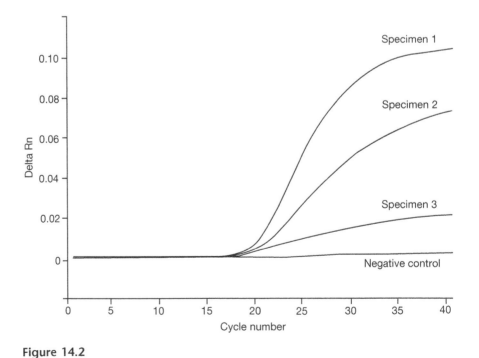

Figure 14.2

Fluorescent signal for three representative positive specimens in the RSV MGB probe assay.

sequenced and probe targets compared to the RSV MGB TaqMan® probe sequence. For this investigation, the same primers used in the respiratory syncytial virus assay were used to sequence the forward and reverse DNA strands of the PCR products.

The sequencing results showed that specimen 1, which provided optimal fluorescence, had no mismatches with the RSV MGB probe. In contrast, specimens 2 and 3, which provided lower fluorescent signal, each had a single mismatch (bold underlined) with the probe. The mismatched bases in each of these specimens were at different positions on the probe and resulted in varying fluorescent signals.

RSV MGB probe	TCAATACCAGCTTATAGAAC
Specimen 1	TCAATACCAGCTTATAGAAC
Specimen 2	TC**T**ATACCAGCTTATAGAAC
Specimen 3	TCAATACCAGCTTA**C**AGAAC

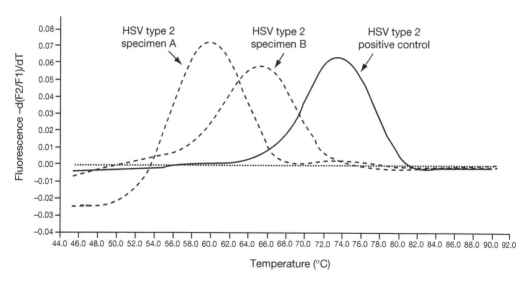

Figure 14.3

LightCycler® melting curve analysis for two herpes simplex virus type 2 positive specimens showing uncharacteristic melting temperatures compared to a herpes simplex virus type 2 positive control. The results were obtained using a hybridization probe-based assay targeting the DNA polymerase gene.

Troubleshooting guide 14.3: Uncharacteristic herpes simplex virus type 2 melting curves

Two herpes simplex virus (HSV) type 2 positive specimens providing uncharacteristic melting curves were further investigated to identify the cause of the shift in melting temperature. Briefly, PCR products from these specimens were sequenced and the probe target sequence was compared to the Espy hybridization probe sequences, which should be fully complementary to HSV type 2 DNA. For this investigation, it was unnecessary to design primers outside the Espy primer targets, as the probe target sequences were of primary interest. Thus, the primers utilized in the Espy HSV assay (*Protocol 16.2*) were used to sequence the PCR products in forward and reverse direction.

The sequencing results showed that both of the HSV type 2 positive specimens had mismatches with the Espy hybridization probes. For specimen A there were three mismatches (bold-underlined) compared to the Espy probe 2 sequence and for specimen B there were two mismatches (bold-underlined) compared to the Espy probe 1 sequence.

Probe 1	GTACATCGGCGTCATCTGCGGGGGCAAG
Strain A	GTACATCGGCGTCATCTGCGGGGGCAAG
Strain B	GTACATCGGCGT**T**AT**T**TGCGGGGGCAAG
Probe 2	TGCTCATCAAGGGCGTGGATCTGGTGC
Strain A	TGCTCAT**T**AAGGGCGT**C**GA**C**CTGGTGC
Strain B	TGCTCATCAAGGGCGTGGATCTGGTGC

Troubleshooting guide 14.4: Underestimation of BK viral load due to sequence variation

The specimen providing discrepant BKV viral load results between the BK-qLC and BKV-Tag-qPCR assays was further investigated. Briefly, the DNA segment of the VP2 gene targeted by the BK-qLC PCR assay was sequenced using primers external to those used in the assay. The sequence was then compared to the primer and probe sequences used in the BK-qLC assay.

The results showed a single base mismatch (bold-underlined) with the forward primer was responsible for the change in amplification characteristics in the BK-qLC assay. It should be noted that the annealing temperature used in the BK-qLC assay was 55°C whereas the Tm of this primer (calculated using Primer Premier version 5; Premier Biosoft International, CA, USA) was only 51.8°C assuming no mismatches and 47.6°C if the 3' cytosine base was excluded. Thus, the low Tm of this primer was likely to have compounded the effects of sequence mismatch.

Forward primer   GTAAAAGACTCTGTAAAAGACTCC
Specimen   GTAAAAGACTCTGTAAAAGACTC**C̲G̲**

Solid organ transplant monitoring

15

Omaima M. Sabek

15.1   Introduction

Over the past several decades, many factors have led to higher rates of patient and graft survival in organ transplantation. These include enhanced immunosuppression, infectious disease prophylaxis, advances in surgical technique, better patient selection criteria, optimal timing of transplantation, and better options for treating complications (Hariharan *et al.*, 2000; Adam and Del Gaudio, 2003). As a result of these advanced immunosuppressive treatments, graft loss due to acute rejection is now rare among solid-organ transplants. The optimal strategy for immunosuppressive treatment in the future will tailor low-dose immunosuppression to the patient's needs with avoidance (or early withdrawal) of corticosteroid and lower doses of calcineurin inhibitors. Drugs aimed at tolerance induction and inhibition of adhesion molecule expression might also lead to better outcomes for many recipients in the near future (Ortiz *et al.*, 2003; Heeger, 2003). However, these substantial gains have not prevented graft loss so much as they have delayed it. To improve transplant outcomes, we need a better understanding of predisposing immunologic and non-immunologic factors. With more sensitive analytical methods, it is becoming increasingly clear that efforts to prevent chronic rejection and avoid inappropriate immunosuppression must include and effective means to proximally diagnose acute cellular rejection (ACR), including sub-clinical forms of injury. The role of concomitant viral infection is also being recognized as increasing germane, and these methods may also assist in more precisely identifying the origin and the 'intent' of immune cell infiltrates. Molecular techniques have become a mainstay for most biomedical research. In particular, sensitive methods for gene transcript detection have played an important role in our understanding of the basic mechanisms promoting allosensitization and adaptive immune regulation. It is becoming increasingly clear that these techniques will aid to better understand human transplant biology, and, more importantly, guide clinical decision making with mechanistically based information. By allowing for critical analysis of the blood and graft microenvironment, these assays are providing early mechanistic data aiding in the correlation of basic and clinical scientific concepts. This chapter will discuss the use of real-time polymerase chain reaction in gene transcript quantification, gene polymorphism analysis, viral load detection and quantification. Clinical correlations will be presented as examples of how these techniques may have clinical relevance.

15.2 Real-time quantitative PCR

Real-time quantitative PCR allows for a rapid and precise relative quantification of gene transcripts. Real-time PCR has been shown to exceed the sensitivity of northern blots and RNase protection assays and allows for quantitative study of multiple transcripts with high reproducibility which is a major strength of this method. The reproducibility of the results obtained by real-time PCR is another. Moreover, because the chemistry used is standard and the primers and probes are designed to be compatible, there is consistency. The advantage for using the real-time PCR in a clinical setting, however, is the speed by which the test can be performed. The turnaround time for the assay from time of sample collection is approximately 4 to 6 hour. This is important clinically because surveillance testing of transplant patients using this methodology can be performed without alteration in clinic or hospital procedure and within the already established clinical pathways for posttransplant care.

15.3 RNA normalization

With the increased interest in gene marker based diagnosis, the choice of housekeeping genes for the various assays becomes an important consideration that could affect sensitivity, specificity, and across laboratory reproducibility of results. In our Laboratory, we performed a detailed examination of 11 available housekeeping genes in peripheral blood of healthy volunteers before selecting a suitable housekeeping gene for our assay (Sabek et al., 2002). Expression of the following housekeeping genes was evaluated: 18S rRNA (18S), acidic ribosomal protein (PO), beta-actin (βA), cyclophilin (CYC), glyceraldehyde-3–phosphate dehydrogenase (GAPDH), phosphoglycerokinase (PGK), β2-microglobulin (β2m), β-glucuronidase (GUS), hypoxanthine phosphoribosyl transferase (HPRT), transcription factor IID, TATA-binding protein (TBP), transferring receptor (TfR). The results of indicated that 18S, β-actin, and β-glucuronidase have the least variability of expression over time and are more reliable for normalizing our data. Surprisingly, other housekeeping genes used in previous human studies, such as GAPDH and cyclophilin, had higher variability between specimens and generated irreproducible results. The lack of reliability of GAPDH as a housekeeping gene has been observed by others as well (Tricarico et al., 2002; Bustin, 2000). Hoffmann et al. (2003) has shown that 18s RNA was an appropriate candidate housekeeping control for transcript normalization since its expression levels were not found to vary among different tissues, cell types, with or without stimulation, or from experimental treatments. In contrast GAPDH and β-actin expression varied widely, exhibiting C_t fluctuations and nearly 10-fold differences in some activated cell populations. In addition, cyclophilin is not a practical calibrator for transplantation-related studies due to its direct role in the calcineurin pathway affected by standard immunosuppressive therapies. GUS was found to be a fairly constant and reliable expression target. However, the level of GUS transcript expression may compete with several immune targets and, therefore, does not allow for multiplex assays and could invalidate quantitative precision. Based on these findings 18s

RNA was regarded as the most reliable internal housekeeping control for analyses. However, it is vital to test for the effect of different drugs used on the expression levels of the housekeeping genes if we are to incorporate this technique as standard clinical tool for monitoring allograft dysfunction and guiding individualized immunotherapeutic interventions. Further discussion of the choice of a normalizer gene is presented in Chapter 4.

15.4 Immunologic monitoring in solid organ transplantation

Improvement of patient and graft survival rates and the post transplantation quality of life depends on many pre- and post-transplant factors, such as tissue compatibility, immunosuppressive therapy, as well as the early detection and treatment of both acute and chronic graft rejection. Allograft antigens presented to the host immune system in association with the major histocompatibility complex (MHC)-encoded molecules initiate lymphocyte activation and proliferation of the primed lymphocytes leading to both cellular and/or humoral rejection. The cytokines are essential for the differentiation, proliferation, and amplification of the T-cell responses (Schwartz, 1992). The most important one is interleukin-2 (IL-2), which is essential for activated T-cell proliferation. Monitoring intragraft cytokines such as IL-2, IL-4, IL-10 and interferon-γ using real-time PCR (Flohe *et al.*, 1998; Hahn *et al.*, 2001), although invasive, is simple and has been shown to be effective in characterizing and predicting graft rejection and tolerance. Koop *et al*, (2004) showed that differentiation between chronic rejection and chronic cyclosporine toxicity in renal transplant patients may be more reliably detected using the molecular technique of a real-time PCR quantitative analysis of the renal cortical mRNA levels of laminin β2 and extracellular matrix regulation molecule transforming growth factor-β. Using a less invasive technique to monitor intracellular T-cell cytokines such as (Il-2, IL-4,IL-10, TNF, RANTES, and IFNγ) levels in lymphocyte subsets (CD3/CD4 and CD3/CD8) as well as activation markers CD25 (IL-2R), CD40 CD69, CD71, CD103 and cytotoxic effector molecules such as perforin, granzyme B and FasL in peripheral blood and urine cytokines was shown to be predictive (Li *et al.*, 2001; Ding *et al.*, 2003; Muthukumar *et al.*, 2003; Dadhania *et al.*, 2003; Tatapudi *et al.*, 2004; Shoker *et al.*, 2000; Satterwhite *et al.*, 2003; Sabek *et al.*, 2002). Simon *et al.* (2003) reported that renal rejection diagnosis using perforin and granzyme B gene expression measurements was possible 2–30 days before traditional clinical diagnosis (i.e. elevation of serum creatinine) and that gene expression measurement is a useful tool for the recognition of graft rejection in its earliest stages. Findings from these studies suggest that serial measurements could be implemented as a monitoring system to highlight patients at higher risk of rejection, making them candidates for biopsy or pre-emptive anti-rejection therapy. Such early intervention may be more effective than therapy initiated after organ damage has already occurred.

15.5 Pharmacogenetics in solid organ transplantation

Many pharmacogenomic predictors of drug response are now available, and include both drug metabolism-disposition factors and drug targets. It is well recognized that different transplant recipients respond in different ways to immunosuppressive medication (Zheng *et al.*, 2005). The inter-individual variations are greater than the intra-individual variations, a finding consistent with the notion that inheritance is a determinant of drug responses. In the general population, it is estimated that genetics accounts for 20 to 95% of the variability in drug disposition and effects (Kalow *et al.*, 1998). Many other non-genetic factors, such as organ function, drug interactions, and the nature of the disease, probably influence the effects of medication. The recent identification of genetic polymorphisms in drug-metabolizing enzymes and drug transporters led to the hypothesis that genetic factors may be implicated in the inter-individual variability of the pharmacokinetic or pharmacodynamic characteristics of immunosuppressive drugs, major side effects, and immunologic risks. The promising role of pharmacogenetics and pharmacogenomics in the effort to elucidate the inherited basis of differences between individual responses to drugs, lies in the potential ability to identify the right drug and dose for each patient. As an example, Mycophenolate mofetil (MMF) is an immunosuppressant widely used for the prophylaxis of organ rejection in renal, pancreas, and liver transplantation (Cai *et al.*, 1998; Detry *et al.*, 2003). It inhibits the inducible isoform of the enzyme inosine-monophosphate dehydrogenase (IMPDH II). IMPDH II is necessary for *de novo* purine synthesis in activated lymphocytes, whereas other cells generate purine nucleotides via the salvage pathway (Allison and Eugui, 2000). The use of MMF in liver transplantation (Detry *et al.*, 2003), can be used to reduce the dosage of calcineurin inhibitors and steroids (Moreno *et al.*, 2003). However, MMF administration may be associated with bone marrow and gastrointestinal toxicity, thus requiring therapeutic drug monitoring. The quantitation of gene transcription of the IMPDH inducible isoform using quantitative real-time PCR is feasible and reliable and may represent an important alternative to enzyme activity measurements based on liquid chromatographic methods. Therefore, the capability to discriminate between increased expressions of the inducible versus the constitutive isoform appears to be of critical importance. Moreover, a quantitative PCR assay may be used to analyze the drug target, thus reflecting MMF efficacy and tolerability, allowing a personalized immunosuppressive therapy.

15.6   Cytokine gene polymorphism analysis

While a wide range of factors contributes to allograft survivability, routine screening of cytokine gene polymorphisms may have important clinical relevance and therefore should be considered in the design of both pre- and post-treatment regimens. Most cytokines have been demonstrated to be transcriptionally controlled. Cytokines influence the local activation of cells and play a critical role in the regulation of immune responses. While functional affects have been attributed to cytokine gene variants (Hoffmann *et al.*, 2001; Louis *et al.*, 1998; Turner *et al.*, 1997), their role in allograft rejection remains controversial. The level of production of many of these cytokines may be important in accelerating or slowing the rejection process. The inheritance of genetically determined polymorphisms has been implicated in the development of both acute and chronic renal allograft rejection (Hutchinson, 1999; Hutchinson *et al.*, 2000) and peripheral tolerance (Burlingham *et al.*, 2000). Indeed, some studies have shown allelic variations within cytokine genes might act on the microenvironment of responding T cells and APCs, and thereby regulate allograft survival. A more clear definition of cytokine genotypes may indicate how recipients will respond to their transplants and guide both optimal immunotherapy, organ selection, or enable selection of patient subgroups for individualized clinical trials (Hoffmann *et al.*, 2001, Sankaran *et al.*, 1999).

15.6.1 Recipient and donor gene polymorphisms

In a recent study by Hoffmann *et al.* (2003) where they compared patient and donor allelic genotypes in renal transplant patients and established phenotypes to allograft status as defined by Banff criteria in serial protocol biopsies (Racusen *et al.*, 1999). They showed that high production alleles for either IL-10 or IFN-gamma were dominant in recipients with Banff criteria for chronic changes compared to a control group. They also showed that in the rejections patients that were resistant to initial steroid therapy alone, those individuals were of the high IFN-gamma producer phenotype. Not only recipient but donor related factors, in particular the age and ethnicity of the donor can play a role in an early rejection episode (Marshall *et al.*, 2001; Swanson *et al.*, 2002; Verran *et al.*, 2001). Donor brain death and allograft ischemia are a traumatic process that can affect cytokine production (Pratschke *et al.*, 2000). The impact of high donor IFN-gamma and IL-10 production was shown to be significant (Hoffmann *et al.*, 2003) on recipients presenting with sub-clinical rejection, ACR or chronic changes when compared to those donors with stable recipients and normal allograft histology. These results suggest that donor-derived production may play an essential role in inflammatory processes initiating rejection. Given the shortage of suitable donor organs, the acceptance of organs from marginal donors is rising, warranting an assessment of donor specific variables (Gjertson *et al.*, 2002; Randhawa, 2001).

15.6.2 Ethnicity and cytokine gene polymorphism

Ethnicity clearly is associated with dramatic differences in cytokine polymorphism distributions. The disparity between graft survival rates among African–American and other populations suggests that the cytokine polymorphisms described by many laboratories, including our own (Lo *et al.*, 2004; Egidi and Gaber, 2003; Hartwig *et al.*, 1993; Hardinger *et al.*, 2001; Lo *et al.*, 2001; Ilyas *et al.*, 1998), may play an incremental role in ethnic-based survival rates and subsequent differences in immune reactivity. Indeed, there is a documented deficit in long-term allograft survival for African–Americans as compared to Caucasian, Hispanic, and Asian populations (USRDS, 2000). Poor graft survival among African–Americans had been attributed in part to socioeconomic factors and to inferior HLA matching. African–American donor kidneys also have been associated with a worse graft survival (Yates *et al.*, 2003). Studies examining potential effects of polymorphisms on pharmacokinetics and immune responsiveness have also been performed showing that peripheral blood cells from healthy adult

African–Americans express significantly more B7 costimulatory molecules (CD80, CD86) than Caucasians and mount more vigorous immune responses to mitogens and antigens *in vitro* (Hutchings *et al.*, 1999; Lindholm *et al.*, 1992; Tornatore *et al.*, 1995). Factors such as these could represent additional key factors for racial variation in allograft loss (Cox *et al.*, 2001; Kerman *et al.*, 1991; Gaston *et al.*, 1992). There are striking differences in the distribution of cytokine polymorphisms among ethnic populations (Hoffmann *et al.*, 2002; Cox *et al.*, 2001). Blacks, Hispanics and Asians have marked differences in their inheritance of IL-6 alleles and IL-10 genotypes that result in high expression when compared with Caucasian. High IL-6 producing individuals have previously been shown to be at heightened risk for ACR and/or some forms of end-stage renal disease (Casiraghi *et al.*, 1997; Hoffmann *et al.*, 2003). The high production of IL-6 in concert with pro-inflammatory cytokines, may result in an increase risk for allograft rejection in the African–American population, or at least increase the impact of alterations in immunosuppressive non-compliance. While, Asians exhibit IFN-gamma genotypes that result in diminished cytotoxic effects thereby a decrease risk of acute graft rejections. This observation suggests that a population of recipients more prone to ACR may be predicted through study of cytokine gene polymorphisms.

15.7 Viral infection in transplant patients

Viral pathogens have emerged as the most important microbial agents having deleterious effects on solid organ transplant recipients. Antiviral chemoprophylaxis involves the administration of medications to abort transmission of, avoid reactivation of, or prevent progression to disease from, active viral infection. Real-time PCR amplification techniques are currently used to determine the viral load in clinical samples for an increasing number of targets, and in a short turn-around time, and to determine whether variants relevant for antiviral resistance are present (see, for example, Chapter 14). Technological improvements, from automated sample isolation to real-time amplification technology, have given the ability to develop and introduce systems for most viruses of clinical interest, and to obtain clinical relevant information needed for optimal antiviral treatment options. Treatment initiation is based not only on the viral load itself but also on a combination of difference criteria, including the kinetics of the viral load and the patient's immune condition. Cytomegalovirus (CMV) is the major microbial pathogen having a negative effect on solid organ transplant recipients. CMV causes infectious disease syndromes, and is commonly associated with opportunistic super infection. CMV has also been implicated in the pathogenesis of rejection. The ability to detect persistent low-level CMV replication after termination of antiviral therapy as well as the viral load at the completion of antiviral therapy is also of interest and may lead to treatment adjustment and prevent the risk of emergence of resistant strains (Boivin *et al.*, 2001). The viral load at the completion of antiviral therapy has been shown to predict the risk of relapsing (Sia *et al.*, 2000). But, most importantly, a negative result from a reliable sensitive assay can rule out CMV replication and may prevent unnecessary treatment. Herpes simplex virus (HSV) reactivation in solid organ transplant recipients is invasive, more frequent, takes longer to heal, and has greater potential for dissemination to visceral organs than it does in the immunocompetent host. Epstein–Barr virus (EBV) has its most significant effect in solid organ transplant as the precipitating factor in the development of post-transplant lymphoproliferative disorders (PTLD) as well as rejection (Jabs *et al.*, 2004). Over the past few years, PCR assay has shown that a rapid increase in peripheral blood EBV DNA load is predictive of PTLD (Qu *et al.*, 2000). Identifying patients at high risk or in an early stage of PTLD enables EBV infection to be pre-emptively or promptly dealt with by administering anti-CD20 monoclonal antibody or EBV-CTL, or simply by reducing the immunosuppressive therapy (Wagner *et al.*, 2004; Kuehnle *et al.*, 2000). In conclusion, EBV reactivation, increase in viral-load detection by quantitative real-time PCR, can be managed simply by modulating immunosuppression. Hepatitis B virus (HBV) and hepatitis C virus (HCV) are important pathogens in the solid organ transplant population as indications for transplantation. Prophylaxis for recurrent HBV and HCV after liver transplantation is controversial, suppressive rather than preventive, and potentially lifelong. Influenza infection after solid organ transplant is acquired by person-to-person contact. During epidemic periods of influenza, transplant populations experience a relatively high frequency of infection, and influenza may affect immunosuppressed solid organ transplant recipients more adversely than immunocompetent individuals. As we carry all of the above knowledge to the clinic, immunologic monitoring should start prior to transplantation and should continue for as long as immunosuppressive therapy is needed.

References

Adam R, Del Gaudio M (2003) Evolution of liver transplantation for hepatocellular carcinoma. *J Hepatol* **39**(6): 888–895.

Allison AC, Eugui EM (2000) Mycophenolate mofetil and its mechanisms of action. *Immunopharmacology* **47**(2–3): 85–118.

Boivin G, Gilbert C, Gaudreau A, Greenfield I, Sudlow R, Roberts NA (2001) Rate of emergence of cytomegalovirus (CMV) mutations in leukocytes of patients with acquired immunodeficiency syndrome who are receiving valganciclovir as induction and maintenance therapy for CMV retinitis. *J Infect Dis* **184**(12): 1598–1602.

Burlingham WJ, O'Connell PJ, Jacobson LM, Becker BN, Kirk AD, Pravica V, Hutchinson IV (2000) Tumor necrosis factor-alpha and tumor growth factor-beta1 genotype: partial association with intragraft gene expression in two cases of long-term peripheral tolerance to a kidney transplant. *Transplantation* **69**(7): 1527–1530.

Bustin SA (2000) Absolute quantification of mRNA using real-time reverse transcription polymerase chain reaction assays. *J Mol Endocrinol* **25**(2): 169–193.

Cai TH, Esterl RM, Jr, Nichols L, Cigarroa F, Speeg KV, Halff, GA (1998) Improved immunosuppression with combination tacrolimus (FK506) and mycophenolic acid in orthotopic liver transplantation. *Transplant Proc* **30**(4): 1413–1414.

Casiraghi F, Ruggenenti P, Noris M, Locatelli G, Perico N, Perna A, Remuzzi G (1997) Sequential monitoring of urine-soluble interleukin 2 receptor and interleukin 6 predicts acute rejection of human renal allografts before clinical or laboratory signs of renal dysfunction. *Transplantation* **63**(10): 1508–1514.

Cox ED, Hoffmann SC, DiMercurio BS, Wesley RA, Harlan DM, Kirk AD, Blair PJ

(2001) Cytokine polymorphic analyses indicate ethnic differences in the allelic distribution of interleukin-2 and interleukin-6. *Transplantation* **72**(4): 720–726.

Dadhania D, Muthukumar T, Ding R, Li B, Hartono C, Serur D, Seshan SV, Sharma VK, Kapur S, Suthanthiran M (2003) Molecular signatures of urinary cells distinguish acute rejection of renal allografts from urinary tract infection *Transplantation* **75**(10): 1752–1754.

Detry O, de Roover A, Delwaide J, Meurisse M, Honore P (2003) The use of mycophenolate mofetil in liver transplant recipients. *Expert Opin Pharmacother* **4**(11): 1949–1957.

Ding R, Li B, Muthukumar T, Dadhania D, Medeiros M, Hartono C, Serur D, Seshan SV, Sharma VK, Kapur S, Suthanthiran M (2003) CD103 mRNA levels in urinary cells predict acute rejection of renal allografts. *Transplantation* **75**(8): 1307–1312.

Egidi MF, Gaber AO (2003) Outcomes of African–American kidney-transplant recipients treated with sirolimus, tacrolimus, and corticosteroids. *Transplantation* **75**(4): 572; author reply 573.

Flohe S, Speidel N, Flach R, Lange R, Erhard J, Schade FU (1998) Expression of HSP 70 as a potential prognostic marker for acute rejection in human liver transplantation *Transpl Int* **11**(2): 89–94.

Gaston RS, Hudson SL, Deierhoi MH, Barber WH, Laskow DA, Julian BA, Curtis JJ, Barger BO, Shroyer TW, Diethelm AG (1992). Improved survival of primary cadaveric renal allografts in blacks with quadruple immunosuppression. *Transplantation* **53**(1): 103–109.

Gjertson DW, Dabrowska DM, Cui X, Cecka JM (2002) Four causes of cadaveric kidney transplant failure: a competing risk analysis. *Am J Transplant* **2**(1): 84–93.

Hahn AB, Kasten-Jolly JC, Constantino DM, Graffunder E, Singh TP, Shen GK, Conti DJ (2001) TNF-alpha, IL-6, IFN-gamma, and IL-10 gene expression polymorphisms and the IL-4 receptor alpha-chain variant Q576R: effects on renal allograft outcome. *Transplantation* **72**(4): 660–665.

Hardinger KL, Stratta RJ, Egidi MF, Alloway RR, Shokouh-Amiri MH, Gaber LW, Grewal HP, Honaker MR, Vera S, Gaber AO (2001) Renal allograft outcomes in African American versus Caucasian transplant recipients in the tacrolimus era. *Surgery* **130**(4): 738–745.

Hariharan S, Johnson CP, Bresnahan BA, Taranto SE, McIntosh MJ, Stablein D (2000) Improved graft survival after renal transplantation in the United States, 1988 to 1996. *N Engl J Med* **342**(9): 605–612.

Hartwig MS, Hall G, Hathaway D, Gaber AO (1993) Effect of organ donor race on health team procurement efforts. *Arch Surg* **128**(2): 1331–1335.

Heeger PS (2003) What's new and what's hot in transplantation: basic science ATC 2003. *Am J Transplant* **3**(12): 1474–1480.

Hoffmann SC, Pearl JP, Blair PJ, Kirk AD (2003) Immune profiling: molecular monitoring in renal transplantation. *Front Biosci* **8**: e444–e462.

Hoffmann SC, Stanley EM, Cox ED, DiMercurio BS, Koziol DE, Harlan DM, Kirk AD, Blair PJ (2002) Ethnicity greatly influences cytokine gene polymorphism distribution. *Am J Transplant* **2**(6): 560–567.

Hoffmann SC, Stanley EM, Darrin Cox E, Craighead N, DiMercurio BS, Koziol DE, Harlan DM, Kirk AD, Blair PJ (2001) Association of cytokine polymorphic inheritance and in vitro cytokine production in anti-CD3/CD28–stimulated peripheral blood lymphocytes. *Transplantation* **72**(8): 1444–1450.

Hutchings A, Purcell WM, Benfield MR (1999) Peripheral blood antigen-presenting cells from African–Americans exhibit increased CD80 and CD86 expression. *Clin Exp Immunol* **118**(2): 247–252.

Hutchinson IV (1999) The role of transforming growth factor-beta in transplant rejection. *Transplant Proc* **31**(7A): 9S–13S.

Hutchinson IV, Pravica V, Sinnott P (2000) Genetic regulation of cytokine synthesis: Consequence for acute and chronic organ allograft rejection. *Graft* 56: 281–286.

Ilyas M, Ammons JD, Gaber AO, Roy S, 3rd, Batisky DL, Chesney RW, Jones DP, Wyatt RJ (1998) Comparable renal graft survival in African–American and Caucasian recipients. *Pediatr Nephrol* 12(7): 534–539.

Jabs WJ, Maurmann S, Wagner HJ, Muller-Steinhardt M, Steinhoff J, Fricke L (2004) Time course and frequency of Epstein–Barr virus reactivation after kidney transplantation: linkage to renal allograft rejection. *J Infect Dis* 190(9): 1600–1604.

Kalow W, Tang BK, Endrenyi L (1998) Hypothesis: comparisons of inter- and intra-individual variations can substitute for twin studies in drug research. *Pharmacogenetics* 8(4): 283–289.

Kerman RH, Kimball PM, Van Buren CT, Lewis RM, Kahan BD (1991) Stronger immune responsiveness of blacks vs whites may account for renal allograft survival differences. *Transplant Proc* 23(1) Pt 1: 380–382.

Koop K, Bakker RC, Eikmans M, Baelde HJ, de Heer E, Paul LC, Bruijn JA (2004) Differentiation between chronic rejection and chronic cyclosporine toxicity by analysis of renal cortical mRNA. *Kidney Int* 66(5): 2038–2046.

Kuehnle I, Huls MH, Liu Z, Semmelmann M, Krance RA, Brenner MK, Rooney CM, Heslop HE (2000) CD20 monoclonal antibody (rituximab) for therapy of Epstein–Barr virus lymphoma after hemopoietic stem-cell transplantation. *Blood* 95(4): 1502–1505.

Li B, Hartono C, Ding R, Sharma VK, Ramaswamy R, Qian B, Serur D, Mouradian J, Schwartz JE, Suthanthiran M (2001) Noninvasive diagnosis of renal-allograft rejection by measurement of messenger RNA for perforin and granzyme B in urine. *N Engl J Med* 344(13): 947–954.

Lindholm A, Welsh M, Alton C, Kahan BD (1992) Demographic factors influencing cyclosporine pharmacokinetic parameters in patients with uremia: racial differences in bioavailability. *Clin Pharmacol Ther* 52(4): 359–371.

Lo A, Egidi MF, Gaber LW, Shokouh-Amiri MH, Nazakatgoo N, Fisher JS, Gaber AO (2004) Observations regarding the use of sirolimus and tacrolimus in high-risk cadaveric renal transplantation. *Clin Transplant* 18(1): 53–61.

Lo A, Stratta RJ, Egidi MF, Shokouh-Amiri MH, Grewal HP, Kizilisik AT, Alloway RR, Gaber AO (2001) Outcome of simultaneous kidney-pancreas transplantation in African–American recipients: a case control study. *Transplant Proc* 33(1–2): 1675–1677.

Louis E, Franchimont D, Piron A, Gevaert Y, Schaaf-Lafontaine N, Roland S, Mahieu P, Malaise M, De Groote D, Louis R, Belaiche J (1998) Tumour necrosis factor (TNF) gene polymorphism influences TNF-alpha production in lipopolysaccharide (LPS)-stimulated whole blood cell culture in healthy humans. *Clin Exp Immunol* 113(3): 401–406.

Marshall SE, McLaren AJ, McKinney EF, Bird TG, Haldar NA, Bunce M, Morris PJ, Welsh KI (2001) Donor cytokine genotype influences the development of acute rejection after renal transplantation. *Transplantation* 71(3): 469–476.

Moreno JM, Rubio E, Gomez A, Lopez-Monclus J, Herreros A, Revilla J, Navarrete E, Sanchez Turrion V, Jimenez M, Cuervas-Mons V (2003) Effectiveness and safety of mycophenolate mofetil as monotherapy in liver transplantation. *Transplant Proc* 35(5): 1874–1876.

Muthukumar T, Ding R, Dadhania D, Medeiros M, Li B, Sharma VK, Hartono C, Serur D, Seshan SV, Volk HD, Reinke P, Kapur S, Suthanthiran M (2003) Serine proteinase inhibitor-9, an endogenous blocker of granzyme B/perforin lytic pathway, is hyperexpressed during acute rejection of renal allografts. *Transplantation* 75(9): 1565–1570.

Ortiz AM, Troncoso P, Kahan BD (2003) Prevention of renal ischemic reperfusion injury using FTY 720 and ICAM-1 antisense oligonucleotides *Transplant Proc* **35**(4): 1571–1574.

Pratschke J, Wilhelm MJ, Kusaka M, Beato F, Milford EL, Hancock WW, Tilney NL (2000) Accelerated rejection of renal allografts from brain-dead donors. *Ann Surg* **232**(2): 263–271.

Qu L, Green M, Webber S, Reyes J, Ellis D, Rowe D (2000) Epstein–Barr virus gene expression in the peripheral blood of transplant recipients with persistent circulating virus loads. *J Infect Dis* **182**(4): 1013–1021.

Racusen LC, Solez K, Colvin RB, Bonsib SM, Castro MC, Cavallo T *et al.* (1999) The Banff 97 working classification of renal allograft pathology. *Kidney Int* **55**(2): 713–723.

Randhawa P (2001) Role of donor kidney biopsies in renal transplantation. *Transplantation* **71**(10): 1361–1365.

Sabek O, Dorak MT, Kotb M, Gaber AO, Gaber L (2002) Quantitative detection of T-cell activation markers by real-time PCR in renal transplant rejection and correlation with histopathologic evaluation. *Transplantation* **74**(5): 701–707.

Sankaran D, Asderakis A, Ashraf S, Roberts IS, Short CD, Dyer PA, Sinnott PJ, Hutchinson IV (1999) Cytokine gene polymorphisms predict acute graft rejection following renal transplantation. *Kidney Int* **56**(1): 281–288.

Satterwhite T, Chua MS, Hsieh SC, Chang S, Scandling J, Salvatierra O, Sarwal MM (2003) Increased expression of cytotoxic effector molecules: different interpretations for steroid-based and steroid-free immunosuppression. *Pediatr Transplant* **7**(1): 53–58.

Schwartz RH (1992) Costimulation of T lymphocytes: the role of CD28, CTLA-4, and B7/BB1 in interleukin-2 production and immunotherapy. *Cell* **71**(7): 1065–1068.

Shoker A, George D, Yang H, Baltzan M (2000) Heightened CD40 ligand gene expression in peripheral CD4+ T cells from patients with kidney allograft rejection. *Transplantation* **70**(3): 497–505.

Sia IG, Wilson JA, Groettum CM, Espy MJ, Smith TF, Paya CV (2000) Cytomegalovirus (CMV) DNA load predicts relapsing CMV infection after solid organ transplantation. *J Infect Dis* **181**(2): 717–720.

Simon T, Opelz G, Wiesel M, Ott RC, Susal C (2003) Serial peripheral blood perforin and granzyme B gene expression measurements for prediction of acute rejection in kidney graft recipients. *Am J Transplant* **3**(9): 1121–1127.

Swanson SJ, Hypolite IO, Agodoa LY, Batty DS, Jr, Hshieh PB, Cruess D, Kirk AD, Peters TG, Abbott KC (2002) Effect of donor factors on early graft survival in adult cadaveric renal transplantation. *Am J Transplant* **2**(1): 68–75.

Tatapudi RR, Muthukumar T, Dadhania D, Ding R, Li B, Sharma VK, Lozada-Pastorio E, Seetharamu N, Hartono C, Serur D, Seshan SV, Kapur S, Hancock WW, Suthanthiran M (2004) Noninvasive detection of renal allograft inflammation by measurements of mRNA for IP-10 and CXCR3 in urine. *Kidney Int* **65**(6): 2390–2397.

Tornatore KM, Biocevich DM, Reed K, Tousley K, Singh JP, Venuto RC (1995) Methylprednisolone pharmacokinetics, cortisol response, and adverse effects in black and white renal transplant recipients. *Transplantation* **59**(5): 729–736.

Tricarico C, Pinzani P, Bianchi S, Paglierani M, Distante V, Pazzagli M, Bustin SA, Orlando C (2002) Quantitative real-time reverse transcription polymerase chain reaction: normalization to rRNA or single housekeeping genes is inappropriate for human tissue biopsies. *Anal Biochem* **309**(2): 293–300.

Turner D, Grant SC, Yonan N, Sheldon S, Dyer PA, Sinnott PJ, Hutchinson IV (1997) Cytokine gene polymorphism and heart transplant rejection. *Transplantation* **64**(5): 776–779.

USRDS (2000) Annual data report. The National Institutes of Health, The National Institute of Digestive Diseases and Kidney Diseases, Bethesda, MD.

Verran DJ, deLeon C, Chui AK, Chapman JR (2001) Factors in older cadaveric organ donors impacting on renal allograft outcome. *Clin Transplant* **15**(1): 1–5.

Wagner HJ, Cheng YC, Huls MH, Gee AP, Kuehnle I, Krance RA, Brenner MK, Rooney CM, Heslop HE (2004) Prompt versus preemptive intervention for EBV lymphoproliferative disease. *Blood* **103**(10): 3979–3981.

Yates CR, Zhang W, Song P, Li S, Gaber AO, Kotb M, Honaker MR, Alloway RR, Meibohm B (2003) The effect of CYP3A5 and MDR1 polymorphic expression on cyclosporine oral disposition in renal transplant patients. *J Clin Pharmacol* **43**(6): 555–564.

Zheng HX, Zeevi A, McCurry K, Schuetz E, Webber S, Ristich J, Zhang J, Iacono A, Dauber J, McDade K, Zaldonis D, Lamba J, Burckart GJ (2005) The impact of pharmacogenomic factors on acute persistent rejection in adult lung transplant patients. *Transpl Immunol* **14**(1): 37–42.

Protocol 15.1: Monitoring solid organ transplant recipients for rejection

A. URINE, BLOOD, AND BIOPSY PROCESSING

1. Urine

1.1   Urine collection

1.1.1   Urines are to be placed on ice and/or into the refrigerator immediately after obtaining sample.

1.1.2   Time collected needs to be noted on the specimen.

1.1.3   Samples need to be processed up at our earliest opportunity.

1.1.4   Log volume and appearance (cloudy, clear).

1.2   Urine processing (process within 1 h of collection).

1.2.1   Spin the total amount of urine for 5 min at 2,200 rpm.

1.2.2   Remove 1 ml urine supernatant and transfer to a 2 ml freezing vial labeled with the accession number only and identify it as urine supernatant. Freeze at –80 in box provided for viral detection study.

1.2.3   Resuspend the pellet in 1 x PBS and transfer to a 1.5 ml capped microcentrifuge tube. Spin for 2.5 min at 3,400 rpm.

1.2.4   Following the wash, remove the supernatant and resuspend in 0.5 ml 1 x PBS.

1.2.5   Mix 20 µL of this suspension, 20 µL Trypan Blue and 60 µL 1 x PBS. Count the number of leukocytes present. Note the presence of epithelial cells, bacteria, etc.

1.2.6   Pellet the remaining suspension by spinning at 1,000 rpm, aspirate supernatant, either add 0.5 ml RNA later and freeze at 20°C or proceed with RNA extraction as discussed later.

2. Whole blood

2.1   Whole blood collection
2.1.1 Three 7 ml ACD tubes are to be drawn, mixed immediately, and kept at room temperature until picked up.
2.1.2 Check tubes for a complete name, date drawn, and second identifier, such as DOB.
2.1.3 If there is any danger of the specimen reaching temperatures less than 60°F, pack the tubes inside a double container, preferably Styrofoam.

2.2   Specimen logging
2.2.1 Note the temperature of the specimen if other than RT when received.
2.2.2 Record total volume prior to spinning; note the packed cell volume (PCV) and any hemolysis after spinning.

2.3   Specimen processing
2.3.1 Pour all blood into a 50 ml conical tube. Note the volume then fill to the 50 mark with RPMI with 2% FCS. Mix briefly and spin at 2,200 rpm for 5 min with the brake in the OFF position.
2.3.2 Aspirate the supernatant to within 1/4 inch of the buffy coat. Note PCV prior to doubling the volume in increments of 10 with RPMI with 2% FCS. Mix well by rocking to break up any leukocyte clumps.
2.3.3 Overlay 5 ml Ficoll with 10 ml red cell suspension in a 15 ml conical tube. Spin for 30 min at 1,400 rpm with the brake on LOW.
2.3.4 Remove the mononuclear interface to a 15 ml conical tube containing RPMI as soon as the centrifuge stops. Top the tube off with RPMI and spin at 1,600 rpm for 5 min. Repeat this wash.
2.3.5 Resuspend the pellet in 2 ml cold 1 x PBS and count. Specimens containing a large number of platelets should be rewashed. Note viability on the worksheet.
2.3.6 Pellet the cells by spinning for 2.5 min at 3,400 rpm aspirate PBS and either resuspend in 0.5 ml RNA Later and freeze at −20°C or proceed with the RNA extraction.

3. Biopsy   Tissue is either stored in RNA Later at −20°C
                                 until it is processed or processed immediately

B. RNA ISOLATION, AND DNASE TREATMENT PROTOCOL

Using RNAqueous™ kit and the manual provided by Ambion

Buffer preparation: 64% ethanol: Add 38.4 ml 100% ACS grade
                                ethanol to the bottle that contains 21.6 ml
                                nuclease-free water (provided in the kit) for a final
                                concentration of 64% ethanol.
                                Wash solution 2/3: dilute the solution with 64%
                                ml 100% ACS grade ethanol before use

1. Rinse cells once in PBS by gentle resuspension
    and centrifugation, to remove any RNA Later
    (Qiagen) solution
2. Remove PBS and add 300 uL lysis/binding
    solution per 10^6 to 10^7 cells. *If the lysate is
    extremely viscous, further dilute with
    lysis/binding solution*
3. Add an equal volume of 64% ethanol and
    mix well by repeated pipetting
4. Insert an RNAqueous™ filter cartridge
    (Ambion) into one of the RNase-free
    collection tubes supplied. Apply the
    lysate/ethanol mixture to the filter. *700 μL is
    the maximum volume that can be applied at
    one time. It is not recommended to exceed
    1,800 μL lysate/ethanol mixture per filter*
5. Centrifuge for 1 min, discard the flow-through,
    and reuse the tube for the washing steps.
6. Wash the filter cartridge with 700-μL wash
    solution #1 (provided in kit), centrifuge for 1
    min, discard the flow-through, and reuse the
    tube for the subsequent washes
7. Wash the filter cartridge twice with 500-μL
    wash solution #2/3, centrifuge for 1 min, and
    discard the flow-through, centrifuge for an
    extra 2 min to remove any traces of wash
    solution
8. Transfer filter cartridge to fresh collection
    tube. Add 25 μL of elution solution to the
    center of the filter. Incubate the tube with
    cartridge in a heat block set at 65–70°C, for
    10 min. Recover the elute by centrifuging for
    1 min
9. To maximize recovery of RNA, repeat step 7,
    except do not transfer the cartridge to a fresh
    collection tube

10. To measure optical density, take 10 μL of the eluted RNA and add 90 uL of DEPC water. Measure the absorbance at 260 nm using the GeneQuant Pro spectrophotometer (Amersham Pharmacia Biotechnology)

DNase treatment using DNA-free™ kit and the manual provided by Ambion.

Use individually wrapped tubes and tips.
1. Add 0.1 volume** of 10X DNase 1 buffer and 1 uL of DNase 1–(2 units) to the RNA. Mix gently and incubate at 37°C for 20–30 min
2. Add 0.1 volume or 5 μL, whichever is greater, of the resuspended DNase inactivation reagent to the sample and mix well
3. Incubate for 2 min at room temperature.
4. Centrifuge the tube for 1 min to pellet the DNase inactivation reagent
5. RNA should be stored at –80°C until used
** Volume means 0.1X the amount of RNA

C. REAL-TIME PCR PROTOCOL

1. TaqMan® probes and primers.
   a. Probes and primers for the gene of interest are designed using the Primer Express® software (PE Applied Biosystems, CA, USA).
   b. All primer pairs were designed to produce amplicons smaller than 150 bp
2. Preparing *TaqMan® probes* from ABI already resuspended at 100 μM.
   a. A dilution of 1/20 is made to give a 5μM solution
   b. Aliquot the probe solution and store in the dark at 20°C. It is not recommended to thaw and freeze more than twice
3. The *primers* arrive lyophilized with the amount given on the tube in pmols (such as 150.000 pmol which is equal to 150 nmol).
   a. If X nmol of primer is resuspended in XμL of H_2O, the resulting solution is 1 mM
   b. Freeze this stock solution in aliquots
   c. When the 1 mM stock solution is diluted 1/100, the resulting working solution will be 10 μM
   d. To get the recommended 50–900 nM final primer concentration in 50 μL

reaction volume, 0.25–4.50 µL should be used per reaction (2.5 µL for 500 nM final concentration)

The pre-developed (**PDAR**) **primers and probes** are supplied as a mix in one tube. It is recommended to use 2.5 µL in a 50 µL reaction volume.

4. Setting up one-step TaqMan® reaction

    Using the TaqMan® EZ RT-PCR kit (PE Applied Biosystems) and the manual provided

    a) Prepare a reagent mix containing all the PCR components except target RNA.
        Preparing a reagent mix is recommended in order to increase the accuracy of the results

Reagent mix	Volume for one sample	Final concentration
RNase-free H_2O	14.5 uL [18 uL For PDAR 18S MIX]	
5X TaqMan® EZ buffer	10.0 µL	1X
Manganese acetate (25 mM)	6.0 µL	3 mM
dATP (10 mM)	1.5 µL	300 µM
dCTP (10 mM)	1.5 µL	300 µM
dGTP (10 mM)	1.5 µL	300 µM
dUTP (20 mM)	1.5 µL	600 µM
Primer F (10_M) *	2.5 µL	500 nM
Primer R (10_M) *	2.5 µL	500 nM
TaqMan® Probe (5 µM)*	1.0 µL	100n M
rTth DNA Polymerase (2.5 U/µL)	0.25 µL	0.1 U/µL
AmpErase UNG (1 U/µL)	0.5 µL	0.01 U/µL
Total mix	45 µL	
Target RNA	5 µL	

* If a PDAR is used, 2.5 µL of primer + probe mix used

    b) Amplification of the target genes tested and the 18S RNA is to be performed in duplicate in the same plate
    c) In each plate a non-template control (for each reagent mixture) will be run to test for any primer-dimmer or contamination
    d) In the case of using an absolute standard curve or relative standard curve, the standard must be run in duplicate in each plate
    e) The program consists of heating at 50°C for 2 min, 60°C for 30 min, and 95°C for 5 min, followed by 40 cycles of a two-

stage temperature profile of 94°C for 20 sec and 62°C for 1 min

f) Accumulation of the PCR products is detected by directly monitoring the increase in fluorescence of the reporter dye

g) Data points collected in this manner are analyzed at the end of thermal cycling

5. Analyzing and interpreting the results

a) Save the run before it starts by giving it a name (not as untitled). Also at the end of the run, first save the data before starting to analyze

b) The choice of dye component should be made correctly before data analysis. Example; if the probe is labeled with FAM and VIC is chosen there will be some result but the wrong one

c) When analyzing the data ensure that the default setting for baseline is 3–15. If any C_t value is <15, the baseline should be changed accordingly (the baseline stop value should be 1–2 smaller than the smallest C_t value)

d) A threshold for the amplification of each gene of interest is set by drawing a line that intersects the exponential phase of the logarithmic amplification curves for all samples being analyzed for expression of target gene (A threshold line should be drawn above the background noise and below the plateau region). The cycle number at which the threshold line intersects the linear curve for each sample is used to determine the threshold cycle (C_t) value

e) Rn^+ is the Rn value of a reaction containing all components; Rn^- is the Rn value of an unreacted sample (baseline value or the value detected in NTC). ΔRn is the difference between Rn^+ and Rn^-. It is an indicator of the magnitude of the signal generated by the PCR

Final quantification can be done by any method. For the purposes of renal transplant monitoring relative quantification methods are satisfactory (see Chapters 2, 3, 6, and 7).

Real-time PCR applications in hematology

16

Anne M. Sproul

The success of treatment of patients with leukemia is judged by several criteria. For the patient and clinician the most desirable outcome is sustained remission from disease and long-term survival. Unfortunately it is unlikely that complete eradication of leukemic cells is ever achieved. Hematological remission is defined as fewer than 5% blast cells in the bone marrow as determined by morphology. Disease levels below this threshold detected by more sensitive methods such as flow cytometry, cytogenetics (including FISH), and PCR are referred to as minimal residual disease (MRD). The sensitivity of PCR has extended the detection limit to one leukemic cell in one million normal cells.

Although qualitative or endpoint PCR has played a major role in the diagnosis and monitoring of leukemia, it is now accepted that quantitative data is more informative. Several large trials including molecular monitoring have shown that some patients still have PCR detectable disease during and at the end of treatment, for example the *AML1ETO* transcript may persist in AML patients in long-term remission (Jurlander *et al.*, 1996). Whereas these studies have been able to predict poor outcome for some patient groups, e.g. *PMLRARA* positivity at the end of treatment indicates impending relapse (Lo Coco *et al.*, 1999), it has been more difficult to gain accurate prognostic information from PCR results, e.g. childhood acute lymphoblastic leukemia (ALL) (van Dongen *et al.*, 1998) or chronic myeloid leukemia (CML) (Hochhaus *et al.*, 2000). However by measuring the MRD levels it has been shown that it is possible to stratify risk groups on the basis of MRD level at specific time points in treatment (Biondi *et al.*, 2000).

Real-time PCR is the only truly quantitative method available. It offers other advantages over conventional PCR, for example more rapid turnaround as no post PCR processing is necessary and since closed systems are used the risk of contamination is abolished. These are important in diagnostic laboratories. This chapter refers to real-time PCR using hydrolysis or TaqMan® probes; this reflects the author's personal experience not bias. The same analyses can be performed using hybridization probes and should be regarded as interchangeable. However the reagents referred to have been designed for assays using probes and are not validated for the use of DNA-binding dyes such as SYBR® Green 1.

16.1 Specimens

Bone marrow aspirates are the most common specimen for MRD monitoring in acute leukemia. Peripheral blood is also used but the sensitivity is at least ten-fold lower than in marrow. For patients in the early stages of treatment, blood may be used; however to achieve maximal sensitivity bone marrow should be analyzed wherever possible. Peripheral blood monitoring is preferred for patients with CML.

16.2 Specimen quality

It is sometimes difficult to obtain good quality specimens for molecular analysis; there are often conflicting demands on the clinician obtaining the specimen to provide specimens for morphology, flow cytometry, and cytogenetics in addition. This may result in clotted or small specimens. The amount of nucleic acid isolated may be very low. Specimens may be posted to a central laboratory for analysis and be more than 48 hours old. These factors must be borne in mind when addressing the matter of sensitivity; even the best assay is reliant on the quantity of input material; one cannot assume a sensitivity of 10^{-5} if only 10^4 molecules were analyzed.

16.3 Template preparation

Sensitive accurate quantitative PCR relies on the use of high-quality templates, the protocols below work and relatively simple and inexpensive; however there are several other methods that work equally well.

16.4 DNA isolation

The choice of DNA extraction method is determined by several factors; DNA yield and quality, convenience, cost and individual preferences. Although PCR can yield results from very poor specimens, e.g. forensic material, the use of poor quality specimens will compromise accurate and sensitive diagnosis. The exception to this is the use of paraffin embedded tissues for the extraction of nucleic acids. This valuable diagnostic resource can yield DNA suitable for analysis; however, the quality of the isolated DNA is dependent on the treatment of the tissue prior to and during fixation. The DNA will be of lower molecular weight, however due to the smaller amplicons suited to real-time PCR, good results can be obtained. Many commercially available kits produce high-quality DNA preparations that are good templates for real-time PCR, however one drawback is the low concentration of the DNA produced. Whilst these amplify very efficiently; there may be too few DNA molecules to ensure maximal sensitivity.

16.5 PCR inhibition

Some DNA specimens may be contaminated with inhibitors of PCR. This becomes apparent when values obtained from standard curves to not correlate with DNA concentration calculated from optical density readings. Moppet *et al.* (2003) have demonstrated that the addition of bovine serum albumin (0.04% final concentration) to the PCR mix can overcome some degree of inhibition.

16.6 RNA isolation

Solvent based methods using commercially available reagents such as Tri-Reagent® (Sigma) or Trizol® (Invitrogen) are simple and adaptable to varying cell numbers, but involve the use of toxic, organic solvents. Column based kits, from suppliers such as Qiagen and Promega, circumvent the use of organic solvents but are far less flexible in terms of increasing cell numbers. If centrifugation of the columns is performed the process becomes tedious for more than a few specimens and cross contamination is more likely. If large numbers of specimens are to be processed the use of vacuum compatible columns is recommended. As with DNA preparation for MRD studies, the final RNA concentration may not be sufficiently high.

DNA contamination is a frequent problem of RNA preparation. Although most applications will design the real-time PCR primers and probes to span exons (some genes are single exons and this is not possible), cDNA synthesis and subsequent PCR may be less efficient and sensitivity reduced due to lower input of cDNA. Removal of contaminating DNA by DNase digestion is easily carried out by utilizing RNase-free DNase followed by the Rneasy Minelute Kit (Qiagen). Our laboratory uses Tri-Reagent® for routine isolation of total RNA from blood and marrow cells.

Specimens for RNA based MRD analysis should be processed as soon as possible to prevent decrease in sensitivity due to RNA degradation. Blood specimens may be drawn directly into PAXgene tubes (Qiagen), however these are expensive, but are of great value as blood specimens may be shipped at room temperature without RNA degradation.

16.7 cDNA synthesis

The choice of primer for cDNA synthesis should reflect the downstream PCR requirements. Diagnostic real-time RT-PCR will require amplification of at least one control transcript and the specific test sequence. Therefore the use of specific reverse PCR primers to prime cDNA synthesis is not recommended as separate cDNA reactions would be required and subsequent analysis of other sequences would require further cDNA synthesis. However, if maximum sensitivity is required, cDNA synthesis using the downstream PCR primer will increase the yield of sequence specific PCR product.

Oligo d(T) priming utilizing the poly(A) tail present at the 3' end of most (but not all) mRNA molecules is used in many applications. It has the major advantage of priming cDNA synthesis almost exclusively from mRNA. However, for detection of fusion transcripts in leukemia, this method results in lower sensitivity, as many of the fusions occur towards the 5' end of the mRNA molecule and oligo d(T) primed cDNA is heavily biased towards 3' ends. It is also less tolerant of slightly degraded RNA templates.

Random priming using random hexamers is the method of choice for most applications in leukemia diagnosis and MRD monitoring. The cDNA produced is representative of the entire expressed RNA population; this does include the major component of cellular RNA the ribosomal RNA (rRNA), however this does not appear to compromise the sensitivity of real-time PCR. The Europe against Cancer (EAC) program designed and validated primer and probe combinations for the identification and monitoring of fusion transcripts in leukemia (Gabert *et al.*, 2003). Optimal conditions for cDNA synthesis were also investigated and the recommended protocol is used in our laboratory.

16.8 Relative versus absolute quantitation

There are strong arguments for the use of standard curves and absolute quantitation; however most clinicians want to know how an individual patient is responding to treatment. There has been much discussion of the most meaningful units for expression of MRD results; copies/μL of blood (most MRD monitoring for acute leukemia is performed on bone marrow); copies/μg of RNA (this is only an indication of the RNA concentration and cDNA synthesis efficiency). Perhaps the most useful is to express the ratio of test gene to a control gene expression level (i.e. the target amount is divided by the control gene amount to obtain a normalized target value), however this is an expensive option. Most diagnostic laboratories insist on the inclusion of standard curves with each batch of test samples. Centers dealing with leukemia patient samples require to provide quick turnaround time; may have limited budgets; have access to a shared instrument only, and perform multiple assays per run; therefore multiple standard curves prove expensive for small numbers of specimens. Larger centers with high throughput use absolute quantitation; the reproducibility achieved by the best instruments is such that it is not necessary to run a standard curve with each batch of specimens. However, the principal investigator must ensure that the standard curve data are robust and not subject to individual operator variation; it is vital that the low control is included in each run if sensitivity is being reported. *Figure 16.1* illustrates the use of absolute quantitation of *BCRABL* and *ABL* and *Figure 16.2* illustrates the expression of the results as a percentage ratio of *BCRABL/ABL* of an individual patient.

Relative quantitation using the diagnostic specimen as the calibrator may be the method of choice. Results can be expressed as log reduction from the time of diagnosis. This is informative to the clinician and patient. Reports must state clearly that the transcript level is relative to the diagnostic level, not the sensitivity of the assay as $\Delta\Delta C_t$ can result in MRD levels of <1.00E-07. Once validated, relative quantitation assays are convenient to run as multiple assays can be performed on one plate. Cell lines may be used as sources of calibrator samples; however the data obtained are relative to an arbitrary calibrator and do not reflect the percentage of malignant cells directly, this approach is useful for examining serial specimens from patients.

The final choice of method will depend on the circumstances of the individual patient and laboratory. Our laboratory uses both absolute quantitation expressing results as a ratio of test gene to control gene and relative quantitation using the individual patient's diagnostic sample as calibrator for their follow-up specimens. However the transcript ratio is established for the diagnostic sample by absolute quantitation and regular quality control is performed using standard curves.

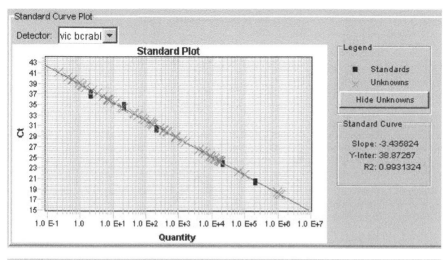

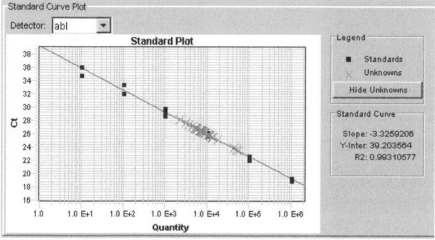

Figure 16.1

Standard curves for absolute quantitation of *BCRABL* in patients undergoing treatment for CML using *cABL* as control gene.

16.9   Control genes for MRD in leukemia

Whether absolute or relative quantitation is carried out, the choice of control gene is very important. Older publications refer to sequences such as β-actin, GAPDH or 18S ribosomal RNA for gene expression assays. The first two have pseudogenes and inaccuracies may result from DNA contamination, the latter is expressed at very high levels and is not a suitable control for single copy genes without limiting primer concentrations. Both test and control genes must amplify with the same efficiency for the $\Delta\Delta C_t$ method to be valid. Another important consideration is that the expression of the control gene may be influenced by the treatment the patient is

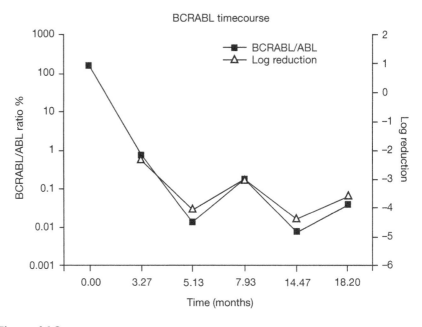

Figure 16.2

Tracking response to therapy in a patient with CML.

undergoing and should remain stable during the course of treatment. The Europe Against Cancer (EAC) project addressed the choice of control genes (Beillard *et al.*, 2003); *ABL*, *GUS*, and *B2M* (β-2- microglobulin) emerged as the most suitable. For genomic DNA targets recommended control genes include *B2M*, *ALB* (albumin), and *HBB* (β-globin).

16.10   Controls for real-time PCR

The use of appropriate controls is vital for accurate diagnostic PCR. If absolute quantitation is performed the most convenient standard is a plasmid containing the target sequence. Kits for cloning of PCR products are available from many sources, but some expertise in manipulation of plasmids and bacteria is required. Plasmid standards for many of the common fusion transcripts are available commercially (for example, from Ipsogen, Marseille, France). cDNA produced from *in vitro* transcribed RNA can also be used to generate standard curves; however it cannot be assumed that the efficiency of reverse transcription is equal to that of the test specimens; also the RNA is subject to degradation.

Plasmid DNA should be isolated using high-quality commercial kits such as those manufactured by Qiagen or Promega. The concentration of the purified DNA should be determined accurately. It is vital that the size (assuming a molecular weight of 660 kD for a base pair) of the plasmid and the inserted sequence are known in order to calculate the number of molecules per microlitre using Avogadro's number (see Chapter 2). As DNA can stick to the walls of test tubes; it is advisable to use carrier DNA. We prepare our plasmid dilutions in a solution of 10 µg/ml lambda phage DNA in 0.1x TE (Tris-EDTA) buffer. The choice of carrier DNA should be considered carefully to ensure that there is no cross reactivity with PCR primers and probes. The range of standard concentrations should reflect the dynamic range of the assay and the values expected from patient samples; in practice this is often from 10^6 to less than 10 molecules per microlitre. Standards should be aliquoted and stocks stored at –20°C, working aliquots should be stored at 4°C. These should be discarded when the reproducibility of the lower copy number replicates deteriorates (usually between one to six months).

In addition to positive controls negative controls must also be incorporated; ideally these should include reagent blanks for all stages of processing and PCR master mix containing probes and primers but no template control (NTC). In practice we include a specimen processed in the same batch as the test samples but known to be negative for the target being analyzed (NAC; no amplification control).

16.11   Assay design

Design of primers and probes has been described elsewhere (see Chapters 5 and 7). The value of standardized reagents cannot be stressed highly enough (Gabert et al., 2003). However, where this is not possible, sequencing and design of patient specific reagents can be successful. Sufficient template should be added to ensure adequate sensitivity. The number of replicates required varies with application; however most laboratories would consider three to be the minimum for accurate quantitation and removal of outliers. The definition of replicate also varies; some centers process separate aliquots of cells, but it is more common to process one batch and set up multiple wells adding DNA or cDNA to the individual wells. If only one replicate is positive with a high C_t, the result should be confirmed using more replicates.

16.12   Laboratory precautions

The measures necessary to prevent PCR contamination in real-time PCR are broadly similar to those required for qualitative PCR. Due to the nature of the PCR process, contamination is a major potential problem. Contamination with PCR products may lead to false positive results. Unexpected positive PCR may result in the patient requiring re-sampling and cause unnecessary anxiety. There should be separate laboratories for specimen preparation, reagent preparation, PCR set-up and post-PCR processing. Although one of the major advantages of real-time PCR is that there is no need for post-PCR processing, many laboratories carry out qualitative PCR for the same target sequences. If it is not possible to have separate laboratories, careful consideration should be given to the design and utilization of the available space. Personal experience has shown that the majority of contamination problems occur during specimen processing where large numbers of specimens are being handled at one time.

16.13 PCR reaction set-up

PCR reactions should be set up in a dedicated laboratory adhering to clinical diagnostic laboratory rules. It may be recommended that a laminar flow cabinet with means of ultra-violet sterilization should be used. Opinion is divided on the necessity of laminar flow conditions; the operator may be lulled into a false sense of security and not heed the basic rules for setting up PCR.

1. Use aerosol resistant tips
2. Use dedicated micropipettes only. Reproducibility between runs will be affected by the use of different pipettes
3. If possible reagents, such as primer and probe stocks, and homemade master, should be prepared and aliquoted on an entirely separate site. Use the smallest aliquots that are practical. Discard empty containers; do not refill. Always discard an aliquot if contamination is suspected
4. Protect reagents for real-time PCR from light. Store aliquots in amber tubes
5. Wear gloves at all times and change frequently
6. Wear disposable gowns or keep a set of laboratory coats for PCR set-up. If different laboratories are used, change laboratory coats between areas
7. Be aware of aerosol generation. Centrifuge specimen tubes to ensure that there is no DNA/cDNA on the caps that may contaminate your gloves
8. Isolate known positive specimens from expected low level MRD specimens
9. Include NTC and NAC controls
10. Use dUTP and UNG containing master mixes

The use of plasmids as positive controls may be unavoidable; however concentrated plasmid DNA solutions contain very high numbers of target molecules. If possible plasmid DNA isolation and standard preparation should be carried out in a separate location. If possible, seal the PCR vessels containing test samples and negative controls before pipetting positive controls whether they are plasmid or cell line cDNA/DNA.

16.14 Interpretation and quantitation

A sample is considered positive if the C_t value is less than the number of cycles performed. However this does not mean that quantitation is possible in every case. Accurate quantitation is only possible within the reproducible range of the assay, i.e. higher or equal to the reproducible sensitivity. For plasmid controls this is usually between five and ten copies.

'Negative specimens' are those where no amplification has occurred and the control gene amplification is satisfactory (see below). A sample can also be considered negative if the lowest C_t is within the range of the NAC controls or if all of its C_t values are more than 4 cycles apart from the highest C_t of the maximal sensitivity control.

Arbitrary cut-off values for control gene levels may be set. For RT-PCR in our laboratory we do not report a test sequence as 'undetected' if the *cABL* copy number is less than 10^4 copies per microlitre of cDNA or if the mean C_t is greater than 30. Specimens that fail to achieve the above criteria would be re-purified and concentrated prior to reverse transcription, as the most common reasons for such failures are DNA contamination, low RNA concentration or RNA degradation. Examination of RNA integrity can be performed using the Agilent 2100 Bioanalyser (Agilent Technologies) or gel electrophoresis. However RNA degradation is rare in fresh samples properly processed. If the sample was more than 48 hours old at the time of processing some RNA turnover/degradation is inevitable, clinical reports of negative findings should include comments regarding time to receipt of specimens.

16.15   Sensitivity

The sensitivity that can be obtained by real-time PCR analysis depends on several variables; some target independent and specimen dependent, i.e. specimen sensitivity; others dependent on the nature of the target, i.e. assay sensitivity. The assay sensitivity can be established by performing dilution experiments, these can be dilutions of patient material, cell lines or plasmids. The potential presence of fusion genes in healthy individuals means that MRD levels of less than 10^{-6} should be interpreted with caution. Sensitivity of allele-specific oligonucleotide (ASO) assays for rearranged antigen receptor genes are usually established by performing dilution of the patient's leukemic cell DNA in mononuclear cell DNA from several donors.

However, the values obtained do not reflect the sensitivity obtained for the actual test sample. This is dependent on the quality and quantity of the template used. If absolute quantitation has been used, specimen sensitivity can be estimated by dividing the maximal reproducible sensitivity of the test gene by the copy number of the control gene, e.g. in an assay where the 10 copy/l standard is the limit of reproducibility and the *ABL* gene copy number is $2.3 \times 10^4/\mu l$; the sensitivity is $10/2.34 \times 10^4$ or $4.27 \times 10^{-4.}$

For DNA based ASO approaches the test DNA is usually quantified by amplification of a control gene such as *B2M* or *ALB* in an absolute quantitation assay using genomic DNA of known concentration to prepare standards.

16.16   Targets for detecting MRD

16.16.1 Fusion transcripts

The recurrent chromosome translocations in leukemia are excellent targets for the tracking of residual disease in patients undergoing treatment. The fusion genes that result from the chromosome rearrangements are specific (in the main) for leukemic cells. PCR assays that use primers complimentary to the fusion partners can be highly sensitive due to very low or absent background. However, some leukemia specific fusion transcripts, such as *BCRABL*, are found in a significant proportion of healthy individuals (Bose *et al.*, 1998). Sensitivities equivalent to one positive cell in one hundred thousand negative cells can be achieved by nested qualitative RT-PCR reactions. RT-PCR is applicable to the detection of fusion transcripts due to the great distances spanned by the introns in the partner genes. The use of the spliced RNA transcript not only overcomes the problem of long PCR; it increases the sensitivity due to the presence of multiple copies of the fusion transcript rather than the single copy of the rearranged gene. However, the labile nature of RNA is a major disadvantage in the use of RT-PCR assays as diagnostic tests. Not only does RNA degrade as cells lose viability; purified RNA is vulnerable to the ubiquitous presence of ribonucleases in the laboratory environment. The problems associated with the use of RNA can be minimized with a few sensible precautions in the laboratory and the optimization of specimen collection and delivery.

As quantitative gene expression has been described in great detail in Chapter 7, another complete protocol is not presented. A brief outline of our protocol for the quantitation of *BCRABL* in patients with CML treated with imatinib is given as an example.

- We perform gene expression TaqMan® assays using 10 µl assays per well. This is possible as 1 µl of cDNA may be dispensed accurately (however it is difficult to dispense such small volumes of concentrated genomic DNA)
- Do not dilute cDNA from patients on imatinib therapy
- Prepare large stocks of diluted probe and primers and store in small aliquots in amber tubes at –20°C

Constituent	Volume per well	Volume per 96-well plate
Molecular biology grade H_2O	3 µl	315 µl
10 µM forward primer	0.3 µl	31.5 µl
10 µM reverse primer	0.3 µl	31.5 µl
5 µM TaqMan® probe	0.4 µl	42 µl

- Each well should contain:

2X Master mix	5 µl
Probe and primer mix	4 µl
Template (cDNA or plasmid)	1 µl

- Standard curves for both *BCRABL* and *ABL* are used on each run. We use an in-house plasmid construct that contains a full-length b3a2 *BCRABL* transcript, permitting the same plasmid to be used for both. We use the plasmid standards from Ipsogen to calibrate our own preparations. However, both *BCRABL* and *ABL* standards must be purchased, as they are different constructs
- Use triplicate reactions for all tests and controls
- Carry out 50 cycles of PCR to ensure that samples with high C_t values represent true amplification
- Analyze results and examine replicates
- Export results to Microsoft Excel
- Express results as *BCRABL*/total *ABL* × 100
- Express individual specimen sensitivity as last reproducible point on standard curve/*ABL* copy number

16.16.2 Rearrangements of immunoglobulin/TCR genes in lymphoid neoplasia

The rearrangement of multiple gene segments to form functional immunoglobulin or T-cell receptor genes occur early in lymphocyte development and each lymphocyte has a particular combination of variable (V), diversity (D) and joining (J) segments. The junctional regions V-(D)-J are clonal markers of lymphoid neoplasia and can be used as targets for diagnosis and disease tracking. However the 'uniqueness' of each rearrangement means that it is not possible to use standardized real-time PCR reagents; in most instances individual clonal rearrangements are sequenced and primers and/or probes are designed. As probes are the most costly component in a real-time PCR assay, it would be prohibitively expensive to design patient specific probes. Therefore, several groups have designed assays using one ASO primer and a germline primer and probe (*Figure 16.3*). This reduces the complexity of the assays as only one forward primer be designed, although in practice at least three versions of the ASO are designed. These approaches are discussed in depth by (van der Velden *et al.*, 2003). The use of ASO primers in lymphoid malignancies is highly specialized and requires experience in design and testing of reagents to produce the highly sensitive and

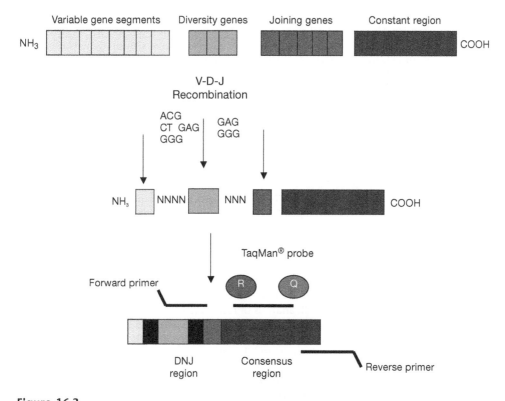

Figure 16.3

Generation of allele specific primers for MRD quantitation.

specific assays required for monitoring and adapting treatment. However, due to the background of similar rearrangements and limited repertoire at some loci, these assays do not routinely achieve the sensitivity and specificity of those using chromosome translocations as targets.

References

Beillard E, Pallisgaard N, Van der Velden VH, BI W, Dee R, Van der Schoot E, Delabesse E, Macintyre E, Gottardi E, Saglio G, Watzinger F, Lion T, Van Dongen JJ, Hokland P, Gabert J (2003) Evaluation of candidate control genes for diagnosis and residual disease detection in leukemic patients using 'real-time' quantitative reverse-transcriptase polymerase chain reaction (RQ-PCR) – a Europe against cancer program. *Leukemia* **17**(12): 2474–2486.

Biondi A, Valsecchi MG, Seriu T, D'Aniello E, Willemse MJ, Fasching K, Pannunzio A, Gadner H, Schrappe M, Kamps WA, Bartram CR, Van Dongen JJ, Panzer-Grumayer ER (2000) Molecular detection of minimal residual disease is a strong predictive factor of relapse in childhood B-lineage acute lymphoblastic leukemia with medium risk features. A case control study of the International BFM study group. *Leukemia* **14**(11): 1939–1943.

Bose S, Deininger M, Gora-Tybor J, Goldman JM, Melo JV (1998) The presence of typical and atypical BCR-ABL fusion genes in leukocytes of normal individuals: biologic significance and implications for the assessment of minimal residual disease. *Blood* **92**(9): 3362–3367.

Fehse B, Chukhlovin A, Kuhlcke K, Marinetz O, Vorwig O, Renges H, Kruger W, Zabelina T, Dudina O, Finckenstein FG, Kroger N, Kabisch H, Hochhaus A, Zander AR (2001) Real-time quantitative Y chromosome-specific PCR (QYCS-PCR) for monitoring hematopoietic chimerism after sex-mismatched allogeneic stem cell transplantation. *Journal of Hematotherapy & Stem Cell Research,* **10**(3): 419–425.

Gabert J, Beillard E, Van der Velden VH, Bi W, Grimwade D, Pallisgaard N, Barbany G, Cazzaniga G, Cayuela JM, Cave H, Pane F, Aerts JL, De Micheli D, Thirion X, Pradel V, Gonzalez M, Viehmann S, Malec M, Saglio G, Van Dongen JJ (2003) Standardization and quality control studies of 'real-time' quantitative reverse transcriptase polymerase chain reaction of fusion gene transcripts for residual disease detection in leukemia – a Europe Against Cancer program. [Review]. *Leukemia* **17**(12): 2318–2357.

Hochhaus A, Weisser A, La Rosee P, Emig M, Muller MC, Saussele S, Reiter A, Kuhn C, Berger U, Hehlmann R, Cross NC (2000) Detection and quantification of residual disease in chronic myelogenous leukemia. [Review]. *Leukemia* **14**(6): 998–1005.

Jurlander J, Caligiuri MA, Ruutu T, Baer MR, Strout MP, Oberkircher AR, Hoffmann L, Ball ED, Frei-Lahr DA, Christiansen NP, Block AM, Knuutila S, Herzig GP, Bloomfield CD (1996) Persistence of the AML1/ETO fusion transcript in patients treated with allogeneic bone marrow transplantation for t(8;21) leukemia. *Blood* **88**(6): 2183–2191.

Langerak AW, Szczepanski T, Van der Burg M, Wolvers-Tettero IL, Van Dongen JJ, (1997) Heteroduplex PCR analysis of rearranged T cell receptor genes for clonality assessment in suspect T cell proliferations. *Leukemia* **11**(12): 2192–2199.

Lo Coco F, Diverio D, Falini B, Biondi A, Nervi C, Pelicci PG (1999) Genetic diagnosis and molecular monitoring in the management of acute promyelocytic leukemia. [Review]. *Blood* **94**(1): 12–22.

Maas F, Schaap N, Kolen S, Zoetbrood A, Buno I, Dolstra H, De Witte T, Schattenberg A, Van de Wiel-van Kemenade E (2003) Quantification of donor and recipient hemopoietic cells by real-time PCR of single nucleotide polymorphisms. *Leukemia* **17**(3): 630–633.

Moppett J, Van der Velden VH, Wijkhuijs AJ, Hancock J, Van Dongen JJ, Goulden N (2003) Inhibition affecting RQ-PCR-based assessment of minimal residual disease in acute lymphoblastic leukemia: reversal by addition of bovine serum albumin. *Leukemia* **17**(1): 268–270.

Pongers-Willemse MJ, Seriu T, Stolz F, D'Aniello E, Gameiro P, Pisa P, Gonzalez M, Bartram CR, Panzer-Grumayer ER, Biondi A, San Miguel JF, Van Dongen JJ (1999) Primers and protocols for standardized detection of minimal residual disease in acute lymphoblastic leukemia using immunoglobulin and T cell receptor gene rearrangements and TAL1 deletions as PCR targets: report of the BIOMED-1 CONCERTED ACTION: investigation of minimal residual disease in acute leukemia.[see comment]. *Leukemia* **13**(1): 110–118.

Van der Velden VH, Hochhaus A, Cazzaniga G, Szczepanski T, Gabert J, Van Dongen JJ (2003) Detection of minimal residual disease in hematologic malignancies by real-time quantitative PCR: principles, approaches, and laboratory aspects. [Review]. *Leukemia* **17**(6): 1013–1034.

Van der Velden VH, Wijkhuijs JM, Jacobs DC, Van Wering ER, Van Dongen JJ (2002) T cell receptor gamma gene rearrangements as targets for detection of minimal residual disease in acute lymphoblastic leukemia by real-time quantitative PCR analysis. *Leukemia* **16**(7): 1372–1380.

Van der Velden VH, Willemse MJ, Van der Schoot CE, Hahlen K, Van Wering ER, Van Dongen JJ (2002) Immunoglobulin kappa deleting element rearrangements in precursor-B acute lymphoblastic leukemia are stable targets for detection of minimal residual disease by real-time quantitative PCR. *Leukemia* **16**(5): 928–936.

Van Dongen JJ, Seriu T, Panzer-Grumayer ER, Biondi A, Pongers-Willemse MJ, Corral L, Stolz F, Schrappe M, Masera G, Kamps WA, Gadner H, Van Wering ER, Ludwig WD, Basso G, De Bruijn MA, Cazzaniga G, Hettinger K, Van der Does-Van der Berg A, Hop WC, Riehm H, Bartram CR (1998) Prognostic value of minimal residual disease in acute lymphoblastic leukemia in childhood. *Lancet* **352**(9142): 1731–1738.

Verhagen OJ, Willemse MJ, Breunis WB, Wijkhuijs AJ, Jacobs DC, Joosten SA, Van Wering ER, Van Dongen JJ, Van der Schoot CE (2000) Application of germline IGH probes in real-time quantitative PCR for the detection of minimal residual disease in acute lymphoblastic leukemia. *Leukemia* **14**(8): 1426–1435.

Protocol 16.1: Post-transplant MRD monitoring

OUTLINE

This is a complex technique involving several steps. The major processes are:

1. Identify an individual patient's monoclonal rearrangements and perform sequencing
2. Design and order clone specific forward primers for at least 2 separate loci
3. Prepare control DNA from the pooled mononuclear cells up at least 10 normal donors
4. Prepare dilution series of patient's diagnostic DNA in control DNA
5. Test clone specific primers for sensitivity and specificity. Re-design and re-test if necessary
6. Test remission specimens from the individual patient and calculate the level of MRD relative to the diagnostic specimen using the standard curve

Reagents and consumables

PCR primers for chosen antigen receptor rearrangements
Polyacrylamide gel solution (Sigma, 40% Stock, 29:1 bis: acrylamide)
TRIS borate EDTA electrophoresis buffer
DNA Sequencing Kit
Germline reverse primers and TaqMan® probes (Verhagen et al., 2000; van der Velden et al., 2002; van der Velden et al., 2002)
Control gene TaqMan® assay, e.g ALB (Verhagen et al., 2000)
2 x TaqMan® universal master mix (Applied Biosystems)
TE Buffer (10mM TRIS, 1mM EDTA, pH 8.0)
Molecular biology grade H_2O
96-well PCR plates
Optical quality adhesive covers or caps

Equipment

Thermal Cycler
Real-time PCR instrument such as ABI 7500

Genespec Micro spectrophotometer (Hitachi)
Vertical polyacrylamide electrophoresis apparatus
(SE 600 Ruby, GE Healthcare)
PCR micropipettes
Waterbath at 37°C
Microcentrifuge
Benchtop centrifuge with microplate carriers

Specimens required

- DNA from patient at presentation (marrow or blood should contain >90% leukemic cells). Adjust concentration to 100 ng/µl
- DNA from pooled mononuclear cells from at least 10 healthy donors (concentration as above). This will be the non-amplified control (NAC) and will also be used as diluent for dilution series of patient's diagnostic DNA. Leukemic cell DNA in a background of polyclonal lymphocyte DNA mimics the DNA isolated from the patient in remission. It will also be used to prepare dilutions for control gene standard curve
- DNA from patient during treatment at 100 ng/µl

Identify clonal rearrangements

Analyze the patient's diagnostic DNA using PCR assays for IgH, TCRδ, TCRγ and kappa deleting element (KDE) as described (Pongers-Willemse *et al.*, 1999) and identify suitable rearrangements by heteroduplex analysis (Langerak *et al.*, 1997). Select monoclonal rearrangements with no evidence of heteroduplex formation. If possible at least two rearrangements from different loci should be selected, e.g. IgH or TCR. Where there has been sufficient resolution of clonal bands they may be excised from the gel and sequenced. Analyze the sequence using software available at http://www.ncbi.nlm.nih.gov/igblast, http://imgt.cines.fr/textes/vquest, or http://vbase.mrc-cpe.cam.ac.uk/vbase1/dnaplot2.php to designate the gene segments involved in the junctional region

Design of clone specific forward primer

Forward patient specific primers are designed (Primer Express®, Applied Biosystems) to lie over the area of clone specific rearrangement. In the case of IgH, this should be the DNJ junction as opposed to the VND junction, to minimize the risk of clonal evolution. For TCRγ targets the primer should lie over the VND or DND junction, and for γ deleting elements KDE the VNKDE join. For maximal specificity avoid including anymore than 3–5 bases in the germline sequence at the 3' end. In practice this can be very difficult to achieve with some delta and many kappa targets.

Due to the more limited diversity of TCR rearrangements, the germline probe assay sensitivity is dependent upon the rearrangement present and the number of insertions present. Rearrangements containing >12 insertions and deletions provide useful markers whilst those with <12 are unlikely to be specific (van der Velden *et al.*, 2002).

Criteria for an acceptable forward primer (in addition to the rules listed in Chapter 7, Table 1)

- Tm of 57–60°C on Primer Express® using Nearest Neighbor algorithm
- Choose shortest primer with the least number of germline nucleotides
- Check for the presence of 3' forward/forward and forward/reverse primer-dimmers and for possible 3' hairpin formation
- In practice, with some rearrangements it is impossible to meet these criteria, however, primers with >3 G/C at the 3' end do not work well
- If Primer Express® will not identify a primer with an appropriate Tm in the region of interest, (particularly common with IgH), primers may be designed by eye and using the Wallace rule (Tm = 4(G + C) + 2(A + T) aim for Tm = 55 – 60°C), using a 2–3 base overhang into consensus at the 3' end
- Design three candidate primers per locus and order from a reliable supplier

Control PCR

A suitable control PCR must be run. The control PCR is used to confirm the amount and quality of amplifiable DNA in each sample analyzed for MRD. It is thus simultaneously a check for the accuracy of DNA quantitation, accuracy of pipetting and for the presence of PCR inhibitors.

As all test PCRs use 500 ng of test DNA, the control standard curve needs to cover this range only. The UK Childhood Leukaemia Network use *ALB*, other European groups use *B2M*, our laboratory uses *ALB*. Other control genes may be used but their chromosomal location should be considered in order to avoid loci that may be deleted or amplified in a particular leukemia.

A standard curve should be constructed from duplicate analysis of a dilution series made from pooled normal control DNA. Dilute the DNA in TE. Only the 10^0, 10^{-1}, 10^{-2}, and 10^{-3} points need be analyzed.

Generating standard curves for assessment of probe sensitivity and measurement of MRD

A dilution series in which leukemic DNA is diluted with pooled normal blood mononuclear cell DNA is used to generate the standard curve.

Preparation of patient DNA dilution series for TaqMan® MRD analysis

Dilution point	Quantity of leukemic (target) DNA	Volume of target DNA	Volume of control DNA
10^0	1000 ng	–	
10^{-1}	100 ng	10 µl	90 µl
10^{-2}	10 ng	10 µl	90 µ
10^{-3}	1000 pg	10 µl	90 µ
10^{-4}	100 pg	10 µl	90 µ
10^{-5}	10 pg	10 µl	90 µ
10^{-6}	1 pg	10 µl	90 µ

1. Adjust concentration of presentation DNA (containing >90% leukemic cells) to 100 ng/ml. Warm in 37°C waterbath for 1 h to ensure that the solution is homogenous, vortex briefly and spin down
2. 10^{-1} Point: 1000 ng (10 µl of 100 ng/µl solution) of target DNA is added to 90 µl of pooled normal donor peripheral blood mononuclear cell DNA (100 ng/µl), making a final solution of 10 ng/µl target DNA. 5 µl of this solution is then used in TaqMan® PCR representing 50 ng of target (= 10^{-1} point)
3. 10^{-2} Point: Leave 10^{-1} point in water bath for 1 h, vortex and spin down. 10 µl of 10^{-1}

solution is added to 90 µl of pooled normal DNA (100 ng/µl), to make 10^{-2} point

4. Repeat step 2 four more times to make 10^{-3}, 10^{-4}, 10^{-5}, and 10^{-6} points

TaqMan® PCR

The following master mix should be used for all PCR reactions, with 20 µl added to each well, 5 µl of sample DNA (i.e. 500 ng DNA) is added.

Constituent	Vol/reaction
TaqMan® Universal Master Mix	12.5 µl
H₂O	5 µl
Forward primer (10 µM)	2.25 µl
Reverse primer (10 µM)	2.25 µl
Probe (5 µM)	0.5 µl
DNA (100 ng/µL)	5 µl
Total volume	25 µl

Initial clone specific PCR and *ALB* PCR conditions are:
2 min at 50°C
10 min at 95°C
50 cycles of 15 sec at 95°C and 1 min at 60°C

Sensitivity and specificity testing

The clone specific primers must be tested for efficiency and specificity before using to analyze patient specimens. An acceptable primer must reach a sensitivity of 10^{-4} and not have significant reaction with pooled normal DNA control (NAC).

Method

1. Amplify the 10^{-2} dilution point of the leukemic (target) DNA in duplicate and 6 NAC (normal mononuclear cell DNA) control wells. Use 500 ng DNA per well
2. Examine C_t of 10^{-2} point. For an efficient primer the C_t should be less than 30
3. Examine C_t of NAC. Ideally this should 50, or at least 11 greater than C_t of 10^{-2} if sensitivity of 10^{-4} is to be achieved
4. Choose the primer that best fits these 2 criteria. If none fit the first, then primer redesign is required
5. If none fit the second, then incremental increases in the annealing temperature can be attempted (63, 66, and 69°C recommended)
6. If these steps do not result in a primer that fits the criteria, then that particular rearrangement is not suitable as an MRD marker

7. The sensitivity of a primer/probe combination is defined as the lowest dilution point that has a C_t value at least 3 less than that for NAC for TCR δ, IgK and TCR γ or at least 6 less for IgH. See below for definitions of sensitivity

Reproducible sensitivity   Lowest dilution of standard curve with reproducible C_t values (C_t values that differ by less than 1.5) that is at least 3 less than that for the NAC (at least 6 for IgH)

Maximal sensitivity   The lowest dilution of standard curve giving specific but non-reproducible C_t values, which differ at least 1.5 C_t from C_t value of previous 10-fold dilution or NAC

Quantitation of MRD in remission specimens

1. Adjust DNA concentration to 100 ng/μl in TE
2. The patient's dilution series is used as standard curve for MRD measurement
3. The dilution series of pooled normal mononuclear cell DNA is used as standard curve for control gene for quantity and quality of test DNA
4. The clone specific forward primer, germline reverse primer and probe are used to amplify; the patient's leukemic dilution series in duplicate (the 10^{-6} is usually not included as such sensitivity is not achieved); duplicate NAC (pooled normal) 500 ng DNA per well; duplicate NTC controls; the test sample in triplicate with 500 ng DNA per well
5. The control PCR is used to amplify; normal control dilution series in duplicate; duplicate NTC; duplicate test samples

Both assays can be performed on same plate unless the clone specific PCR assay requires an increased annealing temperature as control gene assays may not work at higher temperatures. See Figure 16.4 for plate set-up

6. Perform 50 cycles of PCR and analyze the data
7. The slope of the clone specific standard curves should be in the range –3.3 to –4.0. The correlation coefficient of curves >0.95. The maximum C_t variation between replicates of a point <1.5. This defines the reproducible range

	1	2	3	4	5	6	7	8	9	10	11	12
A	$L10^{-1}$	$L10^{-1}$	Test 1	Test 1	Test 1	$C10^{0}$	$C10^{0}$	$L10^{-1}$	$L10^{-1}$	Test 1	Test 1	Test 1
B	$L10^{-2}$	$L10^{-2}$	Test2	Test2	Test2	$C10^{-1}$	$C10^{-1}$	$L10^{-2}$	$L10^{-2}$	Test2	Test2	Test2
C	$L10^{-3}$	$L10^{-3}$	Test 3	Test 3	Test 3	$C10^{-2}$	$C10^{-2}$	$L10^{-3}$	$L10^{-3}$	Test 3	Test 3	Test 3
D	$L10^{-4}$	$L10^{-4}$				$C10^{-3}$	$C10^{-3}$	$L10^{-4}$	$L10^{-4}$			
E	$L10^{-5}$	$L10^{-5}$				NTC	NTC	$L10^{-5}$	$L10^{-5}$			
F	NAC	NAC				Test 1	Test 1	NAC	NAC			
G	NAC	NAC				Test2	Test2	NAC	NAC			
H	NTC	NTC				Test 3	Test 3	NTC	NTC			

Columns 1–5 and 8–12 are the clone specific PCR experiments, columns 6 + 7 the *ALBUMIN* control.

KEY: each well contains 500 ng of DNA except NTC which contain TE

L10 = leukemic dilution series

C10 = normal DNA dilution series

Test = test samples

NAC = normal mononuclear control DNA

NTC = no template control

Adapted with permission of Dr J. Hancock, University of Bristol

Figure 16.4

Clone specific TaqMan® MRD plate set-up for 2 markers.

8. Curves that do not fit these criteria are not acceptable, as they will result in inaccurate MRD quantitation

9. DNA concentration (and amplifiability) of the test samples should be determined from the control gene results. If the test sample DNA is within the range 400–600 ng, the MRD result should not be adjusted; however values between 250–400 ng should be adjusted. If test samples give a value significantly lower than the supposed quantity of DNA put into the control PCR, inhibition is the likely cause. Inhibition may be overcome by using BSA in the PCR master mix (Moppett *et al.*, 2003) (a 2% w/v solution is prepared, filter sterilized through a 0.2 µm filter and 5 µl is used in each 25 µl master mix)

10. Quantitative results should be reported: if the replicates are with 1.5 C_t and within the reproducible range of the assay; if at 2/3 wells

have a reproducible C_t below the maximal sensitivity of the assay. If both of these conditions are satisfied the results calculated from the standard curve should be reported

11. Positive but unquantifiable results are reported if; there are non-reproducible C_t values outside reproducible range but outside range of NAC; if at least 2/3 wells have a reproducible C_t above the maximal sensitivity but outside the range of NAC. Report the data as $<10^{-x}$ (where X = lower limit of reproducible range)

12. Report as not detected: if C_t values are 50; if C_t values are within the range of NAC

Monitoring post-stem cell transplantation chimerism

The monitoring of patients who have undergone allogeneic stem cell transplantation has relied on semi-quantitative methods, such as short tandem repeat (STR) PCR with fluorescent primers or Southern blotting. Real-time PCR based on the detection of single nucleotide polymorphisms has been developed in some centers and is under evaluation, however encouraging results are emerging and sensitivities of 0.1–0.01% may be possible (Maas *et al.*, 2003).

Use of Y chromosome sequences

Male patients with female donors can be monitored using Y chromosome specific assays (Fehse *et al.*, 2001), these are highly sensitive (0.001%) and there may always be a threshold of positive cells, therefore serial analysis is more useful, with any increase in the level of male cells being regarded as suspicious. Investigation of mixed chimerism should always be done in tandem with MRD analysis where suitable disease markers are available. Our laboratory uses an in-house assay based on detection of the *TSPY* repeats on Y chromosome. The use of a reiterated target makes this assay very sensitive; due to the variation in number of *TSPY* repeats in individual males, the $\Delta\Delta C_t$ method is used. Although this is not an expression assay, the values obtained are sufficiently accurate to reflect fluctuation in the level of male cells against a female background. The control gene used is *HBB*, but other genomic sequences may be used. Pre-transplant DNA or buccal swab DNA from the recipient is used as the calibrator sample.

Protocol 16.2: Use of TSPY repeats on Y-chromosome in chimerism studies

REAGENTS AND CONSUMABLES

Optical Adhesive Cover (Applied Biosystems)
96-well real-time PCR plate (Applied Biosystems)
1.5 mL amber tubes (Axygen)
Aerosol-resistant pipette tips (Axygen)
Universal TaqMan® Master Mix (Applied Biosystems)
Molecular Biology Grade Sterile H_2O
Oligonucleotide Primers (10 µM in molecular biology grade H_2O)
TaqMan® probes (5 µM in molecular biology grade H_2O, in amber tubes)

Primers and probes for TSPY PCR

TSPY 1139-F	5' TGT CCT GCA TGT TGG CAG AGA
TSPY 1204-R	5' TCA AAA AGA TGC CCC AAA CG
TSPY 1162-T	
TaqMan® Probe	5' JOE-CCT TGG TGA TGC CGA GCC GC-3'TAMRA
Beta globin 509-F	5' CTG GCT CAC CTG GAC AAC CT
Beta globin 587-R	5' CAG GAT CCA CGT GCA GCT T
Beta globin 541-T	
TaqMan® Probe TAMRA	5' FAM-TGC CAC ACT GAG TGA GCT GCA CTG TG-3'

EQUIPMENT

7500 real-time PCR instrument (Applied Biosystems)
Genespec 1 micro spectrophotometer (Hitachi)
Centrifuge with microplate adapters
Micropipettes

METHOD

1. Purified DNA is the preferred template material; however this assay can be performed using white blood cells isolated by red cell lysis techniques or ficoll gradient separation

2. Prepare master mixes for *TSPY* and *HBB* (or preferred control gene). Set up triplicate reactions for both. Remember that each male patient requires his own individual calibrator specimen. If possible, include a known female DNA isolated at the same time as the test specimens (NAC) and NTC

Constituent	Vol./reaction TSPY	Vol./reaction HBB
TaqMan® Universal Master Mix	12.5 µL	12.5 µL
H$_2$O	3.5 µl	5.0 µl
Forward primer (10 µM)	0.75 µl	0.75 µl
Reverse primer (10 µM)	2.25 µl	0.75 µl
Probe (5 µM)	1.0 µl	1.0 µl
DNA (100 ng/µL)	5 µL	5 µL

3. Dispense 20 µL of mix to appropriate wells of a 96-well plate (use adjacent sets of columns)
4. Add 5 µL of the test specimens, NAC and NTC controls, before adding the pre-transplant calibrator specimens
5. Seal the plate carefully with an optical adhesive cover
6. Spin briefly
7. Set up the real-time instrument as per manufacturer's instructions. Use the default Applied Biosystems thermal cycling protocol of 95°C for 10 min followed by 50 cycles of 15 sec at 95°C and 60 sec at 60°C
8. Save the data and export the raw C$_t$ data to Microsoft Excel
9. Open the data in Excel. Sort by specimen and detector and compare the replicate values. Discard outliers and calculate the mean C$_t$ for each set of triplicates
10. Calculate the ΔC_t for each specimen thus:
ΔC_t = Mean C$_t$ (*TSPY*) – Mean C$_t$ (*HBB*)
11. Calculate the $\Delta\Delta C_t$ for each test specimen thus:
$\Delta\Delta C_t = \Delta C_t$ Test – ΔC_t Calibrator
12. Calculate the relative *TSPY* copy number using the formula $2^{-\Delta\Delta C_t}$
13. The sensitivity of this assay is such that low level *TSPY* sequences will be detected post stem cell transplant (assuming sufficient DNA is analyzed) in patients with complete donor chimerism. This is probably due to contamination with male cells during sampling.

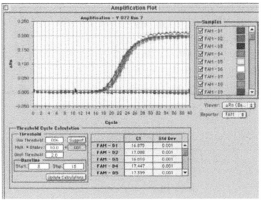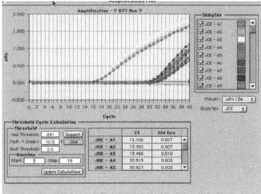

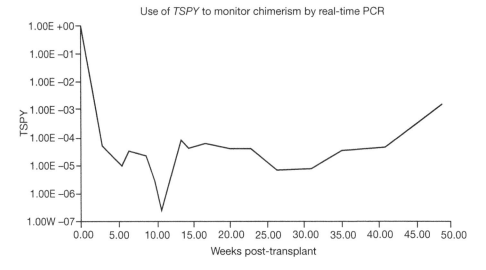

Use of *TSPY* to monitor chimerism by real-time PCR

Figure 16.5

Use of *TSPY* to monitor chimerism by real-time PCR.

However, it is more important to establish a threshold for each patient. Increasing levels should be investigated where possible with MRD specific assays, the underlying disease should be considered; acute leukemia patients may have at least a tenfold higher level in bone marrow compared to peripheral blood if the *TSPY* level is an indicator of MRD

Serial analysis of peripheral blood from a young male patient transplanted for high risk ALL is shown in *Figure 16.5*; he became MRD positive at fifteen weeks post-transplant and relapsed clinically at week 48.

Real-time PCR for prenatal diagnosis of monogenic diseases caused by single nucleotide changes – the example of the hemoglobinopathies

17

Joanne Traeger-Synodinos, Christina Vrettou and Emmanuel Kanavakis

17.1 Introduction to prenatal diagnosis (PND) in clinical genetics

In developed countries, genetically determined disorders account for up to one third of admissions to pediatric wards and are a significant cause of childhood deaths. Although the Human Genome Project and related advances in molecular biology promise means for the long-term curative treatment of many severe genetic disorders, the current approach for controlling these disorders remains prevention, including the application of prenatal diagnosis (PND), recognized as an important reproductive option.

PND aims to provide an accurate, rapid result as early in pregnancy as possible. A prerequisite involves obtaining fetal material promptly and safely, and current methods include trophoblast sampling (also known as chorionic villi sampling or CVS) or amniocentesis. Fetal cells and free fetal DNA are also present in the circulation of the pregnant mother and provide a potential source for 'non-invasive' fetal sampling, but reliable protocols have yet to be established for clinical application (Hahn and Holzgreve, 2002). Termination of affected pregnancies is the major disadvantage of PND, and there is still an on-going debate about its ethical and social implications.

Preimplantation genetic diagnosis (PGD) represents a 'state-of-the-art' procedure which potentially avoids the need to terminate affected pregnancies through identification and transfer of only unaffected embryos established from in-vitro fertilization (IVF) (Handyside *et al.*, 1990).

Although there are several technical, practical, and ethical issues associated with performing PGD, the procedure is progressively being incorporated within the available option of clinical services for preventive genetics, and is considered particularly appropriate for couples with an unsuccessful reproductive history (Kanavakis and Traeger-Synodinos, 2002; Traeger-Synodinos *et al.*, 2003; Sermon *et al.*, 2004).

17.2 Classic mutation detection methods for prenatal diagnosis of monogenic diseases and best practice guidelines

17.2.1 Classic mutation detection methods

The hemoglobinopathies were the first genetic diseases to be characterized at the molecular level and have been used as a prototype for the development of almost every new technique of mutation detection for more than 20 years. Consequently there are numerous PCR-based techniques described in the literature that can be used for PND of the globin gene mutations. The most commonly used techniques include: allele specific oligonucleotide (ASO) hybridization or dot blot analysis, reverse dot blot analysis, the amplification refractory mutation system (ARMS), denaturing gradient gel electrophoresis (DGGE), gap PCR, restriction fragment length polymorphism (RFLP) analysis and direct DNA sequencing (Old *et al.*, 2000; Kanavakis *et al.*, 1997). Each technique has its relative advantages and disadvantages in terms of simplicity, speed, and cost. The choice of techniques used by any laboratory will be influenced by the infrastructure and expertise available in the laboratory, the budget, the sample throughput, and the spectrum of mutations in the target population. In any laboratory providing PND, it is recommended that at least two alternative methods are available for validating each mutation in every prenatal sample.

Without doubt, the provision of PND requires the highest standards in laboratory practice to minimize sources of potential errors and ensure an accurate result. Based on more than two decades of experience, a general consensus has been reached on the precautionary measures to minimize risk of diagnostic errors (Old *et al.*, 2000; Traeger-Synodinos *et al.*, 2002).

17.2.2 Best practice guidelines for prenatal diagnosis

Best practice guidelines, which are applicable to PCR-based prenatal genetic testing for any monogenic disorder, include:

1. To divide the original CVS or amniotic fluid samples to two aliquots, so the fetal DNA can be analyzed in duplicate. In addition when CVS is the source of fetal genetic material, to perform careful microscopic dissection (see below in the protocols for preparation of fetal samples in 17.5).
2. To obtain fresh parental blood samples along with the fetal sample in order to confirm parental mutations and also as a source of control DNA.
3. To analyze parental DNA samples and other appropriate controls for the relevant disease-causing mutations alongside the fetal DNA.

4. To repeat the fetal DNA analysis to double check the result, ideally by using a second independent method.

5. To exclude presence of contaminating maternal DNA in the fetal sample through analysis of appropriate hypervariable genetic markers (e.g. microsatellite loci) in the parental and fetal DNA; the fetal DNA should demonstrate inheritance of one maternal and one paternal allele for any loci tested.

6. Finally, the PND report should detail the methods of DNA analysis used and state the risk of misdiagnosis based on the literature and the audit results of the diagnostic laboratory.

17.3 Sources of fetal samples for prenatal diagnosis

The aim of PND is to provide a result as early in the pregnancy as possible, and thus it is preferable if the fetal sample is obtained within the first trimester by CVS sampling between 10–12 weeks of pregnancy. The alternative is to obtain an amniocentesis sample from about 14–16 weeks of pregnancy. Overall CVS samples provide more DNA and are less prone to co-sampling of maternal cells, which can lead to contamination of the fetal DNA.

The yield of DNA extracted from a trophoblast sample weighing around 20–25 mg is typically 30–40 μg. DNA can be prepared from amniotic fluid cells directly or after culturing. The quantity of DNA extracted directly from 15 ml of non-cultivated amniotic fluid cells (~5 μg) is sufficient for PND based on PCR methods in most cases.

The risk of contamination with maternal DNA can be ruled out in most cases by the presence of one maternal and one paternal allele following the amplification of 'informative' polymorphic repeat markers, as mentioned above in the best practice guidelines.

The analysis of single-cells within the context of PGD involves a different procedure in that the single-cell is simply lysed in the eppendorf tube in which the cell was initially placed. This releases the cellular DNA content and the reagents for the initial processing are added directly to the tube; these are usually for the first-round PCR reaction, although some protocols start with whole genome amplification or multiple displacement amplification. A protocol for lysing single cell samples is outlined within the protocol described for single-cell genotype analysis (see 'Cell Lysis' in 17.5).

17.4 Real-time PCR protocols for PND and PGD applied to the hemoglobinopathies: background and design of protocols

17.4.1 Real-time PCR and allele discrimination using the LightCycler® (system 1.0 or 1.5)

Real-time PCR integrates microvolume rapid-cycle PCR with fluorometry, allowing real time fluorescent monitoring of the amplification reaction for quantitative PCR and/or characterization of PCR products for rapid genotyping, precluding any post-PCR sample manipulation. The LightCycler® (Roche Molecular Biochemicals) is one such system. The detection of potential sequence differences for genotyping applications (usually Single Nucleotide Polymorphisms, SNPs) employs the use of two fluorescent probes which hybridize to adjacent internal sequences within target amplified DNA, one of which covers the region expected to contain the mutation(s). Close proximity of annealed probes facilitates fluorescence resonance energy transfer (FRET) between them. The probes are designed to have different melting temperatures (Tm) whereby one with the lower Tm spans the mutation site(s). Monitoring of the emitted fluorescent signals as the temperature increases will detect loss of fluorescence (F) as the lower Tm probe melts off the template. A single base mismatch under this probe results in a Tm shift of 5–10°C, allowing easy distinction between wild-type and mutant alleles. The ability to detect base mismatches under the low Tm probe (mutation detection probe) and the use of two different colored probes (LightCycler® system 1.0 or 1.5) allows more than one mutation to be screened in a single PCR reaction.

17.4.2 Molecular basis of β-hemoglobinopathies

The β-globin gene is a relatively small gene (<2000 bp) located in the short arm of chromosome 11. Although more than 180 causative mutations have been reported for β-thalassemia syndromes (http://www.globin.cse.psu.edu), in any given population, there are a limited number of common mutations and a slightly larger number of rare mutations. The majority of common mutations tend to cluster within neighboring gene regions, and this facilitates the use of a small number of probe sets for mutation detection.

17.4.3 Principles behind design of LightCycler® probe sets and assays in the β-globin gene (appropriate for Systems 1.0 and 1.5)

All the allele-specific probes were designed to be complementary for the wild-type sequence, allowing the detection of any sequence variation compared to normal under the region of probe-hybridization (Vrettou *et al.*, 2003*)*. This precluded the need to use multiple independent assays using separate mutation-specific detection probes. This potentially allows the distinction of any allele with a nucleotide variation located under the length of the probe, minimizing costs and time required to screen a large spectrum of mutations, as is necessary for most populations where β-thalassemia is common. Although most mutations have a distinct melting profile and can be detected by comparison with controls, the definitive characterization of each mutation can be achieved by a second method such as an ARMS-PCR assay (Old *et al.*, 1990). As stated in the best practice guidelines, any laboratory performing DNA diagnostics should have more than one available mutation detection method.

The protocol that we designed for the Greek population, which is also suitable for most other populations, where the β-hemoglobinopathies are prevalent, uses just four combinations (or sets) of probes for mutation detection, on two alternative PCR amplicons (*Figure 17.1*). The LightCycler® PCR primers (sets LC1 and LC2) were designed with the aid of computer software (Amplify version 2.0, Bill Engels, 1992–1995) to amplify two

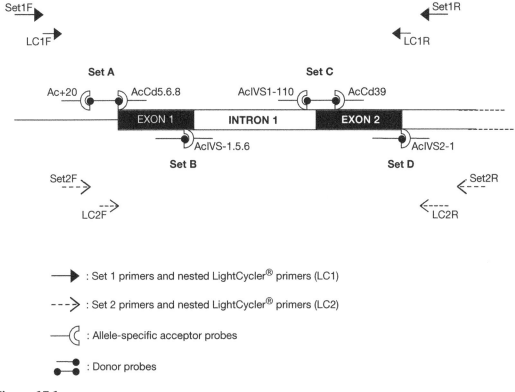

Figure 17.1

The position of the β-globin gene primers and LightCycler® hybridization probe sets appropriate for prenatal diagnosis and preimplantation genetic diagnosis protocols. The primers illustrated include two alternative sets for use in the first-round of PCR relevant only for single cell genotyping (Set 1 or Set 2) and two alternative sets of primers for use in the real-time PCR genotyping assay on the LightCycler® (sets LC1 or LC2). F = forward primer; R = reverse primer.

alternative regions of the β-globin gene surrounding the majority of the most common β-thalassemia mutations in all world populations, along with the HbS mutation (*Table 17.1* and *Figure 17.1*).

Design of LightCycler® mutation detection probe sets took into account secondary structure properties (e.g. hairpin-loops) and potential primer-dimer formation between the probes themselves and the PCR primers, evaluated by computation prior to synthesis (TIB MOLBIOL, Berlin, Germany). In addition, design of mutation detection probes avoided the regions of the β-globin gene known to contain common sequence variants such as codon 2 (CAC>CAT). The LightCycler® systems 1.0 and 1.5 can detect 2 fluorescent labels (LightCycler® Red 640 [LC Red 640] and LightCycler® Red 705 [LC Red 705]) as well as SYBR® Green. The choice of fluorescent label for each probe will depend upon the relative frequency of mutations in the population under study and the potential requirement for multiplexed assays when more than one mutation is investigated within a

Table 17.1 LightCycler® mutation detection probe sets

Beta-gene mutation	Probe set	Acceptor probe name and sequence	Donor probe name and sequence
CAP +20 (C>T)* CAP+22 (G>A)	Set A	Ac +20: 5'-tc tga cac aac tgt gtt cac tag ca 3' **LC Red****	Donor set A: **FITC** 5' cct caa aca gac acc atg gtg cac c-3' **FITC**
HbS (Cd 6 A>T) Cd5 (-CT) Cd6 (-A) Cd8 (-AA) Cd 8/9 (+G)		Ac Cd5.6.8: **LC Red**** 5' gac tcc tga gga gaa gtc tgc-3' **P*****	
IVSI-1 (G>A) IVSI-1 (G>T) IVSI-2 (T>G) IVSI-2 (T>C) IVSI-2 (T>A) IVSI-5 (G>A) IVSI-5 (G>C) IVSI-5 (G>T) IVSI-6 (T>C)	Set B	Ac IVSI-1.5.6: **LC Red**** 5' tgt aac ctt gat acc aac ctg ccc a-3' **P*****	Donor set B: 5' tgc cca gtt tct att ggt ctc ctt aaa cct gtc-3' **FITC**
IVSI-110 (G>A) IVSI-116 (T>G)	Set C	Ac IVSI-110: 5'-tct gcc tat tgg tct att ttc cc-3' **LC Red****	Donor set C: **FITC** 5'-ccc tta ggc tgc tgg tgg tc-3' **FITC**
Cd39 (C>T) Cd37 (TGG>TGA) Cd41/42 (delTTCT)		Ac Cd39: **LC Red**** 5'-acc ctt gga ccc aga ggt tct t-3' **P*****	
IVSII-1(G>A)	Set D	AcIVS2–1: **LC Red**** 5' tct cag gat cca cgt gca gct tg-3' **P*****	Donor set D: 5' gtc cca tag act cac cct gaa g-3' **FITC**

The LightCycler® PCR reactions also include a set of (β-globin gene specific PCR primers, either set LC1 or set LC2. **LC1** Forward (F): 5'-GCT GTC ATC ACT TAG ACC TCA-3'; **LC1** Reverse®: 5'-CAC AGT GCA GCT CAC TCA G-3'; **LC2** Forward (F) 5'-CAA CTG TGT TCA CTA GCA AC-3'; **LC2** Reverse® 5'-AAA CGA TCC TGA GAC TTC CA-3'.
FITC: Fluorescein.
* = Polymorphism linked with the IVSII-745 (C>G) mutation.
** = LC Red: The fluorescent label used for each probe will depend upon the relative frequency of mutations in the population under study and the potential requirement of multiplexed assays (Red640 or Red705).
***P= Phosphorylated.

single sample. For example in the Greek population, IVSI-110 G>A is the most common mutation, so the probes for most other mutations encountered in Greece are labeled with the opposite fluorescent marker to that used for IVSI-110 G>A.

More specifically, two of the probe combinations (named set A and set C) include 2 acceptor (mutation detection) probes with one central donor probe (*Figure 17.1*) foreseeing the use of one or both acceptor probes of the set according to the needs of any genotyping assay. Each of the acceptor probes in sets A and C are labeled with different acceptor fluorophores and the central donor probe, designed to span the distance between the two acceptor probes, is labeled with a fluorescein (F) molecule at both 5' and 3' ends. Set B was designed to screen for several neighboring mutations with use of a single mutation detection (acceptor) probe and D was designed to detect a single mutation, each in combination with a donor probe which is labeled with fluorescein only at the end adjacent to the acceptor probe (*Table 17.1, Figure 17.1*). In all sets the mutation-screening (acceptor) probes were designed to have a lower Tm relative to the donor probes, thereby ensuring that the fluorescent signal generated during the melting curve, is determined only by the mutation probe.

17.4.4 Additional considerations in design of single-cell genotyping for PGD using real-time PCR

In clinical PGD, the diagnosis has to be achieved on a single cell, with only one initial PCR reaction possible. In addition, the result has to be available within 24 hours. PCR-based diagnosis of single-cells is susceptible to PCR failure, allele drop-out (when one of the alleles fails to amplify to detectable levels) and contamination. PCR failure means no result, which is undesirable, but allele drop-out and contamination are dangerous as they can lead to an unacceptable misdiagnosis. Thus PCR protocols have to be optimized to minimize PCR failure and allele drop-out, and contamination must be stringently avoided and ideally monitored in each individual sample (by the parallel analysis of hypervariable microsatellite loci) to preclude misdiagnosis. Finally when confronting a relatively common monogenic disorder such as the β-thalassemias and hemoglobinopathies, ideally the method should be flexible to facilitate detection of a wide spectrum of disease-associated mutation-interactions, precluding the design and standardization of case-dependant protocols each time.

Thus the overall strategy we designed (Vrettou *et al.*, 2004) involved a first-round multiplex PCR (approximate time 1.5 h), containing β-globin gene first-round PCR primers (Set1 or Set2) for the amplification of a region of the β-globin gene surrounding the case-specific mutations along with fluorescently-labeled primers for amplification of two polymorphic microsatellite markers GABRB3 (at the GABAA receptor b3 locus in the Angelman/Prader–Willi region on chromosome 15) and D13S314 (*Table 17.2*). The first-round multiplex PCR was followed by: 1) real-time nested PCR with hybridization probes for analysis of β-globin gene alleles (approximate time 40 min), and in parallel, 2) size analysis of the two microsatellite markers on an automatic sequencer (approximate time 35 min) for monitoring and precluding contamination. Genomic DNA from the parents in each cycle was amplified in parallel along with blank samples from the IVF unit and PCR premix blanks.

17.4.5 Potential advantages of real-time PCR protocols for PND and PGD

Advantages of real-time PCR protocols for PND

The ability to carry out rapid DNA analysis for genotype characterization has become an increasingly important requirement for the clinical diagnostic laboratory. A wide variety of methods exist for detecting point mutations in a DNA molecule, including ASO, PCR-RFLP, ARMS, DGGE,

Table 17.2 Sequences of primers included in the first-round multiplex PCR for PGD protocol

Primers	Sequence of primers	Product size, bp
β gene		
Set 1*		
F	5'-GAA GTC CAA CTC CTA AGC CA-3' (70373–70393)**	
R	5'-CAT CAA GGG TCC CAT AGA CTC-3' (71041–71062)**	689
β gene		
Set 2*		
F	5'-CAA CTG TGT TCA CTA GCA AC-3' (70560–70580)**	
R	5'-TAA AAG AAA CTA GAA CGA TCC-3' (71176–71197)**	637
GABRB3		
F	5'-CTC TTG TTC CTG TTG CTT TCA ATA CAC -3'	
R***	5'-CAC TGT GCT AGT AGA TTC AGC TC -3'	191–201
D13S314		
F***	5'-GAG TGG AGG AGG AGA AAA GA-3'	
R	5'-GTG TGA CTG GAT GGA TGT GA-3'	137–155

* For amplifying the β-globin gene, include either Set1 or Set2 in first-round multiplex PCR, with subsequent use of either LC1 primers or LC2 primers (as appropriate) for mutation analysis on the LightCycler® (see *Figure 17.1*).
** Co-ordinates of primers based on GenBank NG_000007.
*** 5' fluorescently labeled with Cy5.5.
F: forward primer.
R: reverse primer.

and direct sequencing. However, these methods require several hours and sometimes days to complete a diagnosis, and, consequently, there is a need for more rapid, high-throughput assays.

When applied for PND, the LightCycler® real-time PCR method is very rapid and accurate. Once the parental mutations are known, PND requires less than 3 hours for completion, including the stage of DNA extraction from the fetal sample. The distribution of mutations means that many cases require only one PCR reaction (including two LightCycler® mutation detection probe sets if necessary), minimizing both the time and cost of the PND procedure. Furthermore, in cases where the two parental mutations differ but are under the same detection probe, the melting curve of a normal allele is always distinguished when present, thus preventing a misdiagnosis.

Advantages of real-time PCR protocols for PGD

The sensitivity of real-time PCR is expected to maximize the detection of amplified DNA copies, even when present in low numbers. This potentially reduces the frequency of amplification failure and allele drop-out. Furthermore, even if allele drop-out does occur, it is monitored by the use of the nested real-time PCR, which will detect both alleles contributing to the genotype. Embryos with 'impossible' genotypes are considered to have allele drop-out and are not recommended for transfer. Only those embryos in which a normal allele is definitely detected are recommended for transfer.

The protocol as designed for the β-hemoglobinopathies is capable of detecting a wide range of mutations and genotype interactions simply through selecting the appropriate β-globin gene primers and LightCycler® hybridization probe(s) (*Table 17.1* and *Figure 17.1*). This is an important consideration when offering PGD for a common disease with a wide spectrum of mutation interactions, as it precludes the need to standardize case-specific protocols each time a new couple requires PGD.

The incorporation of real-time PCR into the PGD protocol means that the entire procedure for genotype analysis is very rapid, requiring less than 6 hours for completion, and well within the 24–hour time limit between blastomere biopsy and embryo transfer.

References

Hahn S, Holzgreve W (2002) Prenatal diagnosis using fetal cells and cell-free fetal DNA in maternal blood: what is currently feasible? *Clin Obstet Gynecol* **45**: 649–656.

Handyside AH, Kontogianni EH, Hardy K, Winston RM (1990) Pregnancies from biopsied human preimplantation embryos sexed by Y-specific DNA amplification. *Nature* **344**: 769–770.

Kanavakis E, Traeger-Synodinos J, Vrettou C, Maragoudaki E, Tzetis M, Kattamis C (1997) Prenatal diagnosis of the thalassemia syndromes by rapid DNA analytical methods. *Mol Hum Reprod* **3**: 523–528.

Kanavakis E, Traeger-Synodinos J (2002) Preimplantation Genetic Diagnosis in Clinical Practice. *J Med Genet* **39**: 6–11.

Old J, Petrou M, Varnavides L, Layton M, Modell B (2000) Accuracy of prenatal diagnosis for haemoglobin disorders in the UK: 25 years' experience. *Prenat Diagn* **20**: 986–991.

Old JM, Varawalla NY, Weatherall DJ (1990) Rapid detection and prenatal diagnosis of beta-thalassaemia: studies in Indian and Cypriot populations in the UK. *Lancet* **336**: 834–837.

Sermon K, Van Steirteghem A, Liebaers I (2004) Preimplantation genetic diagnosis. *Lancet* **363**: 1633–1641.

Traeger-Synodinos J, Vrettou C, Palmer G, Tzetis M, Mastrominas M, Davies S, Kanavakis E (2003) An evaluation of PGD in clinical genetic services through 3 years application for prevention of beta-thalassemia major and sickle cell thalassemia. *Mol Hum Reprod* **5**: 301–307.

Traeger-Synodinos J, Old JM, Petrou M, Galanello R (2002) Best practice guidelines for carrier identification and prenatal diagnosis of haemoglobinopathies. European Molecular Genetics Quality Network (EMQN). CMGC and EMQN, http://www.emqn.org

Vrettou C, Traeger-Synodinos J, Tzetis M, Malamis G, Kanavakis E (2003) Rapid screening of multiple β-globin gene mutations by real time PCR (LightCycler™): application to carrier screening and prenatal diagnosis for thalassemia syndromes. *Clin Chem* **49**: 769–776.

Vrettou C, Traeger-Synodinos J, Tzetis M, Palmer G, Sofocleous C, Kanavakis E (2004) Real-time PCR for single-cell genotyping in sickle cell and thalassemia syndromes as a rapid, accurate, reliable and widely applicable protocol for preimplantation genetic diagnosis. *Hum Mutat* **23**: 513–521.

Protocol 17.1: Cleaning chorionic villi (CVS) by microscopic dissection

The CVS sample is usually sent from the gynecology clinic in a universal vial in 10–15 ml of saline or culture medium. Sometimes small pieces of maternal tissue are co-biopsied with the villi, and other times a considerable amount of maternal blood is also present in the vial along with the CVS. The villi have a distinct, white frond-like appearance, as opposed to the reddish-pink amorphous appearance of other tissues, which in most cases can be distinguished visually. It is essential that any maternal tissue present in the sample is removed, as the presence of maternal tissue may result in a diagnostic error. There should be enough tissue from most CVS samples for two separate aliquots to allow analysis in duplicate.

Equipment and reagents

1. Bench centrifuge appropriate for eppendorf centrifugation around 3000 g
2. Eppendorf tubes (sterile and DNase/RNase free, 1.5 ml)
3. Sterilized fine forceps
4. Sterile Petri dishes
5. Disposable pipettes
6. Dissecting microscope with ×10 magnification
7. Sterile saline solution (0.9% NaCl)

Method

1. Take two Eppendorf tubes, label clearly and add 200 µl of saline solution to each
2. Pour the contents of the universal vial into a sterile Petri dish, taking care that no tissue remains on the walls of the universal tube
3. Rinse the universal vial with about 5–10 ml saline or culture medium and add to the sample already in the Petri dish
4. Place the Petri dish on a dissecting microscope and remove any maternal tissue adhering to the villi using small clean fine forceps (or two sterile needles)
5. Using forceps place pieces of villi into the two Eppendorf tubes, taking care that the villi do not stick to the forceps ('stirring' the

forceps in the saline solution in the tubes facilitates their release)

6. Once all the villi have been placed into the Eppendorf tubes, close the lids of the tubes (discard the contents of the Petri dish as for any biological sample)

7. Spin the CVS samples at approximately 3000 g in a centrifuge suitable for Eppendorf tubes

8. Carefully tip off the supernatant, wash the villi again with 200 µl of saline solution and again discard the supernatant

The villi are now ready for DNA extraction

Protocol 17.2: Centrifugation of amniotic fluid cell samples to collect amniocytes

The gynecology clinic usually sends approximately 10–15 ml of amniotic fluid sample in a Universal vial. Some laboratories recommend that an aliquot be set up as a backup culture, because occasionally a direct amniotic fluid sample fails to provide sufficient DNA to achieve a diagnosis, or as a source of DNA if maternal contamination is indicated in the direct sample. However, in our laboratory in Athens we do not set up amniocyte cultures as we have rarely failed to achieve a complete prenatal analysis when applying PCR methods, although on rare occasions we have had to request a new amniocentesis sample due to evidence of maternal contamination.

Equipment and reagents

1. Centrifuge appropriate for 10–15 ml tubes and centrifugation at around 3000 g
2. Sterile centrifuge tubes, 10–15 ml capacity
3. Eppendorf tubes (sterile and DNase/RNase free, 1.5 ml) and an appropriate centrifuge for centrifugation around 3000 g
4. Sterile Petri dishes
5. Disposable pipettes
6. Saline solution (0.9% NaCl), sterile

Method

1. Distribute the amniotic fluid sample equally between two separate sterile 10–15 ml centrifuge tubes
2. Spin the samples at approximately 3,000 g to pellet the cells and carefully remove the supernatant
3. Resuspend each pellet in approximately 0.5 ml of saline solution and using a pipette transfer to sterile clearly labeled 1.5 ml eppendorf tubes
4. Repeat step 2

5. Wash the pellet once more with 0.5 ml of saline solution and repeat step 4

The amniocytes are now ready for DNA extraction.

Protocol 17.3: DNA extraction from chorionic villi samples or amniocytes using commercially available DNA purification columns

The kit that we use in the laboratory in Athens is the QIAMP DNA mini kit (Qiagen, Hilden, Germany) and we follow the 'Blood and Body Fluid Spin Protocol'.

1. Resuspend the washed, pelleted villi or amniocytes in 200 μl of saline solution
2. Add 20 μl of proteinase K solution, as described in the directions in the kit
3. Follow the remaining steps exactly as described in the 'Blood and Body Fluid Spin Protocol'

Note: If amniotic fluid cell samples produce a pellet that is only just visible after the cells are initially harvested then use half the volume of elution buffer for the final elution step (100 μl versus 200 μl)

Protocol 17.4: Protocol for rapidly screening multiple β-globin gene mutations by Real-Time PCR: application to carrier screening and prenatal diagnosis of thalassemia syndromes

The method described is based on our publication (Vrettou *et al.*, 2003) and even without modification is suitable for detecting the majority of the most common β-globin gene mutations worldwide.

Equipment and reagents

1. LightCycler® system version 1.0 or 1.5 (Roche)
2. Bench centrifuge for Eppendorf tubes (with well depth approximately 4.5 cm) and appropriate for centrifugation around 3,000g
3. 32 centrifuge adapters in aluminium cooling Block, LightCycler® Centrifuge Adapters (Roche, 1 909 312)
4. LightCycler® glass capillary tubes (20 μl), (Roche, 11 909 339 001)
5. Filter tips for maximum volume of 20 μl and 200 μl along with compatible accurate adjustable pipettes
6. Eppendorf tubes for making the premix
7. A pair of PCR primers selected according to mutations under study (either LC1F plus LC1R or LC2F plus LC2R as shown in *Table 17.1* and *Figure 17.1*)
8. Mutation detection probe sets, appropriate

for mutations under study (see *Table 17.1* and *Figure 17.1*)

9. LightCycler® -D NA Master Hybridization probes Kit (Roche, 2 015 102), which also includes MgCl$_2$ (25 mM) and PCR-grade water

PCR set-up

1. In an Eppendorf tube make a premix for the amplification reactions for a total reaction volume of 20 µL/sample. Each reaction should contain the ready-to-use reaction mix provided by the manufacturer (LightCycler® DNA Master Hybridization Probes) plus MgCl$_2$, a β-globin gene PCR primer pair and LightCycler® fluorescent probe sets for the relevant mutations. A typical PCR reaction for single color detection for one sample is shown in *Table 17.3*

2. When calculating the premix volume, make premix enough for the number of samples being genotyped, a PCR premix blank plus controls for the mutation(s) under investigation. The controls should include a homozygous wild-type sample (N/N), a sample heterozygous for the mutation

Table 17.3 A typical PCR reaction for single color detection for one sample

	Stock Conc*	Final Conc*	µl/sample
H$_2$O (PCR grade)			9.6 µl
MgCl$_2$	25 mM	4 mM	2.4 µl
Beta globin forward primer LC1(F) or LC2(F)	10 µM	0.5 µM	1 µl
Beta globin reverse primer LC1® or LC2®	10 µM	0.5 µM	1 µl
LC Red allele-specific probe (640 or 705)	3 µM	0.15 µM	1 µl
FITC donor	3 µM	0.15 µM	1 µl
Master mix			2 µl
Premix volume			**18 µl**
DNA sample volume			**2 µl**
Total reaction volume			**20 µl**

Conc* = concentration.
The volume of water is always adjusted to give final reaction volume of 20 µl/sample even when more than one probe set is included in the reaction. For example a PCR reaction with dual color detection using 2 allele-specific LC probes (one labeled with Red 640 and the other with Red 705) and a common (central) doubly labeled FITC probe, or even two sets of LC donor-acceptor probes (i.e. 4 probes).

(M/N) and a sample homozygous for the mutation (M/M)

3. Place the appropriate number of LightCycler® glass capillary tubes in the centrifuge adapters in an aluminum-cooling block

4. Distribute accurately 18 μl of premix in all the capillaries

5. Add 2 μl genomic DNA (approximately 50 ng) per sample and controls and 2 μl of double-distilled water to the PCR blank

6. Once the PCR reactions have been set up in the capillaries at 4°C place the caps carefully on each capillary without pressing down yet

7. Remove the capillaries (in their aluminium centrifuge adaptors) from the cooling block and place in a bench centrifuge

8. Spin gently at 300 g for 20 sec to pull the 20 μl reaction volume to the base of the glass capillary

9. Place the glass capillaries carefully into the LightCycler® carousel and simultaneously gently press the cap fully into the capillary and the glass capillary fully down in place in the LightCycler® carousel

10. Place carousel in the LightCycler® and initiate the PCR cycles and melting curve protocols using the LightCycler® software version 3.5.1

Amplification and melting curve analysis

1. Preprogram the LightCycler® software for the following amplification steps: a first denaturation step of 30 sec at 95°C followed by 35 cycles of 95°C for 3 sec, 58°C for 5 sec and 72°C for 20 sec with a temperature ramp of 20°C/s. During the PCR, emitted fluorescence can be measured at the end of the annealing step of each amplification cycle to monitor amplification

2. Immediately after the amplification step, the LightCycler® is programmed to perform melting curve analysis to determine the genotypes. This involves a momentary rise of temperature to 95°C, cooling to 45°C for 2 min to achieve maximum probe hybridization, and then heating to 85°C with a rate of 0.4°C/s during which time the melting curve is recorded

3. Emitted fluorescence is measured continuously (by both channels F2 (640 nm) and F3 (705 nm) if necessary) to monitor the dissociation of the fluorophore-labeled detection probes from the complementary single-stranded DNA (F/T) (F: Fluorescence emitted, T: Temperature). The computer software automatically converts and displays the first negative derivative of fluorescence to temperature vs. temperature ($-dF/dT$ vs. T) and the resulting melting peaks allows easy discrimination between wild-type and mutant alleles (Figure 17.2)

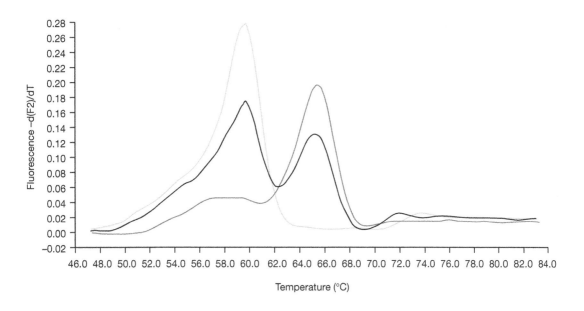

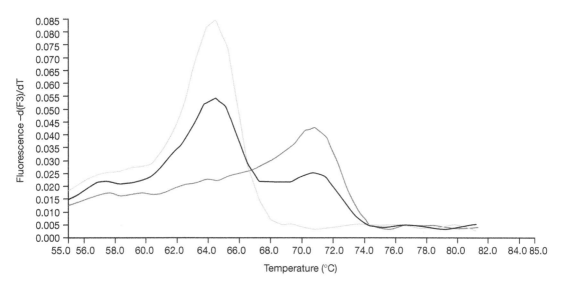

Figure 17.2

Examples of melting curves for two common β thalassemia mutations.
A. Analysis of samples for the IVSI-110 G>A mutation using hybridization probe
Set C with the Ac IVS1–110 and the donor probe set C. **B.** Analysis of samples
for the IVSI-1 G>A mutation using hybridization probe Set B with the Ac IVS-
1.5.6 and the donor probe set B. In both examples, as for all hybridization
probe sets we describe (Vrettou *et al.*, 2003), the right-shifted single peak
represents the melting curve for normal (wild-type) sequences in the region of
hybridization. Any mutation under the region where the allele-specific acceptor
probe hybridizes will cause a reduced melting temperature and thus the melting
curves for homozygous mutant samples will be left-shifted. The samples
heterozygous for a wild-type and mutant allele give a double-peaked melting
curve.

Protocol 17.5: Real-time PCR protocols for single-cell genotyping and PGD

PGD is a multi-step procedure combining expertise both in reproductive medicine and genetic diagnosis. PGD includes embryo biopsy, cell lysis and genotype analysis. For all PGD cycles, precautions against contamination have to be most stringent at all stages. Manipulation of cells and PCR set-up are carried out in separate UV-treated laminar flow hoods. PCR set-up employs dedicated PCR pipettes and pipette tips with filter. In the first round PCR, negative (blank) controls include 2 tubes containing IVF medium and 2 tubes with cell lysis mixture (prepared alongside cell biopsies in the IVF unit), and 1 tube containing PCR mixture alone (prepared in genetics laboratory during PCR set-up). To exclude contamination within the system all negative controls from the first round PCR should be analyzed for the presence of amplified hypervariable microsatellite loci (see below), and additionally subjected to nested PCR analysis in the LightCycler®. It is recommended that one-use disposable gloves, gowns and even face masks are worn by all operators during cell manipulation and PCR set-up.

The protocol described below focuses on the stages of genotype analysis of single cells with the use of real-time PCR.

Equipment for cell lysis and first PCR

1. Dedicated UV-treated laminar flow hood for PCR set-up, and separate laboratory space post-PCR manipulations
2. Filter tips for maximum volume of 20 µl and 200 µl, along with compatible accurate adjustable pipettes (UV treat before each PGD PCR set-up)
3. 200 µl Eppendorf tubes for first round PCR for the parental DNA samples and premix blank. (This size of Eppendorf is also used by IVF centers in which they place the single cells from embryos biopsied)
4. 500 µl or 1.5 ml Eppendorf tubes for making the premix for the first round PCR
5. 1.5 ml Eppendorf tubes for diluting the first-round PCR reactions on the single-cells, prior to LightCycler® PCR set-up
6. Thermal cycler with plate which holds 200 µl Eppendorf tubes

Reagents

1. PCR-grade water
2. PCR-grade solution Proteinase K (Roche, 03 115 887 001), diluted to 150 µg/ml and stored at 20°C in single use aliquots sufficient for a single PGD (average number of embryo-biopsied samples of 10)
3. HotStarTaq® Master Mix (Qiagen, Hilden, Germany)
4. MgCl$_2$ (25 mM)
5. β-globin gene PCR first round PCR primer pair (either Set1F with Set1R or Set2F with Set2R, according to mutations under study, as shown in *Table 17.2* and *Figure 17.1*), and two sets of primer pairs for amplifying the microsatellites GABRB and D13S314 (see *Table 17.2*) labeled with an appropriate fluorophore for detection on the in-house automatic sequencer; in the laboratory in Athens we have a Visible Genetics OpenGene™ System automatic DNA sequencer which can detect Cy5.0 or Cy5.5 fluorescent labels

Cell lysis

1. At the IVF unit each single blastomere (from a 3 day embryo) is placed directly into a 0.2 ml Eppendorf tube in 10 µl of sterile double-distilled sterile water, overlaid with mineral oil (all DNAse and RNase free) and placed at 20°C for at least 30 min
2. Five µl of PCR-grade Proteinase K diluted in sterile double-distilled water were added to the 10 µl of water containing the selected cell, to a final concentration of 50 µg/ml
3. On receipt of single-cell samples at the genetics unit, the Proteinase K is activated by incubation in a thermal cycler at 37°C for 1 h followed by 65°C for 10 min, and finally inactivated by heating to 95°C for 10 min. This treatment lyses the blastomeres

First-round multiplex PCR (to be set up in a UV-treated laminar flow hood)

1. In an Eppendorf tube, make a premix for the amplification reactions for a final reaction volume of 50 µl/sample (taking into account that the single cells have been biopsied and lysed in a volume of 15 µl, i.e. the premix volume for each sample is 35 µl). The contents of a typical single reaction

are shown in *Table 17.4*. Make premix enough for all the embryo samples, the blanks from the IVF unit, the parental DNA samples, and the first-round premix blank

2. Distribute 35 µl of premix to the 15 µl of the single-cell lysate (which is already in 200 µl Eppendorf tubes). Open the lid of each embryo sample individually to add the premix and change gloves in between to prevent cross-contamination between samples

3. For the parental DNA samples add an additional 13 µl double-distilled sterile water and 2 µl genomic DNA to make a final volume of 50 µl, and to the first-round premix blank add 15 µl double-distilled sterile water. These reactions should also be set up in 200 µl Eppendorf tubes and overlaid with mineral oil

4. Immediately following PCR set-up, place the Eppendorf tubes in a thermal cycler. PCR cycling conditions are programmed to include a long initial denaturation time (15 min at 94°C), to activate the hot-start Taq polymerase enzyme (HotStar) and additionally to ensure complete denaturation of genomic template DNA, followed by 18 cycles of 96°C for 30 sec, 60°C for 40 sec, and 72°C for 30 sec, and 18 additional cycles of 96°C for 30 sec, 60°C for 20 sec, and 72°C for 30 sec

Table 17.4 A typical first round multiplex PCR reaction for one cell for single-cell genotyping.

	Stock Conc*	Final Conc*	µl/sample
HotStar Taq® Master Mix			30 µl
Beta globin forward primer:			
Set 1 F or Set 2 F	100 µM	1 µM	0.5 µl
Beta globin reverse primer:			
Set 1 R or Set 2 R	100 µM	1 µM	0.5 µl
D13S314 F	100 µM	0.4 µM	0.2 µl
D13S314 R	100 µM	0.4 µM	0.2 µl
GABRB3 F	100 µM	0.4 µM	0.2 µl
GABRB3 R	100 µM	0.4 µM	0.2 µl
MgCl₂	25 mM	3.3 µM	3 µl
Premix volume			**35 µl**
Cell lysate volume			**15 µl**
Total reaction volume			**50 µl**

Conc* = concentration.

Protocol 17.6: Nested PCR and genotyping of β-globin gene using Real-Time PCR in the LightCycler®

Equipment and reagents

The equipment and reagents used for the nested PCR real-time PCR for genotyping the β-globin gene in single cells are identical to those described in the section 'Protocol for rapidly screening multiple β-globin gene mutations by real-time PCR: application to carrier screening and PND of thalassemia syndromes.' The only additional requirement is double-distilled RNase and DNase-free water in order to make the dilutions of the first-round single cell PCR reactions.

Diluting the first-round PCR reactions and PCR set-up

The nested PCR reactions from the single-cell samples are carried out following the dilution of the first-round PCR reactions to $1/10^6$.

1. To make the dilutions, prepare 2 Eppendorf tubes per cell sample, clearly numbered and each containing 1 ml of double-distilled water. Take a 1 μl aliquot of the first-round PCR and add it to 1 ml of water, mix well, centrifuge and the take a 1 μl aliquot of the $1/10^3$ dilution and add to 1 ml of water, mix well and centrifuge; the final $1/10^6$ dilution is now ready for adding to the nested real-time PCR reaction

2. Nested PCR amplification reactions for μ-globin gene mutation analysis are carried out in LightCycler® glass capillary tubes using 2 μl of diluted aliquot from the first

round PCR amplification, in a total reaction volume of 20 μl exactly as described in the previous section, using PCR primers and the appropriate combination of mutation detection probes are selected according to the parental genotypes (*Table 17.1* and *Figure 17.1*)

Protocol 17.7: Monitoring of contamination by sizing GABRB3 and D13S314 polymorphic microsatellites

The fluorescently (Cy5.0) tagged PCR-generated products of the polymorphic dinucleotide repeat microsatellite markers GABRB3 and D13S314 are generated in the multiplex first round of PCR. The amplicons from the single-cells were sized without diluting the first round PCR reaction; the PCR reactions from the parental DNA samples were diluted approximately 1 to 20 prior to loading on the automatic sequencer. The automatic DNA sequencer presently used in the Athens laboratory is the Visible Genetics OpenGene™ System automatic DNA sequencer with Gene Objects software (Visible Genetics, Evry, France), and each lab will have its own automatic sequencer. Thus we will not describe the protocol or equipment as these will be lab-specific.

The ranges of allele-sizes for each marker are noted in *Table 17.2.*

Protocol 17.8: Technical tips and troubleshooting when performing Real-Time PCR and genotyping using the LightCycler® system 1.0 and 1.5

Handling and storage of PCR primers and probes

All PCR primers to be used on the LightCycler® are diluted to 100 µM, divided into aliquots of convenient volume (e.g. 25 µl) and are stored at −20°C. The primer working solutions (10 µM) are stored at 4°C for up to 3 months.

The LightCycler® hybridization probes are diluted to 3 µM and stored in aliquots of relatively small volume (e.g. 20 µl) at 20°C. A thawed aliquot should not be refrozen, but can be used up to 1 month when stored at 4°C.

Troubleshooting with melting curve analysis

Sometimes heterozygous samples do not give a double peak following melting curve analysis, although the single peak will lie between that of the homozygous wild-type and homozygous mutant samples. This problem tends to occur when there is a relatively high number of samples per run; the maximum number of samples that can be analyzed in a single run using the LightCycler® (system 1.0 and 1.5, software version 3.5) is 32, but we have found that a sample number above approximately 20 may lead to this result (see also Chapter 9 for a discussion of the same problem).

The problem is due to the fact that the temperature increment for the melting curve is too high. During the LightCycler® melting curve analysis the temperature increases form 45°C to 90°C, which requires about 150 sec when using the 'continuous acquisition' mode and a temperature increment of 0.3°C/sec. In continuous acquisition mode the temperature increases continuously, and does not take into account that the measurement of fluorescence between one capillary and the next takes a certain time. If a small number of samples are analyzed, the fluorescence of each will be read more often during the 150 sec of melting curve data acquisition compared to a run with more samples. The result is that there are too few measurement points in the latter situation to calculate a detailed melting curve with two distinct peaks when two alleles are present (as in a heterozygote sample).

There are three possible solutions to this problem:

1. For high sample numbers it is recommended that the temperature increment used for the melting curve is decreased, e.g. to 0.1°C/sec when using continuous acquisition mode, or to 0.4°C/sec when using stepwise acquisition mode
2. If the melting curve still fails to give a satisfactory result, although is not recommended by the manufacturer, we have found that additional melting curves can be performed using other temperature increments (it must be noted that the quality of the melting curves is reduced each time an analysis is performed)
3. In cases when the melting curve still fails to give satisfactory result when more than about 20 samples are analyzed, stop the LightCycler® program following the amplification step and perform melting curve analyses in batches (including the appropriate controls with each melting curve analysis)

Index

Printed and bound by CPI Group (UK) Ltd, Croydon, CR0 4YY

01/11/2024

01782602-0003